Progress in Optical Science and Photonics

Volume 36

Indexed by Scopus

The purpose of the series Progress in Optical Science and Photonics is to provide a forum to disseminate the latest research findings in various areas of Optics and its applications. The intended audience are physicists, electrical and electronic engineers, applied mathematicians, biomedical engineers, and advanced graduate students.

Yesudasu Vasimalla · Santosh Kumar
Editors

Generative AI for Photonic Sensing

Springer

Editors
Yesudasu Vasimalla
Department of Electronics
and Communication Engineering
Centre of Excellence for Nanotechnology
Koneru Lakshmaiah Education Foundation
Guntur, Andhra Pradesh, India

Santosh Kumar
Department of Electronics
and Communication Engineering
Centre of Excellence for Nanotechnology
Koneru Lakshmaiah Education Foundation
Guntur, Andhra Pradesh, India

ISSN 2363-5096 ISSN 2363-510X (electronic)
Progress in Optical Science and Photonics
ISBN 978-981-95-1560-8 ISBN 978-981-95-1561-5 (eBook)
https://doi.org/10.1007/978-981-95-1561-5

This Springer imprint is published by the registered company Springer Nature Singapore Pte Ltd.
The registered company address is: 152 Beach Road, #21-01/04 Gateway East, Singapore 189721, Singapore

Preface

In recent years, the convergence of photonic sensor technology and artificial intelligence (AI) has revolutionized the way we perceive, analyze, and interpret the physical world. This interdisciplinary field combines the high sensitivity and precision of photonic sensors with the adaptive learning and data processing capabilities of AI, particularly generative AI, to create intelligent, efficient, and autonomous sensing systems. This book begins by introducing the fundamentals of photonic sensors, detailing their working principles, types, and key applications across various domains. It then shifts to cover the foundations of generative AI, explaining how models such as GANs, VAEs, and transformers can be trained to simulate, enhance, and interpret complex datasets. The integration of these two cutting-edge domains is a central focus of this work, with dedicated discussions on how AI algorithms are being integrated into photonic sensing frameworks to augment real-time decision-making, noise filtering, pattern recognition, and predictive analysis. Special attention is given to generative AI models tailored for sensor data processing, showcasing their potential in improving data quality, reducing measurement errors, and enhancing the resolution of photonic signals.

Chapters exploring AI-enhanced optical imaging reveal how deep learning improves contrast, clarity, and speed in biomedical and industrial imaging systems. The book further explores the transformative role of AI in optical fiber sensors and sensing networks, where machine learning aids in fault detection, signal demodulation, and network optimization. The application of generative AI in spectroscopy is presented as a powerful tool for spectral reconstruction, data augmentation, and material identification. Moreover, the integration of AI into autonomous systems using LIDAR highlights its importance in robotics, navigation, and smart mobility. The book also delves into environmental and atmospheric sensing, showing how AI can enhance the detection of pollutants, climate variables, and hazardous conditions through advanced photonic techniques. Similarly, the use of AI-driven photonic sensors in medical and biological sensing underscores their significance in diagnostics, real-time monitoring, and personalized health care. A notable chapter is devoted to multi-analyte sensing, where generative AI enables the discrimination and simultaneous detection of multiple targets with high accuracy.

By bringing together these diverse yet interconnected topics, this book serves as a comprehensive resource for researchers, students, and professionals interested in the future of intelligent sensing. The final section addresses emerging trends, ethical considerations, and technical challenges in the development and deployment of generative AI for photonic sensing, emphasizing the need for continued research and collaboration across disciplines. Through this work, readers will gain a deeper appreciation for the immense potential of AI-empowered photonic sensors in advancing science, industry, health care, and environmental stewardship.

Organization of This Book

This book is structured to provide a comprehensive journey through the integration of generative artificial intelligence with photonic sensing technologies.

It begins with Chapter "Understanding the Fundamentals of Photonic Sensors: Principles, Mechanisms and Applications", which lays the foundation by exploring the fundamentals of photonic sensors, detailing their operating principles, underlying mechanisms, and diverse applications in modern science and technology. Chapter "Fundamentals of Generative AI" introduces the core concepts of generative AI, providing readers with a clear understanding of models such as GANs, VAEs, and transformers and their relevance in data generation, interpretation, and enhancement. Building on these foundations, Chapter "AI-Assisted Fiber Optic Sensors for Simultaneous Measurement" presents the implementation of AI-assisted fiber optic sensors for the simultaneous measurement of multiple physical parameters, showcasing the improved precision and functionality made possible through AI integration. Chapter "Generative AI Models for Sensor Data Processing" delves into the design and application of generative AI models specifically tailored for processing sensor data, highlighting their role in noise reduction, pattern recognition, and signal enhancement.

The book continues with Chapter "AI in Optical Fiber Sensors and Sensing Network", which discusses the role of AI in optical fiber sensor networks, emphasizing intelligent data routing, fault detection, and network optimization. Chapter "Generative AI in Spectroscopy" explores the use of generative AI in spectroscopy, demonstrating its ability to reconstruct spectral data, identify substances with high accuracy, and enhance sensitivity. Chapter "Generative AI for Environmental and Atmospheric Sensing" shifts focus to environmental and atmospheric sensing, where AI-driven photonic sensors are used for monitoring air quality, climate patterns, and hazardous pollutants. In Chapter "Medical and Biological Sensing with AI Driven Photonic Sensors", the emphasis moves to medical and biological sensing, where generative AI enhances diagnostics, real-time monitoring, and bio-imaging through smart photonic platforms.

Chapter "Generative AI in Multi-analytes Sensing" investigates multi-analyte sensing, where generative AI enables the detection and differentiation of several biochemical and physical parameters simultaneously with high specificity and efficiency. Finally, Chapter "Future Trends and Challenges in Generative AI for Photonic Sensing" offers a forward-looking perspective on the future trends and challenges in the field, addressing the technical, ethical, and practical aspects of deploying generative AI in photonic sensing systems. Together, these chapters provide a holistic understanding of the transformative synergy between generative AI and photonic sensors, preparing researchers, engineers, and students to explore and innovate in this rapidly evolving domain.

Guntur, India

Yesudasu Vasimalla
Santosh Kumar

Acknowledgments

The authors would like to express their heartfelt gratitude to all those who have contributed to the writing and publication of the book *Generative AI for Photonic Sensing*. Their support and contributions have been instrumental in making this endeavor a success. First and foremost, the authors would like to extend their deep appreciation to Springer Nature Press for their valuable support and for providing the opportunity to publish this book. They would like to specifically thank the dedicated editor who meticulously edited the manuscript, ensuring its quality and readiness for publication. The authors are also indebted to their colleagues for their invaluable input and insights. Whether as direct contributors to the book or through engaging discussions and suggestions, their contributions have enhanced the overall quality of the content. The editors are grateful for their involvement and collaboration throughout the writing process. Dr. Yesudasu Vasimalla and Dr. Santosh Kumar would like to express his sincere gratitude to Shri Koneru Satyanarayana Garu, honorable president of the Koneru Lakshmaiah Education Foundation (deemed to be University), Vaddeswaram, Andhra Pradesh, India, for his immense encouragement and support throughout the writing of this book.

Dr. Yesudasu Vasimalla would also like to express his heartfelt gratitude to his wife, Eedpuganti Hepsibha, and daughter, Susritha Angelina, for their unwavering support, encouragement, and cooperation, as well as to his loving parents.

Dr. Santosh Kumar would also like to express his heartfelt gratitude to his wife, Dr. Ragini Singh, and son, Ayaansh Singh, for their unwavering support, encouragement, and cooperation, as well as to his loving parents.

Yesudasu Vasimalla
Santosh Kumar

Contents

Editors and Contributors

About the Editors

Yesudasu Vasimalla received his bachelor's degree in electronic and communication engineering from Bapatla Engineering College, Guntur, Andhra Pradesh, India, and his master's degree in Communication Systems from the Sardar Vallabhbhai National Institute of Technology, Surat, Gujarat, in 2017 and 2019, respectively. In January 2023, he has obtained a Ph.D. degree in the field of optical biosensors from the Department of Electrical Communication Engineering of the National Institute of Technology, Warangal, Telangana, India. Since May 2022, he is working as Assistant Professor in the Electronics Engineering Department of the Koneru Lakshmaiah Education Foundation, Guntur, Andhra Pradesh, Dhanbad, India. He has published more than 40 peer-reviewed and indexed international SCI Journal papers in the last 5 years. Also, he has published one patent on the detection of human sperm sample using the surface plasmon resonance biosensor. He received the Young Research award from Alpha International Publication in 2023.

Dr. Santosh Kumar is Professor in the Department of Electronics and Communication Engineering at Koneru Lakshmaiah Education Foundation (KLEF), India. He also serves as Lab Head of the Centre of Excellence for Nanotechnology (CoE-NANO), Group Head of the Nanotechnology Research Group (NTRG), and Advisor—R&D at KLEF. He earned his Ph.D. from IIT (ISM) Dhanbad in 2014 and was previously with Liaocheng University, China (2018–2023).

Recognized among the world's top 2% scientists by Stanford University for four consecutive years, Dr. Kumar has supervised 20 M.Tech. dissertations and nine Ph.D. candidates. He has published over 550 research articles in prestigious SCI journals, including 130 IEEE journal papers, with more than 12,000 citations and an h-index of 65. His 100 papers are recognized under various Sustainable Development Goals (SDGs) according to Scopus. His research has been featured in leading journals such as *Biosensors and Bioelectronics*, *Journal of Lightwave Technology*, *Optics Express*, *Optics Letters*, *Applied Physics Letters*, *ACS Applied Nano Materials*, and *Biosensors*.

His expertise spans fiber optic sensors, photonics and plasmonic devices, nano and biophotonics, waveguides, interferometers, and the Internet of Things. He has authored seven scholarly books with CRC Press (Taylor & Francis), IEEE-Wiley, and Springer Nature. Dr. Kumar was recently granted a patent for his groundbreaking optical fiber sensing technology and has published 13 patents on biosensing applications. A highly regarded expert, he has reviewed over 2550 manuscripts for top-tier SCI journals published by IEEE, Elsevier, Springer, OPTICA, SPIE, and *Nature*. His collaborative research extends across institutions in India, China, Portugal, Iraq, Kazakhstan, Brazil, France, and Italy. Dr. Kumar is Fellow of SPIE and IETE, Life Fellow of the Optical Society of India (OSI), and Senior Member of IEEE, OPTICA, and SPIE. He is OPTICA Traveling Lecturer and serves as Chair of the Optica Optical Biosensor Technical Group, USA. He is also Associate Editor for *Scientific Reports (Nature)*, *IEEE Internet of Things Journal*, and*Biomedical Optics Express*. His recent accolades include the Best Performing Associate Editor Award from *IEEE Sensors Journal, USA*, the Best IEEE JLT Paper Award from USA, and the Foreign Distinguished Professor Award from China.

Contributors

R. Adline Freeda Department of Information Technology, KCG College of Technology, Chennai, India

Kawsar Ahmed Department of Electrical and Computer Engineering, University of Saskatchewan, Saskatoon, SK, Canada;
Group of Bio-photomatiχ, Information and Communication Technology, Mawlana Bhashani Science and Technology University, Santosh, Tangail, Bangladesh

Oindrila Banik Department of Biotechnology and Medical Engineering, National Institute of Technology, Rourkela, Rourkela, Odisha, India

Earu Banoth Department of Biotechnology and Medical Engineering, National Institute of Technology, Rourkela, Rourkela, Odisha, India

Nilanjana Basak Group of Bio-photomatiχ, Information and Communication Technology, Mawlana Bhashani Science and Technology University, Santosh, Tangail, Bangladesh

Sayantani Bhattacharya Department of ECE, College of Engineering and Technology, Faculty of Engineering and Technology, SRM Institute of Science and Technology, Kattankulathur, Chennai, Tamil Nadu, India

Francis M. Bui Department of Electrical and Computer Engineering, University of Saskatchewan, Saskatoon, SK, Canada

Abhishek Chaurasia Department of Biotechnology and Medical Engineering, National Institute of Technology, Rourkela, Rourkela, Odisha, India

Debajit Deb Department of ECE, Koneru Lakshmaiah Education Foundation, Vijayawada, India

Koustav Dey Lightwave Technology Laboratory, Department of ECE, Missouri University of Science and Technology, Rolla, MO, USA;
Department of Physics, National Institute of Technology, Warangal, Warangal, Telangana, India

R. Dhaya School of Electrical and Communications Engineering, PNG University of Technology, Lae, Papua New Guinea

Sukanya Ghosh Department of ECE, College of Engineering and Technology, Faculty of Engineering and Technology, SRM Institute of Science and Technology, Kattankulathur, Chennai, Tamil Nadu, India

Praveenkumar Hiremath Mechanics, Materials and Component Design, Lund University, Lund, Sweden

Robin Kalyan Department of High Performance IC Design, Institute of Microelectronics, Agency for Science, Technology and Research, Singapore, Singapore

R. Kanthavel School of Electrical and Communications Engineering, PNG University of Technology, Lae, Papua New Guinea

Dharmendra Kumar Department of Electronics and Communication Engineering, Guru Ghasidas Vishwavidyalaya, Bilaspur, Chhattisgarh, India

Krishna Kumar Dell Technologies, Bengaluru, India

Santosh Kumar Department of Electronics and Communication Engineering, Centre of Excellence for Nanotechnology, Koneru Lakshmaiah Education Foundation, Vaddeswaram, Andhra Pradesh, India

Anamika Kumari Electronics and Telecommunication Engineering Department, COEP Technological University, Pune, Maharashtra, India

Md. Mamun Ali Division of Biomedical Engineering, University of Saskatchewan, Saskatoon, SK, Canada

Rahul Mukherjee ECE Department, School of Engineering and Applied Science, Bennett University, Greater Noida, Uttar Pradesh, India

Debarati Nath Department of ECE, College of Engineering and Technology, Faculty of Engineering and Technology, SRM Institute of Science and Technology, Kattankulathur, Chennai, Tamil Nadu, India

V. Nikhil Department of Physics, National Institute of Technology, Warangal, Warangal, Telangana, India

Sudeep Rathore ECE Department, School of Engineering and Applied Science, Bennett University, Greater Noida, Uttar Pradesh, India

Sourabh Roy Department of Physics, National Institute of Technology, Warangal, Warangal, Telangana, India

Manoj Sharma ECE Department, School of Engineering and Applied Science, Bennett University, Greater Noida, Uttar Pradesh, India

Sneha Sharma Department of Electronics and Communication Engineering, Dayananda Sagar University, Bangalore, India

R. Venkatesh Department of Physics, PSNA College of Engineering and Technology (Autonomous), Dindigul, Tamil Nadu, India

Vibha Vyas Electronics and Telecommunication Engineering Department, COEP Technological University, Pune, Maharashtra, India

Ajay Yadav ECE Department, School of Engineering and Applied Science, Bennett University, Greater Noida, Uttar Pradesh, India

Abbreviations

1D	One-dimensional
ADC	Analog-to-Digital Conversion
AI	Artificial Intelligence
ANN	Artificial Neural Networks
APD	Avalanche Photodiodes
BERT	Bidirectional Encoder Representations from Transformers
BMA	Bayesian Model Averaging
CCD	Charge-Coupled Devices
CLIP	Contrastive language-image pre-training
CNN	Convolutional Neural Network
DAS	Distributed Acoustic Sensing
DBR	Distributed Bragg Reflector
DL	Deep Learning
DM	Diffusion Model
EMI	Electromagnetic Interference
EML	Ensemble Machine Learning
FBG	Fiber Bragg Grating
FOS	Fiber Optic Sensors
GAN	Generative Adversarial Networks
GenAI	Generative AI
GPT	Generative Pretrained Transformers
IR	Infrared
KL	Kullback–Leibler
LASER	Light Amplification by Stimulated Emission of Radiation
LDA	Linear discriminant analysis
LLM	Large language models
ML	Machine Learning
MS	Mass Spectrometry
MSM	Multi-single-multi mode
MZI	Mach–Zehnder Interferometer
NIRS	Near-Infrared Spectroscopy

NLP	Natural Language Processing
NMR	Nuclear Magnetic Resonance
OCT	Optical Coherence Tomography
OTDR	Optical Time-Domain Reflectometry
PCA	Principal Component Analysis
PCF	Photonic Crystal Fiber
PLS	Partial Least Squares Regression
PMT	Photomultiplier Tubes
RF	Random Forest
RL	Reinforcement Learning
RLHF	Reinforcement Learning from Human Feedback
RNN	Recurrent Neural Network
SAM	Segment anything model
SMS	Single-multi-single
SPR	Surface Plasmon Resonance
SVM	Support Vector Machine
SVR	Support Vector Regression
TRM	Transformers Based Model
UV	Ultraviolet
VAE	Variational Auto Encoder
VOC	Volatile Organic Compounds
WDM	Wavelength Division Multiplexing
f	Frequency
λ	Wavelength

Understanding the Fundamentals of Photonic Sensors: Principles, Mechanisms and Applications

R. Venkatesh

Abstract The basic principles and mechanisms of photonic sensors are discussed, including the use of light to sense and quantify associated physical, chemical, or biological changes in their surroundings. Sensor effect in photonic sensors takes place through interaction of light with matter: a phenomenon based on both principles of optics and quantum mechanics, often manifested through absorption, reflection, refraction, and scattering phenomena depending on the interacting materials. It explains how such interactions make them precise for sensing applications by describing how the interactions occur. The various kinds of photonic sensors highlighted in the chapter, including fiber-optic sensors, which due to their ability to send data over long distances with minimal loss, interferometric sensors due to high sensitivity to detect minute changes in phase, and imaging sensors due to their visual representation of physical phenomena. Each type has specific applications where they are most useful-from telecommunications for data transmissions to healthcare, where diagnostics may be diagnosed in real-time and environmental monitoring where the ability to detect pollutants and changes in the conditions within the ecosystem is vital. By discussing these fundamental principles and mechanisms, we hope to broadly explain the operation and importance of photonic sensors in driving measurement technology improvements; ultimately demonstrating the hugely important differences it has made in various sectors of industry.

Keywords Photonic sensors · Light-matter interaction · Fiber-optic sensors · Interferometry · Imaging sensors · Applications · Measurement technology · Electromagnetic waves

R. Venkatesh (✉)
Department of Physics, PSNA College of Engineering and Technology (Autonomous), Dindigul, Tamil Nadu, India
e-mail: sukatesh@gmail.com

Y. Vasimalla and S. Kumar (eds.), *Generative AI for Photonic Sensing*, Progress in Optical Science and Photonics 36, https://doi.org/10.1007/978-981-95-1561-5_1

1 Introduction

1.1 Overview of Photonic Sensors

Photonic sensors are a family of sensors whose operation is based on the interaction of light whether visible, infrared, or any other wavelength of electromagnetic radiation with the environment in order to measure various physical parameters. Unlike the traditional sensors, which may use mechanical or electrical principles, photonic sensors make use of properties of light such as intensity, phase, polarization, or wavelength to detect and quantify changes in the surrounding medium [1]. It functions on the principles that light transmission through or light reflection by the medium changes, depending on one or more conditions such as changes in temperature, pressure, or strain, alteration in chemical makeup, or effects of biological components. The signal is then retrieved using sensitive detectors that convert any observable changes in these light properties (amplitude and phase, perhaps frequency) to measurable signals.

There are several types of photonic sensors, depending on the technology and light properties, such as:

- **Fiber-Optic Sensors**: These make use of optical fibers to transmit light and detect changes in light properties.
- **Surface Plasmon Resonance (SPR) Sensors**: These make use of the interaction of light with surface plasmon waves on metal surfaces to detect chemical or biological changes.
- **Interferometric Sensors**: These make use of interference patterns between light waves to detect tiny changes in physical parameters.
- **Photonic Crystal Sensors**: Utilize engineered structures with periodic properties for highly sensitive measurements.

Photonic sensors have numerous applications in industrial monitoring, medical diagnostics, environmental sensing, and communication systems. They belong to the class of non-contact sensors, showing extremely high sensitivity and excellent reliability, hence proving highly beneficial for use in modern sensor technology for applications in harsh environments [2].

1.2 Historical Context and Development

Photonic sensors have their roots in the early exploration of light and optics. The basis of photonics itself is based on the discovery of the nature of light and its interaction with matter, dating back to ancient civilizations, when scholars like Euclid and Pythagoras first contributed to understanding the properties of light (Fig. 1). However, the science of photonic sensing really took off in the mid-twentieth century.

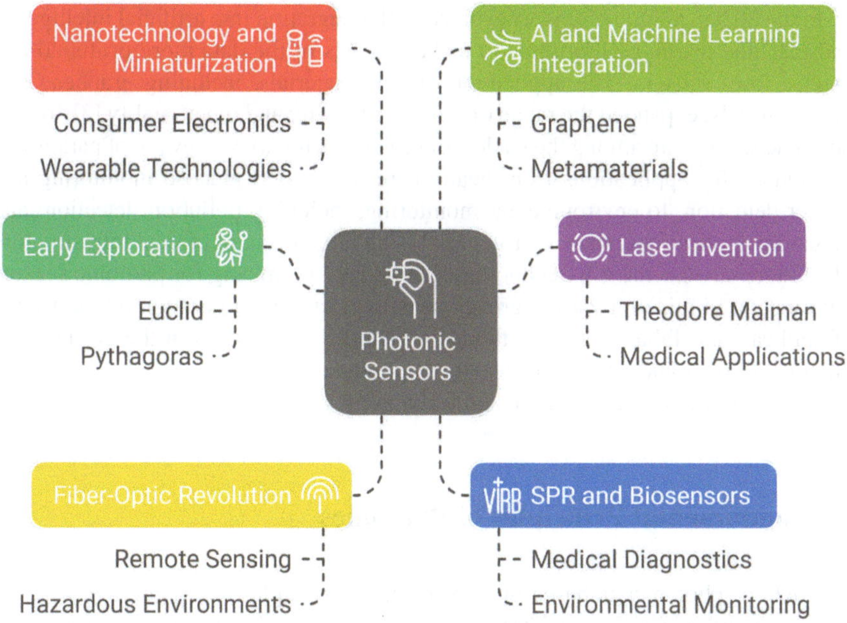

Fig. 1 Evolution of photonic sensors

A significant milestone in the photonic sensors' development occurred in 1960, when Theodore Maiman invented the laser. The laser is an example of a coherent source of light of extremely high coherence. Since the laser is governed by unique control laws, it has opened a wide variety of applications in the sensing and measurement technologies. Most early photonic sensors were concerned with the measurement of physical quantities such as temperature and pressure based on interferometry or fiber-optic technologies. The fiber-optic sensor revolution was born in the 1970s and 1980s with low-loss optical fibers that could transmit light over long distances without significant signal degradation. This development led to fiber-optic sensors in various fields, especially those that require remote sensing capabilities or are hazardous and thus cannot employ traditional electronic sensors.

SPR and other optical biosensors rose in the late 1980s and 1990s. SPR sensors, in particular, revolutionized biological and chemical sensing because they could provide information on molecular interactions without the need for labeling or complex sample preparation. It catalyzed a spate of innovation in medical diagnostics and environmental monitoring and food safety.

During the 2000s, nanotechnology and miniaturization of photonic devices have particularly enabled smaller, cheaper sensors. The development of photonic sensors emerged in consumer electronics, medical devices, and even wearable technologies. Alongside these advancements, development in quantum optics, optical coherence tomography, and similar others accelerated the use of photonic sensors in medicine, material science, and aerospace.

In the 2020s, photonic sensors evolved with integration into artificial intelligence and machine learning algorithms to make analysis of the data more profound and precise from sensors. The application of new materials, including graphene and metamaterials, expanded the range of sensor sensitivity and functionality [3]. Today, photonic sensors are among the leaders of sensing technology; they are of paramount importance for applications from health care, like blood glucose monitoring and cancer detection, to environmental monitoring, including pollution detection, and industrial applications, such as structural health monitoring of critical infrastructure [4]. Future advancement in technology can be very promising for photonic sensors with increased miniaturization, greater sensitivity, and incorporation into a myriad of applications. Table 1 outlines the main historical milestones in the evolution of photonic sensors and how these have made photonic sensors more sophisticated and incorporated into modern science and technology.

2 Fundamental Principles of Photonics

The field of photonics is based on a set of basic principles that guide the behavior of light. In order to understand how photonic sensors work, one needs to grasp these principles and apply them to different technological domains [5].

2.1 Basic Concepts of Light

Light is a kind of electromagnetic radiation that can be described as being both a wave and a particle. It travels through space in the form of an oscillating electric and magnetic field, wherein the electric and magnetic fields are perpendicular to one another and also perpendicular to the direction of propagation [6].

The fundamental properties of light (Fig. 2) include:

- **Wavelength (λ)**: The distance between successive peaks or troughs of the light wave. Wavelength describes the color of visible light and ranges all the way from extremely short wavelengths (gamma rays) all the way to extremely long wavelengths (radio waves).
- **Frequency (f)**: The number of oscillations or cycles per second of the electromagnetic wave. Frequency is inversely proportional to wavelength, and the product of wavelength and frequency equals the speed of light in vacuum: $C = \lambda \cdot f$ where C is the speed of light ($\sim 3 \times 10^8$ m/s).
- **Amplitude**: Wave height that is a direct function of the intensity or brightness of light. The greater the amplitude, the higher the amount of energy that the wave carries.
- **Polarization**: The direction in which the oscillations of the electric field take place. In unpolarized light, the electric field oscillates in all the perpendicular directions

Table 1 Historical development of photonic sensing technology

Time period	Key events and developments	Impact on photonic sensors
Ancient civilizations	Early explorations of light by scholars like Euclid and Pythagoras (around 400–300 BCE) laid the foundations for understanding the properties of light	Initial concepts of light and its behavior, such as reflection and refraction, set the stage for later scientific advancements in optics and photonics
Seventeenth century	Isaac Newton (1666) and others, like Christiaan Huygens, developed theories on light, leading to the wave-particle duality debate	The study of light as both a wave and a particle paved the way for the later exploration of optical phenomena in photonic sensing, especially in interference and diffraction effects
Mid-twentieth century	Discovery of light amplification by stimulated emission of radiation (laser) in 1960 by Theodore Maiman	The invention of the laser provided a coherent, focused light source, revolutionizing photonic sensor technology, especially in interferometry, spectroscopy, and optical communication
1970s–1980s	Development of low-loss optical fibers enabled long-distance light transmission with minimal signal degradation	The advent of fiber-optic technology enabled the development of fiber-optic sensors, ideal for remote and hazardous applications, significantly expanding the use of photonics in sensing
1980s–1990s	Emergence of surface plasmon resonance (SPR) sensors and optical biosensors	SPR sensors provided a breakthrough in biological and chemical sensing, allowing real-time molecular interaction detection without the need for complex sample preparation. This period also saw the application of photonic sensors in medical diagnostics, environmental monitoring, and food safety
Early 2000s	Advancements in nanotechnology and the miniaturization of photonic devices. Growth of optical coherence tomography (OCT) and developments in quantum optics	Nanotechnology enabled the creation of compact, low-cost photonic sensors, leading to their integration into consumer electronics and medical devices. OCT made significant strides in medical imaging, while quantum optics opened new possibilities in sensing

(continued)

Table 1 (continued)

Time period	Key events and developments	Impact on photonic sensors
2010s	Integration of artificial intelligence (AI) and machine learning (ML) into sensor data processing. Exploration of new materials like graphene and metamaterials	AI and ML improved sensor precision and data analysis, allowing for real-time decision-making and predictive analytics. New materials enhanced sensor sensitivity and functionality, enabling sensors to operate in increasingly demanding environments
2020s and beyond	Continued miniaturization and integration of photonic sensors in a range of industries. Advancements in graphene, metamaterials, and quantum sensors	Photonic sensors have become integral to various industries such as healthcare, environmental monitoring, and industrial applications. They are poised to become even more sophisticated, offering greater sensitivity, portability, and integration into systems powered by AI, IoT, and smart technologies

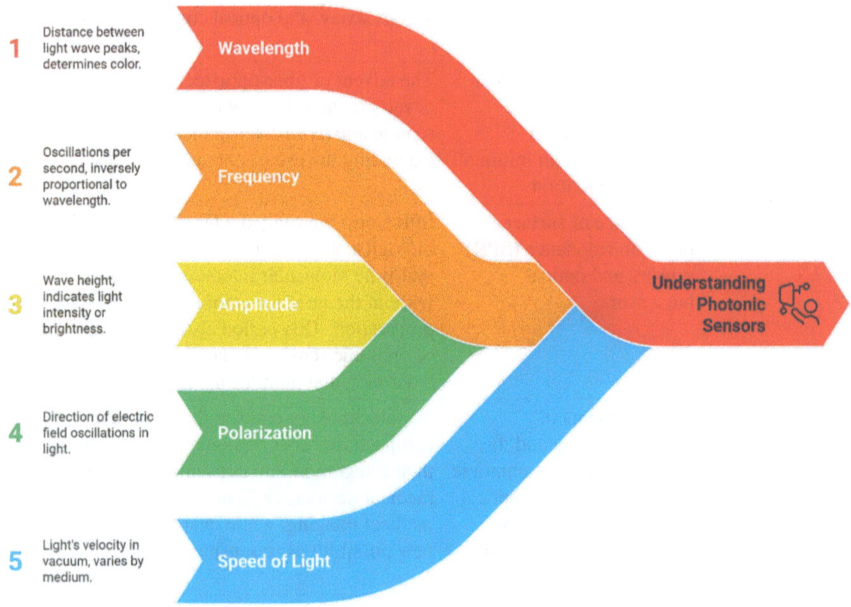

Fig. 2 Fundamental properties of light

of the propagation of light. The polarized light, however, has the oscillation of the electric field in one specific direction.

- **Speed of Light**: The speed at which light travels in vacuum, denoted by c, is approximately 3×10^8 m/s. The speed of light varies with the medium it travels through, such as air, water, or glass.

Such varied kinds of radiations the light in totality entails-low-frequency and long-wave-length radio waves at the extremes of spectrum. But from amongst them all visible light falls but into that section. Mostly used in measurement applications are such small portion portions sensed and recorded in photonic sensors [7].

2.2 *Interaction of Light with Matter*

Photons interact with matter in several different ways, which means that knowing those interactions is integral to how a photonic sensor works. Light and matter are known to have the following principal types of interaction:

- **Reflection**: When light hits a surface and bounces back. The angle of incidence equals the angle of reflection. Reflection can be either specular, like a mirror, or diffuse, like a rough surface.
- **Refraction**: Bending of light while passing from one medium to another with a different refractive index. The degree of bending is determined by Snell's Law:

$$n_1 \sin(\theta_1) = n_2 \sin(\theta_2)$$

  where n_1 and n_2 are the refractive indices of the two media, and θ_1 and θ_2 are the angles of incidence and refraction, respectively.
- **Absorption**: When light is absorbed by a material, the energy from light is transferred to the atoms or molecules of the material. The transferred energy may make the material hot or bring about a chemical or physical change in it. Absorption depends on the wavelength of light and the properties of the material.
- **Transmission**: When light travels through a material. Materials like glass and some plastics are transparent to light at certain wavelengths, so light can pass through. The transmission efficiency depends on the thickness and optical properties of the material.
- **Scattering**: The phenomenon where light is deviated in various directions when it crosses particles or roughness of a medium. Scattering causes the sky to be blue and the sun to be red during sunsets. Scattering is categorized into different types, which are Rayleigh scattering (from molecules and small-sized particles) and Mie scattering (from large-sized particles).
- **Photoluminescence**: It involves a material that absorbs light at one wavelength and re-emits at another. Examples include fluorescence and phosphorescence, in which a material emits light after it is excited by the incident light.
- **Nonlinear Interaction**: In some conditions, especially at high light intensities, light can interact with matter in a nonlinear way, leading to phenomena like second-harmonic generation or self-focusing.

Many photonic sensors rely on these interactions. For example, in the case of fiber-optic sensors, reflection or transmission of light through an optical fiber is used to sense changes in the environment surrounding the sensor, including temperature, pressure, and strain. Surface plasmon resonance sensors work based on the interaction of light with metal surfaces to detect chemical or biological substances [8].

2.3 Wave-Particle Duality

The wave-particle duality concept is the foundation of quantum mechanics, and it's central to the understanding of the behavior of light. It suggests that light has properties of both waves and particles depending on how one observes it.

Wave-Like Behavior: The behavior of light is such that it presents the wave characteristics. Some of these include interference and diffraction. When two waves overlap, the waves can reinforce each other or cancel out depending on the interference: constructive or destructive. Diffraction describes how a light wave bends around obstacles or spreads as it passes through an aperture.

Particle-Like Behavior: In some instances, light behaves as if it consists of individual packets of energy known as photons. The energy of a photon is proportional to the frequency of light. This particle description accounts for the phenomenon that includes the photoelectric effect where the incidence of light on a metal surface causes electrons to be ejected from the material. The photoelectric effect was explained only when the assumption that light is made of particles, which have specific energy, was adopted.

In photonic sensing, wave-like behavior is utilized in techniques that include interferometry, where one measures the differences in phase of light waves; or diffraction, which one can use for determining the characteristics of materials [9]. Often the particle-like behavior is accounted for in processes, like photon counting within imaging sensors, or even within certain quantum sensors operating with ultra-low light sensitivity that measure in terms of photons, discrete, not as waves.

The wave-particle duality of light acts as a useful bridge between classical and quantum descriptions, which can make possible the conception of a multitude of photonic technologies. Such photonic sensors can exploit in various ways either the wave-like or the particle-like properties of light to provide high sensitivity and accuracy in measurements concerning domains ranging from material science up to biological sensing [10]. These basic principles underpin how photonic sensors can detect, manipulate, and analyze light to measure environmental parameters, detect chemical reactions, and many other applications. Understanding these principles is key to advancing sensor technology and applying photonics in real-world systems. Table 2 summarizes the key characteristics of the wave-particle duality of light.

Table 2 Key characteristics: wave-particle duality of light

Characteristics	Wave-like behavior	Particle-like behavior
Nature of light	Light behaves as a continuous wave	Light behaves as discrete packets of energy (photons)
Key phenomena	Interference and diffraction	Photoelectric effect, photon counting
Description of behavior	Light can exhibit constructive and destructive interference; bends around obstacles (diffraction)	Light can be quantized into photons that carry energy proportional to frequency
Mathematical model	Described by wave equations (e.g., Maxwell's equations for electromagnetic waves)	Described by quantum mechanics (e.g., energy of photons is $E = h\nu$, where h is Planck's constant and ν is frequency)
Energy transfer	Energy is spread out over the wave, not localized	Energy is localized in individual photons, each with a quantized amount of energy
Key applications	Interferometry, diffraction techniques, and wave-based photonic sensors	Photon counting, ultra-low light sensors, and quantum sensing applications
Measurement focus	Phase differences, interference patterns, wavelength-based measurements	Detection of individual photons, energy quantization, and particle detection
Examples in technology	Optical interferometers, fiber-optic sensors, spectrometers	Photon detectors, quantum sensors, imaging sensors, light-sensitive devices
Observational dependence	Behavior depends on the experimental setup; wave-like properties emerge in situations involving multiple paths	Behavior depends on detection methods (e.g., when light is observed as discrete quanta in low-light or particle-sensitive setups)

3 Mechanisms of Photonic Sensing

Photonic sensors can be said to work based on the interactions of light itself with a target environment to generate a signal responding to physical, chemical, or biological changes within that environment [11]. The photonic sensing mechanisms draw from light/material interactions as well as the principles of light transfer, detection, and processing involved.

3.1 Light Propagation and Detection

Light Propagation is the way light travels through different media, and it forms the basis for most photonic sensing systems. The way in which light propagates can be controlled based on the type of sensor used, and this control makes photonic sensors highly sensitive and versatile [12].

- **Optical Fibers**: Many photonic sensors have light travel over optical fibers. These guide the light through total internal reflection, wherein the light stays inside the core of the fiber as it is constantly reflected by the walls of the fiber. Therefore, signal is not much degraded as the distance traveled increases significantly, which explains why the sensors based on the fiber optic have been extremely successful in applications like remote sensing.
- **Waveguides**: Photonic waveguides, similar to optical fibers, are structures used for guiding the flow of light along a definite path. Integrated into the chip or a device, such waveguides have found widespread use in an application that goes all the way to integrated photonic circuits, applied in on-chip sensing.
- **Free-Space Propagation**: In some cases of photonic sensors, it relies on propagation via free-space where light transpires into free space specifically, in extraneous environmental measurands are evaluated such as pressure and atmospheric temperature. Light has been forwarded free-space based upon optical-communication and the detection region made of lenses or mirrors and fibers.
- **Mode Propagation**: Light can propagate in various modes depending on the structure and geometry of the sensor. For example, the propagation of light in a fiber-optic sensor may be made to happen in a single mode where only one path exists for light to traverse or multiple modes where more than one path exists within the fiber.

Light detection is an important step in photonic sensing. Photodetectors convert the optical signal into an electrical signal to be processed and analyzed. The types of photodetectors used can be summarized as follows:

- **Photodiodes**: Semiconductor devices that generate current when exposed to light. Photodiodes are commonly used in photonic sensors because of their sensitivity, speed, and efficiency.
- **Photomultiplier Tubes (PMTs)**: These devices detect very weak light signals and are used in the amplification of the current produced by the detection of a photon. These devices are useful for low-light applications.
- **Charge-Coupled Devices (CCDs) and Complementary Metal-Oxide-Semiconductor (CMOS) Sensors**: Used for imaging, these sensors convert photons into electronic charges, which are read out to form an image or signal.
- **Avalanche Photodiodes (APDs)**: Avalanche photodiodes are high sensitivity photodiodes that rely on internal amplification mechanisms to boost the current created by incoming photons, which allows them to sense weak light signals in photonic sensors.

The choice of detector depends upon the wavelength of light being used, the level of sensitivity, and the application [13].

3.2 Sensitivity and Resolution

Sensitivity and resolution refer to the capability of a photonic sensor to measure small variations in the measured quantity and its ability to resolve between two nearly close values.

Sensitivity is defined as the minimum detectable change in the measured quantity, for example, temperature, pressure, chemical concentration, etc. The greater the sensitivity, the more easily slight changes in the environment will cause a detectable change in the sensor output. The sensitivity of a sensor is enhanced by increasing the interaction of light with the target medium. This is done by employing phenomena such as SPR or interferometry.

Resolution is the minimum difference between two values the sensor can resolve. In photonic sensors, it often refers to the resolution or the ability to precisely measure light properties (intensity, phase, wavelength, etc.). High-resolution allows making very fine differences between values, so that even minute changes in measurement value are sensed by the sensor [14]. Table 3 compares sensitivity and resolution in sensors. Sensitivity refers to how sensitive a sensor is to detecting small changes, whereas resolution refers to the smallest difference between values that can be measured.

Many factors determine the sensitivity and resolution of photonic sensors:

1. **Optical Power and Signal-to-Noise Ratio (SNR)**: Higher optical power will enhance sensitivity through the generation of stronger signals; however, the excess noise reduces resolution. Both are maintained when SNR is balanced.
2. **Wavelength and Bandwidth**: Light wavelength selection would influence the light-material interaction. Lower bandwidths might be associated with higher resolution in spectroscopic sensing. A higher bandwidth, conversely, allows for a higher sensitivity due to increased interaction with the medium.
3. **Interferometric and Resonant Techniques**: The interferometry-based technique is interferometry, relying on interference in light waves; resonant sensors, for instance, SPR or optical cavities, which improve the sensitivity and resolution of photonic sensors by employing very minute variations in phase or resonance to measure small environmental variations.
4. **Material and Device Properties**: The choice of material used for optical fibers, waveguides, or detectors can have a significant impact on the performance of the sensor. High-quality materials with low loss and low noise levels improve both sensitivity and resolution.

Table 3 Comparison of sensitivity and resolution in sensors

Factors	Sensitivity	Resolution
Definition	Minimum detectable change in the measured quantity (e.g., temperature, pressure, chemical concentration)	Minimum difference between two values that the sensor can resolve, or the ability to distinguish small variations in measured quantities
Effect on sensor performance	Higher sensitivity allows detection of small environmental changes, improving detection of subtle shifts in the measured quantity	Higher resolution enables precise differentiation between closely spaced values, allowing fine measurements of light properties (intensity, phase, wavelength)
Influence of optical power and SNR	Increased optical power enhances sensitivity by generating stronger signals, but noise reduces sensitivity if not controlled	Resolution depends on a balanced SNR; higher noise reduces the ability to differentiate between closely spaced values, impairing resolution
Influence of wavelength and bandwidth	Longer wavelengths and narrower bandwidths tend to improve sensitivity through greater interaction with the medium (more energy)	Narrower bandwidths typically enhance resolution in spectroscopic sensing (ability to distinguish closely spaced spectral lines)
Interferometric and resonant techniques	Interferometric techniques (e.g., interferometry, SPR) enhance sensitivity by increasing light-medium interaction and detecting small phase shifts	Resonant techniques (e.g., SPR, optical cavities) improve resolution by detecting minute variations in phase or resonance, enabling precise measurement
Material and device properties	High-quality materials reduce noise and losses, enhancing sensitivity by allowing more accurate detection of small changes	Low-loss, low-noise materials improve resolution by enabling finer distinctions between closely spaced light properties or signals
Key applications	Useful in detecting very small changes in the environment, such as small shifts in chemical concentration or temperature	Useful for applications requiring precision measurements, such as high-accuracy spectroscopy or fine displacement measurements
Examples of techniques	Surface plasmon resonance (SPR), fiber optic sensors, chemical and biological sensors	Interferometry, high-resolution spectroscopy, and Fabry–Pérot cavities
Trade-offs	Increased sensitivity may compromise resolution due to noise; higher optical power might worsen resolution if SNR is poor	Higher resolution may lead to reduced sensitivity due to limited signal strength or increased noise sensitivity

(continued)

Table 3 (continued)

Factors	Sensitivity	Resolution
Impact of environmental conditions	Highly sensitive sensors are more affected by environmental noise, requiring careful calibration to avoid false readings	High-resolution sensors may require very stable environmental conditions (temperature, vibration) to accurately resolve minute differences
Key limiting factors	Noise, signal strength, optical power, and light-material interaction efficiency	Bandwidth limitations, signal clarity, and interference between signals

3.3 Signal Processing Techniques

Once light has been captured, signal processing techniques extract meaningful information from the raw data. It plays a vital role in improving photonic sensor performance to develop sensors with very high sensitivity, resolution, and accuracy.

- **Analog-to-Digital Conversion (ADC)**: The photonic sensors yield continuous analog signals that need to be converted into digital signals for ease of processing and interpretation. In modern photonic sensing systems, ADCs play an important role in acquiring data at very high rates.
- **Fourier Transform**: Most light signals are not analyzed in the time domain, but rather in the frequency domain. A Fourier transform is employed to break complex light signals down into their frequency components, and specific spectral components are easier to analyze.
- **Filtering**: Digital or analog filter is used to exclude noise from the signal and to concentrate on the information. Low-pass, high-pass, or band-pass filters can be used, for instance, to isolate a given frequency range of interest and cut out interference.
- **Peak Detection and Tracking**: In certain photonic sensors, for example, SPR or Raman spectroscopy-based sensors, signal peaks are associated with particular environmental conditions or materials. Advanced algorithms can track these peaks and monitor the position, height, or shape of the peaks over time to extract quantitative information.
- **Machine Learning and Artificial Intelligence**: Machine learning and AI have been included in signal processing pipelines to analyze complex sensor data. Such techniques can heighten the recognition of patterns, increase the accuracy of predictions, and automatically adjust and compensate for environmental factors.
- **Real-Time Data Processing**: Many applications are based on the sensors. Be it industrial, or medical sector requires real time process of the information. In all these sensor related applications signal processing techniques can handle high amounts of data to yield decisions promptly along with swift reaction.
- **Calibration and Compensation**: Calibrations often are needed with photonic sensors in order to achieve accuracy measurements. Calibration can be achieved

through comparison of the sensor reading against a known standard, thereby correcting any errors between these two values. Compensation algorithms are sometimes applied for temperature and/or humidity as it impacts sensor behavior.

It provides foundational mechanisms such as light propagation, detection, sensitivity, resolution, and signal processing in developing high-performance sensors over an extremely broad spectrum of applications. Photonic sensors are ideally positioned to provide highly precise and sensitive measurements noninvasively; understanding how the mechanisms of sensing interplay will be central to advancing sensor technologies [15]. The future will see the continued integration of advanced signal processing techniques, along with innovations in materials and detection technologies, further enhancing the capabilities and reach of photonic sensors.

4 Types of Photonic Sensors

Photonic sensors apply properties of light in measuring all possible physical, chemical, or biological parameters. Various photonic sensors have thus come to fruition throughout the years. They depend on specific requirements or applications needed to be satisfied by these devices.

4.1 Fiber-Optic Sensors

Fiber-optic sensors are the most versatile and widely used type of photonic sensors. The transmission of light through an optical fiber is the underlying principle for a number of physical quantities, including temperature, pressure, strain, and even chemical composition [16]. Some key advantages of such fiber-optic sensors are the ability to send light over distances without much signal loss and inability to electromagnetic interference, and are non-invasive in nature.

Types of Fiber-Optic Sensors

- **Intensity-Based Fiber-Optic Sensors**: These sensors detect changes in light intensity while passing through an optical fiber. For example, any change in temperature or pressure on the fiber changes the light intensity, which is then sensed at the output of the sensor. This sensor is relatively simple and inexpensive but is generally less sensitive.
- **Fiber-Optic Strain Sensors**: These sensors use the changes in light transmission to detect mechanical strain. For example, fiber Bragg grating (FBG) sensors utilize a grating structure inscribed within the fiber, which reflects certain wavelengths of light. Changes in the fiber length due to strain alter the reflection wavelength, thereby providing a direct measurement of strain.

- **Distributed Fiber-Optic Sensors**: Distributed sensors measure along the entire length of the fiber whereas point sensors take measurements at only one location on the fiber. Techniques such as optical time domain reflectometry (OTDR), Raman scattering are used in measuring temperature and strain along the length of the fiber. This permits real-time monitoring of large structures like pipelines, bridges, buildings.
- **Chemical and Biological Sensing**: Another application of fiber-optic sensors is for chemical and biological sensing. By coating the fiber with specific functional materials, it can be made to selectively interact with certain chemicals or biological agents, thereby altering the properties of light transmission that can be measured.

Fiber-optic sensors are used in a wide range of applications, including structural health monitoring, environmental sensing, and medical diagnostics, such as in endoscopic imaging.

4.2 Interferometric Sensors

Interferometric sensors measure the changes in the physical parameters, which are displacement, temperature, pressure, and refractive index. It works on the principle of interferometry, comparing the phase difference between two or more light beams. Even small changes in the environment can shift the interference pattern detectably, which means that these measurements can be taken with great accuracy [17].

Types of Interferometric Sensors

- **Michelson Interferometers**: This is the most famous interferometer type; the Michelson interferometer splits a light beam into two arms and then combines it after it travels through two different paths. The interference pattern so obtained contains information about changes in the path length, hence, the distance, temperature, and pressure parameters can be measured.
- **Mach–Zehnder Interferometers**: In these interferometers, light beam is divided into two arms which travel in two different mediums or environments, respectively. Both arms are made to interfere such that interference pattern formed reflects change in the characteristics of the medium such as a change in the refractive index, pressure, or temperature.
- **Fiber-Optic Interferometers**: The use of fiber as the propagation medium for light provides a good and efficient means in interferometry. Fiber-optic interferometers that include Sagnac interferometers, Michelson interferometers, may be applied for detecting rotational motion, pressure variations, temperature, and more can be done even in compact mode and at extreme locations.
- **Optical Coherence Tomography (OCT)**: OCT is a noninvasive imaging procedure that uses the principles of interferometry to give high-resolution images of cross sections of biological tissue. It quantifies the timing delay of returning light from backscattered particles in different layers of tissue with the help of which

it gets a detailed and inner structure images of tissues for use in widespread ophthalmologic, dermatological, and other similar medical fields.

The interferometric sensor has the highest sensitivity and resolution among all sensors, thus recommended for precision applications, including seismic sensing and geophysical survey, as well as in gravitational wave detection.

4.3 Imaging Sensors

Imaging sensors are light-detecting devices that create an image or a map of the light intensity or other properties (e.g., wavelength, polarization) across a scene in two or more dimensions. Imaging sensors are used in a variety of applications, from medical imaging to environmental monitoring to security surveillance.

Types of Imaging Sensors

- **Charge-Coupled Devices (CCDs)**: One of the most popular applications in optical imaging from CCDs is in digital cameras, microscopes, and astronomical telescopes. CCD devices work by collecting incoming light photons as electrical charge, which is then read out from it to create the image. These images have high sensitivity, high resolution, and low noise, but they are a bit slower than CMOS sensors.
- **Complementary Metal-Oxide-Semiconductor (CMOS) Sensors**: CMOS sensors are similar to CCDs but differ in technology that allows for a faster readout and lower power consumption. They are used extensively in consumer electronics, such as in smartphones and cameras, and are also used in scientific and industrial applications. For real-time imaging, CMOS sensors are often favored because of their high processing speed.
- **Near-Infrared (NIR) and Infrared (IR) Imaging**: These detect light in the infrared spectrum; such sensors often use infrared and are applied for thermal imaging, night vision, or surveillance. Using such sensors, the NIR and IR sensors can capture heat signatures or biological processes in terms of monitoring blood flow.
- **Hyperspectral and Multispectral Imaging**: Advanced imaging sensors operating over a much larger spectral range, namely, ultraviolet to infrared wavelengths. They are best suited for applications involving remote sensing, environmental monitoring, and food safety. These offer detailed spectral information for each pixel in the image.
- **Fluorescence Imaging**: In the biological and medical fields, fluorescence imaging involves detecting light emitted by a substance, such as a fluorophore, after it absorbs a particular wavelength of light. It is extremely common for cellular imaging, diagnostic applications, and studying molecular interactions.

Imaging sensors are ubiquitous in medical diagnostics applications, for instance, endoscopy and biopsy imaging, industrial inspection, like quality control, environmental monitoring for levels of pollution in water and air, among others [18].

4.4 Novel Sensor Designs

Technological advances in materials science, nanotechnology, and quantum mechanics have brought novel opportunities for photonic sensing capabilities that are unprecedented and present unfulfilled opportunities for sensing in many fields.

Examples of Novel Sensor Designs

- **Metamaterials-Based Sensors**: Metamaterials are artificial materials engineered to exhibit properties not found in natural materials, often with unique optical properties such as the negative refractive index or unusual dispersion. These materials may be used in the design of sensors that give improved sensitivity and resolution in chemical sensing, biosensing, and optical imaging areas.
- **Graphene-Based Sensors**: Graphene is an amazing, 2D honeycomb of a single layer of carbon atoms. Due to their very remarkable optical and electronic properties, graphene-based photonic sensors are under development for high-speed sensing, chemical detection, and environmental monitoring due to their strong responses to surrounding environments (such as gas or liquid interactions).
- **Quantum Photonic Sensors**: Quantum sensors take advantage of quantum principles like superposition and entanglement to provide the ultimate sensitivity measurements. Quantum interferometers and quantum-limited photodetectors are used in order to detect extremely weak signals with precision up to the classical physics limits.
- **Plasmonic Sensors**: Surface plasmon resonance (SPR) sensors utilize the effect of interaction between light and surface plasmon waves, which represent the collective oscillations of electrons at the interface between a metal and a dielectric medium. Plasmonic sensors are particularly useful in biological and chemical sensing since these sensors can detect molecules without the requirement for labels or other complex sample preparation protocols.
- **Lab-on-a-Chip Sensors**: In these devices, several sensing elements are integrated onto a single microchip, thereby making sensors compact and portable to be used in situ for performing complex analyses. They are applied in point-of-care medical diagnostics and environmental monitoring.
- **Fiber-Optic Plasmonic Sensors**: Fiber-optic sensors combined with plasmonics provide highly sensitive measurements for many applications, such as environmental and biomedical monitoring. Combining these two technologies allows remote sensing with the high sensitivity of plasmonic detection.

New sensor designs are leading photonic sensing toward new horizons and possibilities, making it possible for precision measurement, environmental sensing, biomedical applications, and communications to be further improved. As more of these technologies continue to come to the market, sensors are bound to discover new frontiers in health, energy, and space exploration. Photonics take forms ranging from basic to sophisticated application fields and environments; from fiber optic sensors for distributed sensing, interference sensors for extreme precision, the imaging sensors capable of providing accurate analysis, as well as brand new designs applicable to state of the art and cutting-edge systems [19]. All continue to push to advance and fill new niches in science and industry and then in daily application. The promising results include, among others: improved accuracy for a broad span of applications. Table 4 tabulates the core features, strengths, and usages of four different photonic sensors, giving an overview of their strengths as well as shortcomings for different needs in sensing applications.

5 Applications of Photonic Sensors

Photonic sensors are highly versatile, precise, and non-invasive and are widely applied in many different fields. They can measure the physical, chemical, and biological parameters with high sensitivity and accuracy.

5.1 Telecommunications

Photonic sensors are an essential part of the telecommunications industry. They allow for efficient communication systems and have helped in developing optical networks, fiber-optic communications, and data transmission [20]. This is because photonic sensors can transmit and receive light signals over long distances with minimal loss.

Key Applications in Telecommunications

- **Fiber-Optic Sensing**: Fiber-optic sensors are also widely employed in telecommunication networks to monitor and maintain the health of the network. It integrates sensors inside the optical fibers, which detect the occurrence of faults, temperature fluctuations, strain, and pressure changes along the fiber network. For instance, a distributed fiber-optic temperature sensors (DTS) allows for continuous temperature measurement over long distances and can help identify overheating or damage to cables as potential failures in the network.
- **Optical Time Domain Reflectometry (OTDR)**: OTDR is a method used to determine faults, length of fiber-optic cables, and the status of optical networks. Photonic sensors based on OTDR allow for the determination of breaks or losses in the network and ensure that long-distance optical links operate under

Table 4 Comparison of different types of photonic sensors

Sensor type	Fiber-optic sensors	Interferometric sensors	Imaging sensors	Novel sensor designs
Principle of operation	Uses optical fibers to detect changes in light signals (e.g., intensity, phase, wavelength) as they interact with the environment	Measures phase shifts or interference patterns of light waves, often using interferometers	Captures spatial and temporal information about light to form images, typically with a sensor array	Incorporates new materials, techniques, or hybrid technologies for advanced sensing capabilities
Sensitivity	High sensitivity to temperature, strain, and pressure	High sensitivity to phase shifts, small changes in optical path length	Can be sensitive to variations in light intensity, color, or polarization	Can offer enhanced sensitivity using advanced techniques like resonant interactions or nanostructured surfaces
Resolution	Moderate to high, dependent on the fiber design and detection method (e.g., distributed vs. point sensing)	Very high resolution, capable of detecting minute changes in phase or displacement	Moderate to high, depending on the sensor array and imaging technology used	Typically high, especially with miniaturization and novel material properties
Key applications	Structural health monitoring, environmental sensing, and chemical detection	Precision metrology, optical coherence tomography (OCT), and gravitational wave detection	Medical imaging (e.g., OCT), remote sensing, and surveillance	Next-gen sensors for quantum sensing, environmental monitoring, and medical diagnostics
Advantages	Flexible, remote sensing capability, immune to electromagnetic interference	Extremely precise measurements, high accuracy, and low drift over time	Provides detailed, spatially-resolved data and images	Highly customizable, can leverage cutting-edge materials and hybrid technologies
Limitations	Limited to point or distributed measurements, signal loss over long distances	Requires stable environmental conditions and high-precision optics	Can be complex, requiring advanced processing and high-quality optics	May involve high development cost, complexity, and unproven performance in certain applications

(continued)

Table 4 (continued)

Sensor type	Fiber-optic sensors	Interferometric sensors	Imaging sensors	Novel sensor designs
Examples	Fiber-optic temperature and strain sensors, Distributed Acoustic Sensing (DAS)	Michelson interferometer, Fabry–Pérot interferometer, and Mach–Zehnder interferometer	CCD or CMOS imaging sensors, photon-counting detectors	Plasmonic sensors, meta surface-based sensors, quantum-enhanced devices
Environmental sensitivity	Moderate; sensitive to fiber bending and environmental factors like temperature or pressure	High; sensitive to vibrations, temperature fluctuations, and optical alignment	Moderate; sensitive to environmental factors like light intensity and noise	Varies; some novel designs are highly sensitive to environmental factors but can offer superior performance in specific conditions

optimal conditions. This technique is very important for network maintenance and troubleshooting.

- **Signal Integrity and Quality Monitoring**: In telecommunications, this would mean that the light signal passing through a fiber optic needs to be in integrity to guarantee the quality of data transmission. Photonic sensors can monitor signal quality from the start of dispersion, which affects the amplitude and phase of the optical signals, enabling real-time evaluation of network performance.
- **Wavelength Division Multiplexing (WDM) Systems**: WDM system is a WDM system that would allow multiple signals to be transferred on the same optical fiber; it uses several wavelengths of lights. Photonic sensors are being used for precision measurement and channel monitoring in photonic sensors because they can monitor wavelength shifts in the channels for interference and measures signal strength while avoiding data losses or interference of bandwidth.

Photonic sensors are used to monitor, detect faults, and optimize fiber-optic systems in real time, which makes them an essential part of the network's reliability and efficiency in telecommunication.

5.2 *Environmental Monitoring*

Some of the most potent areas where photonic sensors are significantly making a difference are in the area of environmental monitoring. Some of the primary applications include measures of air and water quality, the monitoring of environmental pollutants, climate change, and several other environmental parameters in both urban and remote settings [21].

Key Applications in Environmental Monitoring

- **Air Quality Monitoring**: Photonic sensors, especially the optical absorption-based and laser spectroscopy-based, are used for detecting pollutants, including carbon dioxide (CO_2), nitrogen dioxide (NO_2), sulfur dioxide (SO_2), and particulate matter (PM). These are used in both outdoor and indoor air quality monitoring systems, thereby enabling cities, industries, and regulatory agencies to monitor pollution in real time.
- **Water Quality Monitoring**: Photonic sensors also measure the quality of water bodies by detecting heavy metals, pH changes, and organic pollutants. Fiber-optic sensors are applied in rivers, lakes, and oceans to measure water temperature, turbidity, and chemical composition. The sensors can be used to monitor the levels of pollution, identify the sources of pollution, and be applied in the management of water.
- **Climate Change and Atmospheric Sensing**: Photonic sensors, in particular, use lidar systems to measure the atmospheric condition including humidity, temperature, and wind speed. This technology of lidar also is used for observing changes in the atmosphere such as greenhouse gas concentration, aerosol content, and clouds, thus providing useful data in climate change studies and modeling.
- **Soil and Agricultural Monitoring**: Photonic sensors can be applied in agriculture for monitoring soil moisture, nutrient content, and crop health. The sensor will help to optimize irrigation techniques, reduce waste water, and improve crop yield through real-time data on soil conditions.

Environmental monitoring systems and photonic sensors are useful in any aspect of natural resource management, mitigating environmental risks for sustainability.

5.3 *Biomedical Applications*

The application of photonic sensors in the biomedical field has significantly transformed diagnostics, medical imaging, and patient monitoring. They provide non-invasive, highly sensitive measurements, thus making them incredibly valuable tools both in clinical practice and in the research environment [22].

Key Photonic Sensing Techniques in Biomedical Applications

- **Optical Coherence Tomography (OCT)**: OCT is a non-invasive imaging method that applies interferometry principles to generate a high-resolution image of cross-sectional biological tissues. It applies in ophthalmology for imaging retina and cornea, amongst many others, that enables an early diagnosis and monitoring process of diseases in the macular degeneration, diabetic retinopathy, and glaucoma. Similarly, in dermatology and cardiology, OCT is applied for imagery purposes in skin and blood vessels.

- **Fluorescence Based Sensing**: Photonic sensors are applied extensively in fluorescence-based assays where light is emitted from a fluorophore after excitation. Fluorescent sensors are applied in a wide range of applications, including detection of specific biomarkers, cellular processes monitoring, and DNA analysis. Fluorescence sensors may be used to identify the presence of tumor cells or biomarkers that signify cancer progression in cancer research.
- **Pulse Oximetry**: This is one of the best examples of how pulse oximetry is widely applied in health care to monitor the oxygen saturation level in a patient's blood. These devices function by transmitting light through the skin (typically at the fingertip or earlobe), which measures light absorption by oxygenated and deoxygenated hemoglobin. The data obtained is used for calculating oxygen saturation levels in patients who suffer respiratory illnesses.
- **Near-Infrared Spectroscopy (NIRS)**: NIRS-based photonic sensors are utilized in critical care and surgery environments to monitor tissue oxygenation. NIRS sensors can measure in real time the oxygen levels of tissues such as the brain and muscles without interfering with the underlying tissue. In neonatal care, this technology monitors the oxygen levels in the brains of preterm infants.
- **Photonic Biosensors**: Photonic sensors are used in biosensors to detect pathogens, allergens, or toxins within biological samples. In most cases, photonic sensors can detect bacteria, viruses, and other microorganisms with great sensitivity and specificity. Photonic biosensors have applications in point-of-care diagnostics, such as detecting infectious diseases or monitoring chronic conditions like diabetes.
- **Fiber Optic-Imaging (Endoscopy)**: Photonic sensors of fiber-optic endoscopes can transmit high-resolution images to the physician, allowing for examination of inner organs, biopsy taking, and diagnosis of a wide range of conditions. Most endoscopic procedures are minimally invasive, and they can be applied across many areas, including gastroenterology, pulmonology, and urology.

Photonic sensors have greatly improved the monitoring, diagnosis, and treatment of medical conditions. They are real-time, precise, and non-invasive solutions in the medical field.

5.4 Industrial Process Control

In an industrial environment, photonic sensors play a vital role in monitoring and controlling parameters that ensure effective, safe, and high-quality operation. Photonic sensors monitor temperature, pressure, humidity, gas concentrations, material properties, and many other parameters in real time.

Key Applications in Industrial Process Control

- **Temperature and Pressure Sensing**: Photonic sensors, including fiber-optic temperature sensors, are used to measure temperature in environments where traditional electronic sensors would not work well, such as high-voltage or high-temperature settings. They are used in aerospace, power generation, and manufacturing industries. They are also used in oil and gas industries to monitor pipeline temperatures and pressures.

- **Gas Sensing**: Photonic sensors are broadly applied for industrial process monitoring involving gases in both safe and hostile environments. Real-time gas sensors using lasers (TDLAS) determine concentrations of carbon dioxide, carbon monoxide, methane, CH_4, and volatile organic compounds (VOCs). The photonic sensors play an important role for safety, to fulfill legal and regulatory requirements, as well as to optimize combustion and monitor emission levels.
- **Quality Control and Material Inspection**: Photonic sensors are used in the manufacturing sector to check quality control and material inspection. Laser-induced breakdown spectroscopy (LIBS) is utilized to check on the material's composition for impurities and contamination, whereas spectroscopic sensors can be used for determining the thickness, refractive index, or chemical composition of materials.
- **Automation and Robotics**: Photonic sensors, such as lidar sensors, are applied in industrial robots and automated systems for the purpose of navigation, object detection, collision avoidance, monitoring of equipment performances, inventory tracking, and optimization of production workflows within smart factories.
- **Remote Monitoring and Diagnostics**: Fiber-optic sensors are increasingly used for the remote monitoring of high-criticality infrastructure such as pipelines, electrical grids, and structural elements in large buildings or bridges. They may be able to detect early warnings of damage, pressure variation, or thermal fluctuation, enabling proactive maintenance, rather than costly downtime.

Photonic sensors help improve efficiency, reduce operational costs, enhance quality of products, and ensure safety in industrial processes. Photonic sensors have found their way to be used as indispensable tools in a wide variety of applications: from telecommunications to environmental monitoring, from biomedical diagnostics to industrial process control. Their sensitivity and precision to detect light makes them ideal for a variety of fields [23]. The future of advancements in photonics is unfolding day by day. It will create further applications that expand the field and offer newer solutions to overcome challenges in the respective industries, health care, and environmental management.

6 Challenges and Limitations

These sensors also have a set of challenges and limitations besides the advantages they will bring to the applications, which lie in their high sensitivity, precision, and versatility. Understanding these challenges is crucial for optimizing sensor performance, ensuring accurate measurements, and overcoming barriers to wide acceptance.

6.1 Sensitivity and Calibration Issues

One of the greatest advantages of photonic sensors is their ability to offer high sensitivity in detecting various parameters. However, achieving and then maintaining this level of sensitivity for a long period of time remains a significant problem, especially under practical conditions.

Challenges

- **Signal Noise and Interference**: Greater sensitivity usually correlates with higher levels of noise and interference. Because even minor environmental fluctuations, fluctuations in sensor materials, or electromagnetic interference can drive small changes in the measured signal, errors tend to creep in during the process. For example, in fiber optic sensors, scattering, back reflections, and modal dispersion distort the signal strength so that differences may not become clearly apparent in such a high-noise, interference-rich scenario.
- **Limited Dynamic Range**: A photonic sensor may sometimes have a limited dynamic range, that is, it might measure the input signals over a given range. If the signal is greater than or smaller than the sensor's dynamic range, it could result in faulty measurements or even saturation of the sensor. Some sensors are extremely sensitive to extremely low light intensity but cannot sense high intensity unless calibrated.
- **Temperature Sensitivity**: Several photonic sensors, especially optical sensors and interferometric sensors, may be susceptible to temperature effects. Changes in temperature can affect refractive indices or cause mechanical stresses in the sensor, and such effects usually manifest as a drift in measurement readings. It requires compensation mechanisms and hence complicates the sensor design further while increasing calibration precision.
- **Calibration Drift**: Photonic sensors would undergo calibration drift arising from aging within the sensor constituent, environmental, and mechanical actions. Calibration drift typically causes the decline in sensor performance accuracy over periods of time with recalibration and thus, particularly difficult for those distributed sensors aimed at scanning large-sized areas or wide structures where partial or total remounting through the length of a sensor is entailed for adjustment.

Solutions

- Applying advanced signal processing algorithms to denoise and achieve better signal-to-noise clarity.
- Thermal and environmental compensation techniques, such as temperature-insensitive fibers or embedded temperature sensors, to counteract temperature effects.
- Calibration routines or even automated calibration systems like self-calibrating sensors may be installed to combat drift with time.

6.2 Environmental Factors

The environment has a large impact on photonic sensors performance, especially for outdoor or rugged environments. Variation in environmental factors, including temperature, humidity, electromagnetic interference, and chemical exposure, may lead to inaccurate photonic sensors performance and lifespan.

Challenges

- **Environmental Interference**: Photonic sensors are susceptible to external environmental influences such as electromagnetic fields, humidity, and vibrations. Electromagnetic interference (EMI) will have a specific impact on industrial sensors, since the machinery or other electrical equipment will generate noise that can interfere with the optical signal of the sensor.
- **Contamination and Physical Damage**: Optical sensors, particularly fiber-optic sensors, may be susceptible to contamination with dust, dirt, and moisture. Surface contamination of the sensor will alter its optical properties and therefore reduce the intensity or distortion of the signal transmitted. The physical damage to a fiber, which can result from a break or a bend, causes loss or failure of the signal.
- **Temperature Extremes**: Extreme temperatures either high or low cause the materials in a photonic sensor to expand or contract. It may result in mechanical stress and deformation or alignment of the components and thus lead to drift or even failure of the sensor. Even the long-term reliability of a photonic sensor can be compromised by thermal cycling, which exposes the device repeatedly to temperature variation.
- **Chemical Sensitivity**: In some environments, photonic sensors may be exposed to corrosive or harsh chemicals that may degrade the optical fibers or coatings used in sensors. Chemical contamination may degrade sensor components and reduce their performance.

Solutions

- Protective coatings and encapsulation can be used on the sensors to protect them against contamination, chemicals, and environmental damage.
- Ruggedized designs and dedicated housings designed for industrial or outdoor applications might also help sensor systems resist environmental conditions.
- Environmental compensation techniques can be utilized, which includes multi-sensor systems and advanced signal processing techniques that allow for filtering effects of environmental interference.
- Self-healing materials, durable fibers tailored for harsh environmental conditions, contribute to photonic sensors' ability to be both strong and more durable.

6.3 Integration with Existing Systems

The photonic sensors are going to be extensively used in practical applications; they need to be incorporated integrally with existing systems and infrastructures. Many times, this integration is difficult, especially in legacy systems or even in requirements-complex environments.

Challenges

- **Compatibility with Legacy Systems**: Most industries and research environments rely on legacy systems that were designed without the consideration of modern photonic sensors. The integration of photonic sensors into these existing systems may require changes in hardware, software, or processes, which can be costly and time-consuming. In some cases, older systems may not support the data formats or communication protocols used by photonic sensors.
- **Data Processing and Analysis**: Photonic sensors often produce vast data, especially in distributed sensing systems. The processing and analysis of such data in real-time can be computationally demanding, necessitating advanced algorithms and processing power. Proper software and hardware infrastructure can be difficult without which it can be a challenge to handle, store, and analyze the vastness of such data produced by these sensors.
- **Signal Transmission and Power Supply**: In some applications, photonic sensors, especially fiber-optic sensors, may need long transmission lines, and the signal might need to be amplified or conditioned for accuracy. The power supply for the sensors and reliable data transmission over long distances can be difficult, especially in remote or inaccessible locations. This is especially true in remote sensing applications where the power source is limited.
- **Standardization and Interoperability**: In contrast, the complete lack of standards in photonic sensors can prevent interoperability across various systems and industries. Various sensor manufacturers may develop different proprietary interfaces or protocols, which can be challenging to integrate into a single system. Compatibility of photonic sensors with different systems and devices is, therefore, extremely essential for its widespread adoption.

Solutions

- Modular designs for sensor development and standard communication protocols to support photonic sensor integration into infrastructure.
- Use edge computing and cloud-based platforms to process data locally or remotely away from the central system, allowing real-time monitoring and analysis with minimal load on the central system.
- Universal data formats and open-source software platforms to ensure interoperability across different sensor types and systems.
- Hybrid sensor systems that embed photonic sensors with other kinds of sensors, such as electronic sensors, to better overcome some of the above-mentioned limitations that exist in signal transmission and power supply.

Despite the marked improvement in the technology of the photonic sensors, several important issues still occur: sensitivity, calibration, sensitivity to environmental influence, and issues with integration of these sensors within already existing systems [24]. To unlock potential of photonic sensors in almost any application of them, each of these obstacles has to be overcome. Overcoming these issues will be innovative material, signal processing, and sensor design combined with new standards and integration frameworks. Reliability, robustness, and usability of photonic sensors will, therefore, benefit from these issues [25]. As photonic sensors take over the functions of industries across the spectrum such as telecommunications and healthcare to environmental monitoring and industrial process control.

7 Future Directions in Photonic Sensor Technology

Photonic sensor technology is quickly evolving because of the continued research, development of new materials in materials science, and advances in optics and nanotechnology. With this in mind, photonic sensors will increasingly be utilized for a vast variety of applications from healthcare and environmental monitoring to industrial automation and telecommunication [26].

7.1 Emerging Trends and Innovations

Emerging trends in the photonic sensing technology are going to bring key advancements in sensors' capabilities and uses. Quantum photonic is found to improve sensitivity by using intrinsic quantum properties-superposition or entanglement-for ultra-sensitivity in measures of gravitational wave detection and certain medical imaging practices. Metamaterials and nano photonics are also miniaturizing biosensors to track trace chemicals as well as biochemical markers, developing biosensing with environmental monitoring more effectively. Integrated photonic circuits are transforming sensor design through efficiency and a reduction in the cost of this technology by collapsing traditional components into compact chips [27]. Miniaturization is also leading toward portable, wearable sensors for both health monitoring and environmental sensing—flexible, stretchable sensors to be embedded into clothing, skin patches. New materials such as graphene and perovskites are improving the performance of sensors, providing even greater sensitivity and efficiency for the applications ranging from biosensing to solar energy harvesting.

7.2 Potential Applications in New Fields

As photonic sensors become more advanced, their applications span various fields with significant benefits to industry and daily life. In smart cities, these sensors are used for air quality monitoring, traffic management, and structural integrity monitoring to maximize energy use while enhancing building safety. In space, they are integrated into lidar systems to create maps of planetary surfaces and find gravitational waves for the advancement of astrophysics. Photonic sensors allow the real-time monitoring of crop health and soil conditions in agriculture, thereby enhancing food safety and farming efficiency [28]. They are also very important in environmental remediation because they enable rapid pollutant detection and thus contribute to pollution control. Additionally, they help monitor ocean health in real time.

7.3 The Role of AI and Machine Learning

The integration of AI and ML with photonic sensors is transforming the way data analysis, decision-making, and optimization of the systems are made. Photonic sensors generate tremendous data, which can be analyzed through AI and ML in the most efficient way possible. For biomedical applications, AI can diagnose at an early stage of disease detection by monitoring trends with wearable sensors, while environmental monitoring uses sensor data to monitor the level of pollution and predict a natural disaster. AI and ML also support real-time decision-making, like in autonomous vehicles, where sensors and AI collaborate to navigate and detect obstacles. In industries, photonic sensors, powered by AI, optimize production, predict equipment failure, and improve system performance. AI is also used for predictive maintenance and smart diagnostics, enabling the anticipation of equipment failures in industrial settings and early detection of diseases in healthcare [29]. The future prospects of photonic sensor technology look bright, based on future developments in quantum photonics, nanotechnology, and integrated circuits, the scenarios set for these sensors will expand. Photonic sensors will be applied to increasingly diverse fields such as agriculture, environmental monitoring, space exploration, and smart city technological development as the new materials and the techniques of sensing continue to evolve. This would also integrate AI and machine learning, making it possible for the full unlocking of photonic sensors to analyze advanced data, provide real-time decisions, and implement predictive maintenance [30]. These will all become a crucial part of solutions that foster innovation and respond to global challenges across different sectors of the industry as these technologies develop.

8 Conclusion

Photonic sensors, a field which has witnessed some phenomenal growth and innovation in the past few decades, continue to advance rapidly today. Such sensors utilize the distinct properties of light for the purpose of detection, measurement, and analysis of different parameters that are fundamentally physical, chemical, and biological in nature and play a key role in multiple industries. From healthcare to environmental monitoring, telecommunications and industrial automation applications, photonic sensors offer ultimate sensitivity, resolution and versatility. Photonic sensors are indeed transformative; they bring many applications that transform the industries and stimulate scientific discovery. As the technology advances, so will photonic sensors become increasingly an integral part of our lives in the medical sciences, environmental studies, space travel, and even industrial automation. While challenges remain, especially in terms of sensor integration, environmental interference, and calibration, these issues are actively being researched and addressed by the ongoing research work and technological development. Emerging trends such as quantum sensing, nano photonics, and AI-driven analysis are making photonic sensors look very promising in the near future. In conclusion, photonic sensors are one of the most promising areas of high-tech advancement in modern science. There is no limit to their potential in augmenting our ability to measure, monitor, and analyze the world around us, making them essential tools in the continuation of scientific exploration, technological innovation, and environmental sustainability.

8.1 Summary of Key Concepts

This chapter covers the fundamental principles of photonic sensors to their applications and future developments. Photonic sensors measure physical phenomena based on light's wave-particle properties. Understanding how light interacts with matter is crucial in designing accurate sensors. These sensors detect changes in the intensity, phase, or frequency of light as it travels through different media, relying on sensitivity, resolution, and signal processing for accurate measurements. There exist fiber-optic sensors (for long-distance measurements), interferometric sensors (to be used in applications requiring high-precision measurements), imaging sensors (to provide visual data), among others, using photonic sensors with advanced materials. The applications of photonic sensors include telecommunications, environmental monitoring, health care, and industrial control with advantages of high sensitivity and non-invasiveness, although challenges in the form of calibration, environmental influences, and system integration need to be addressed. Future technologies in quantum photonics, nano photonics, and AI will boost sensing applications further while being used in smart cities, space research, and food safety.

8.2 The Importance of Photonic Sensors in Modern Science

Photonic sensors are revolutionizing the way of measuring and even interacting with the world. Their precise, non-invasive, and high-sensitivity measurement makes them ideal in a wide scope of applications; from molecular biosensing to monitoring large-scale structural changes. Moreover, they are suitable in fields such as healthcare and environmental monitoring because the data they would collect will be non-intrusive. They are advancing research on areas like quantum physics and material science, thereby fostering breakthroughs into technologies like quantum sensing and nanotechnology. Critical to photonic sensors is also communication, health sciences, and time-sensitive diagnostics of technological innovation in these areas. They also have a direct consequence on the rate of pollution or climate change-a factor aiding human sustainability. Moreover, as the applications of digital technologies, including IoT, AI, and edge computing, increase penetration in society, photonic sensors will bring industries like manufacturing, smart cities, and transportation towards more intelligent, efficient systems.

References

1. M. De, T.K. Gangopadhyay, V.K. Singh, Prospects of photonic crystal fiber as physical sensor: an overview. Sensors **19**(3), 464 (2019). https://doi.org/10.3390/s19030464
2. R.V. Nair, R. Vijaya, Photonic crystal sensors: an overview. Prog. Quantum Electron. **34**(3), 89–134 (2010). https://doi.org/10.1016/j.pquantelec.2010.01.001
3. K. Oh, Optoelectronic and photonic sensors. Front. Sens. **3** (2022). https://doi.org/10.3389/fsens.2022.921795
4. A. Massaro, Photonic crystal waveguides and bio-sensors, in *Photonic Crystals—Introduction, Applications and Theory* (2012). https://doi.org/10.5772/32052
5. N. Tansu, Photonics—advances in fundamental sciences and engineering technologies of light. Photonics **1**(1), 1–8 (2014). https://doi.org/10.3390/photonics1010001
6. M. Nisoli, *Semiconductor Photonics. Principles and Applications* (Società Editrice Esculapio, 2016). https://doi.org/10.15651/978-88-938-5002-5
7. J.-M. Liu, *Principles of Photonics* (2016). https://doi.org/10.1017/cbo9781316687109
8. W. John, F. Nunes, Historical synopsis of light–matter interaction, in *Light–Matter Interaction* (2012), pp. 1–6. https://doi.org/10.1093/acprof:oso/9780198567653.003.0001
9. S. Bi, Discuss wave-particle duality of light, in *Second International Conference on Photonics and Optical Engineering*, vol. 10256 (2017), p. 1025650. https://doi.org/10.1117/12.2260702
10. G. Nemova, Wave-particle duality, in *Field Guide to Light-Matter Interaction* (2022). https://doi.org/10.1117/3.2611513.ch17
11. S. Donati, *Photonic Instrumentation* (2023). https://doi.org/10.1201/9781003391357
12. S.A. Schulz, L. O'Faolain, I. Rey, T.F. Krauss, Understanding propagation loss in slow light waveguides. Adv. Photonics, SLTuB1 (2011). https://doi.org/10.1364/sl.2011.sltub1
13. K. Wada, Light emission and detection for Si photonics, in *2008 IEEE Photonics Global*, Singapore (2008), pp. 1–3. https://doi.org/10.1109/ipgc.2008.4781298
14. J. Miyazaki, T. Kobayahsi, Photothermal microscopy for high sensitivity and high-resolution absorption contrast imaging of biological tissues. Photonics **4**(2), 32 (2017). https://doi.org/10.3390/photonics4020032
15. A. Willner, All-optical signal processing using optical nonlinearities. Adv. Photonics, SPMC1 (2011). https://doi.org/10.1364/sppcom.2011.spmc1

16. M. Yang, J. Dai, Fiber optic hydrogen sensors: a review. Photonic Sens. **4**(4), 300–324 (2014). https://doi.org/10.1007/s13320-014-0215-y
17. Y. Rao, Study on fiber-optic low-coherence interferometric and fiber Bragg grating sensors. Photonic Sens. **1**(4), 382–400 (2011). https://doi.org/10.1007/s13320-011-0042-3
18. Y. Zhuo, B. Cunningham, Label-free biosensor imaging on photonic crystal surfaces. Sensors **15**(9), 21613–21635 (2015). https://doi.org/10.3390/s150921613
19. J. Ma, Y. Yang, Novel designs for quartz tuning fork temperature sensor, in *Proceedings of 2011 6th International Forum on Strategic Technology* (2011), pp. 856–859. https://doi.org/10.1109/ifost.2011.6021155
20. M. Michael, B.E. Caroline, All-optical logic gate using photonic crystals for ultra-fast telecommunication applications, in *Photonic Materials: Recent Advances and Emerging Applications* (2023), pp. 21–42. https://doi.org/10.2174/9789815049756123010005
21. W. Yantasee, Y. Lin, G.E. Fryxell, Electrochemical sensors based on nanomaterials for environmental monitoring, in *Environmental Applications of Nanomaterials* (2007), pp. 401–437. https://doi.org/10.1142/9781860948572_0015
22. A.M. Upadhyaya, P. Sharan, Photonic MEMS sensor for biomedical applications. Comput. Health Inform. Biomed. Appl. 45–66 (2023). https://doi.org/10.1201/9781003331681-3
23. I. Grobelna, Intelligent industrial process control systems. Sensors **23**(15), 6838 (2023). https://doi.org/10.3390/s23156838
24. K. Borzycki, J. Kobelke, P. Mergo, K. Schuster, Challenges in characterization of photonic crystal fibers, in *Optical Sensors 2011; and Photonic Crystal Fibers V, 8073*, 80732M (2011). https://doi.org/10.1117/12.886964
25. R. Khusainov, D. Azzi, I. Achumba, S. Bersch, Real-time human ambulation, activity, and physiological monitoring: taxonomy of issues, techniques, applications, challenges and limitations. Sensors **13**(10), 12852–12902 (2013). https://doi.org/10.3390/s131012852
26. R. Bogue, Quantum dots: a bright future for photonic nano sensors. Sens. Rev. **30**(4), 279–284 (2010). https://doi.org/10.1108/02602281011072143
27. R.I. Epstein, Photonic and electronic cooling, in *2009 International Conference on Emerging Trends in Electronic and Photonic Devices & Systems* (2009), pp. 1–4. https://doi.org/10.1109/electro.2009.5441186
28. L. Daoliang, C. Yingyi, Selected peer-reviewed articles from the smart sensors in modern agriculture. Sens. Lett. **10**(1), 122–123 (2012). https://doi.org/10.1166/sl.2012.1820
29. R. Liang, Y. Sun, An intelligent sensor mobile phone assisting system using AI and machine learning, in *Artificial Intelligence Trends* (2022), pp. 167–178. https://doi.org/10.5121/csit.2022.121004
30. A. Ghosh, A. Pal, N.R. Das, An approach to design photonic crystal gas sensor using machine learning. Optik **208**, 163997 (2020). https://doi.org/10.1016/j.ijleo.2019.163997

Fundamentals of Generative AI

Krishna Kumar

Abstract Generative AI (GenAI) uses machine learning (ML) algorithms to create seemingly new content that can be in the shape of text, images, audio, video, or programming code. GenAI is a revolutionary technology within the field of artificial intelligence that empowers machines to world of GenAI, starting from its definition, historical timeline, and game-changing applications in different domains. First, it goes into the fundamental principles of ML, probabilistic modeling, and the difference between generative and discriminative models. Looking through some of the leading generative models—Variational Autoencoders (VAEs), Generative Adversarial Networks (GANs), diffusion models, and autoregressive models—reveals the variety of techniques that have been harnessed to fuel the advances. The generation of realistic examples by generative models is one of the major achievements in the field of AI (Dash et al. in IEEE Access 12:18330–18357, 2024 [1]; https://www.sciencedirect.com/science/article/pii/S20952635240 0147X#bbib41 [2]; https://medium.com/@nnabwani/full-fine-tuning-peft-prompt-engineering-or-rag-69689ba58a52 [3]). One type of GenAI is found in large language models (LLMs), which use a deep neural network architecture known as the Transformer to capture patterns and information from web-scale training data in a self-supervised manner. These models have achieved human-level quality in generating text, answering questions, and language translation. These abilities have led to a surge of interest in the field and new services. As GenAI continues to evolve and mature, such models will likely become increasingly integrated into people's daily lives (Yadav et al. in Int J Multidiscip Res, 2024 [4]). The chapter goes on to highlight applications of GenAI across diverse sectors—from content creation in media and entertainment to data augmentation and innovation in scientific research. We discuss the challenges (e.g., ethical issues, computational requirements, evaluation difficulties) while emphasizing the strengths and weaknesses of the technology. The market for GenAI models, including LLMs, is growing rapidly with over 100 commercial and open-source offerings with different licensing models, leading to hype but also indicating quick obsolescence of current implementations. Legal and commercial

K. Kumar (✉)
Dell Technologies, Bengaluru, India
e-mail: krishna.kumar8@dell.com

Y. Vasimalla and S. Kumar (eds.), *Generative AI for Photonic Sensing*, Progress in Optical Science and Photonics 36, https://doi.org/10.1007/978-981-95-1561-5_2

risks can be mitigated by selecting the right models and ensuring the right contractual terms and their application (Kılınç and Keçecioğlu in Acad Platform J Eng Smart Syst, 2024 [5]).

Keywords Generative AI (GenAI) · Large language models (LLMs) · Machine learning foundations · Generative models (VAEs, GANs, Diffusion, Transformers) · Applications and use cases · Ethics and challenges in AI

1 Introduction to Generative AI

1.1 What Is Generative AI?

Generative AI (GenAI) is a subset of artificial intelligence that leverages machine learning (ML) techniques to generate new content. This could be in the form of text, images, music, voice, or even new models. The generated output is typically novel and created from random or user-provided data. GenAI (generative AI to be more specific) is a class of artificial intelligence (AI) algorithms that produces new (based on images, text, audio, and more) outputs based on the data on which they have been trained [6]. Generative Adversarial Networks (GANs) were by far the most popular and effective method used in creating the earlier models, such as StyleGAN, which are used for image generation, and TextGAN for natural language generation to generate new content. The current three most known representations of GAI are LLM-based chat and content generation solutions, ChatGPT, Bing, and Google Bard, all of which can engage in human-like conversation and generate new data on the basis of the input data and additional context, if necessary [7].

Unlike traditional AI and ML, which are only trained on a limited dataset to determine mapping abstractions to make pooling inferences, GAI platforms do not just learn from the input dataset but also depend on the relationship of data mixing, modifying, and generating novel content to augment their learning process. As consumer interests parallel, GAIs have recently gained traction as a powerful tool that can be used for various purposes. A prominent example of this is the application of Reinforcement Learning (RL) as a technique for control of robotics, where agents learn from exploration of their actions to find the best way to behave and take action in real-time. Robot becoming humanoid in new motions and movement skills (as GAI interacts with RL) where a human does not need to intervene or provision knowledge. GenAI is made up of two neural networks, a generator to generate new data and a discriminator to evaluate it [8, 9].

Figure 1 shows a high-level GAI architecture, and the paradigm has three large blocks. The "grey" block represents the GAI models, which will be selected and optimized based on the use cases, data, and KPIs. Two examples of GAI models with a large amount of success in real GAI-enabled applications like chatbots, code generation, search engines, and cybersecurity are pretrained LLMs and GANs. The

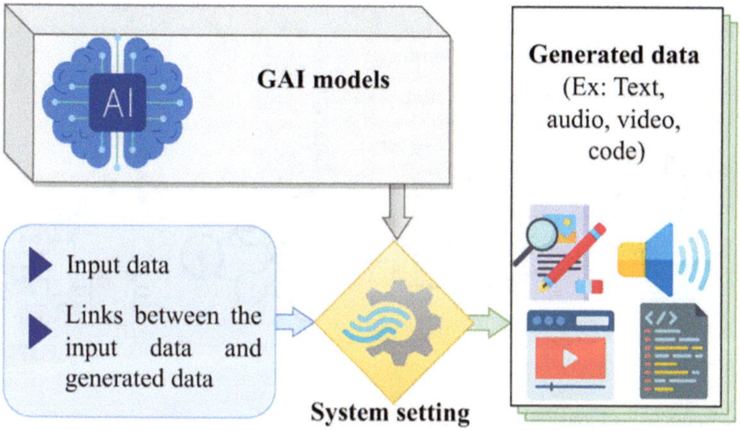

Fig. 1 Illustration of a high-level GAI architecture

"orange" block represents the system settings provided as the input to the GAI models, which help generate the output data. Lastly, the "green" block is for any data produced once they are created by or fed into GAI models based on a prompt, covering data types such as text, audio, and video, as well as code [10, 11].

The three simple steps that describe how using a GAN approach works are shown in Fig. 1.

1. GAN: A neural network that is modeled like how a human brain processes information. Consists of various information or media files like images, text, data, sounds, etc. and forms the basis of the AI.
2. The prompt is a vital input field for the GenAI model to understand the ask/intent.
3. Output: AI uses its neural network to generate new examples like the ones it has been trained on.

1.2 Components of Generative AI

Artificial Intelligence (AI)

Artificial intelligence (AI) is the capability of a machine to perform operations that typically require human intelligence, such as perception, judgment, learning, and problem-solving. The applications of AI are widespread, covering natural language processing, computer vision, robotics, and even gaming [12–14] (Fig. 2).

Use Case: An example of AI working would be chatbots in customer service. These bots utilize NLP (Natural Language Processing) to comprehend customer inquiries and respond accordingly in a more effective and customized manner [15–17].

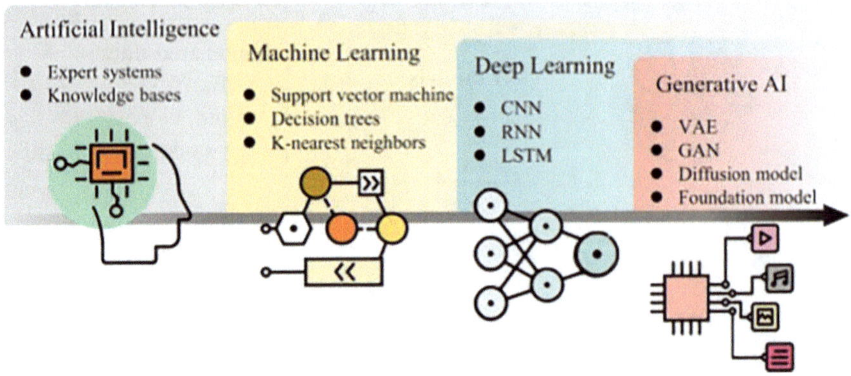

Fig. 2 Components of generative AI

1.3 *The Development and History of Generative Models*

The Development of Generative AI

Chatbots were the first application of GenAI in the 1960s. In this section, we will explore the timeline of significant events that helped GenAI become what it is today.

1966: MIT professor Joseph Weizenbaum creates Eliza, the world's first chatbot that mimics conversations with a psychotherapist. Eliza's ability to respond to users using pattern matching and simple language processing techniques was an early breakthrough in natural language understanding and human–computer interaction.

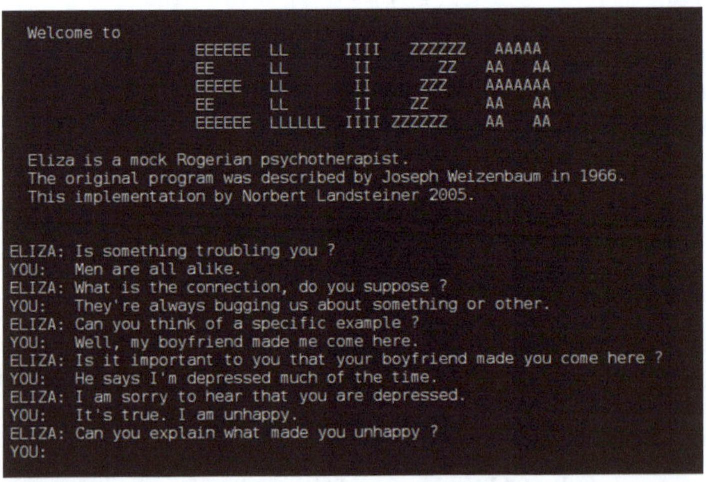

Eliza Conversation

1968: Terry Winograd at MIT created SHRDLU, the first program capable of natural language understanding (in a limited domain). SHRDLU allowed users to manipulate objects through commands spoken in English. The project's success demonstrated how well artificial intelligence can understand and follow a set of specific, detailed directions in real-world scenarios.

1985: Bayesian networks became a popular method in artificial intelligence for representing uncertainties in complex systems and for reasoning about cause-and-effect relationships. Bayesian networks use directed acyclic graphs to represent probabilistic interactions between variables and allow reasoning under uncertainty, being able to perform diagnostics, predictions, and preferred actions.

1989: By introducing Convolutional Neural Networks (CNNs), Yoshua Bengio, Yann LeCun, and Patrick Haffner transformed image recognition. Due to shared weights and convolutions, visual data can also be processed more accurately and efficiently with CNNs than it can with more traditional methods. This advancement formed the foundation for computer vision systems and deep learning applications.

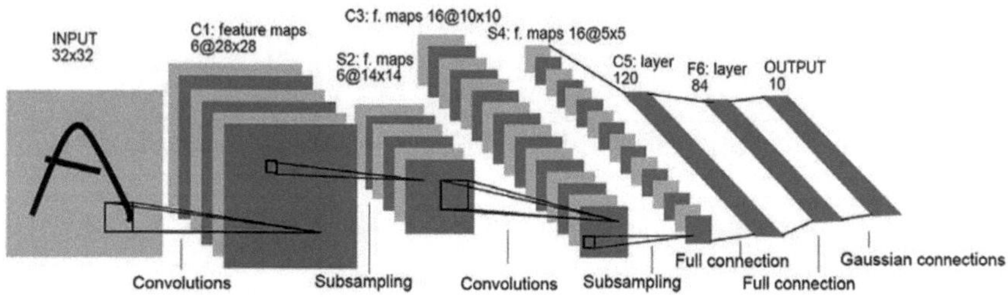

Architecture

2000: Yoshua Bengio et al. publish a paper on neural probabilistic language models, a neural network-based language model, in 2000. Capturing contextual dependencies and learning distributed representations of words improve natural language processing tasks like speech recognition, machine translation, and text generation.

2011: A big moment in consumer AI technology comes with Apple Siri, a voice-activated virtual assistant. With Siri, users can interact with their devices through voice commands, paving the way for a more personalized and intuitive user experience.

2013: Tomas Mikolov develops Word2Vec, a revolutionary method for word embeddings in natural language processing. Instead of one-hot encoding you, Word2Vec uses neural networks to create continuous vectors for each word, capturing the semantic relationships and contextual similarities between words. This leads to a better representation of words, which helps to improve tasks such as sentiment analysis, named entity segregation, document clustering, etc.

2014: Ian Goodfellow and collaborators develop GANs, a new framework for generative modeling. GANs are made up of two neural networks, a generator and a discriminator, that are trained simultaneously in a game-like setting. By employing this technique, it is possible to generate realistic synthetic data, which can be used in applications ranging from image synthesis and style transfer to data augmentation.

2017: "Attention Is All You Need" Vaswani et al. introduce the new structure called transformers, game-changing for the branch of natural language processing. Transformers are, by far, the most used architecture today for many different sequence tasks like machine translation, text summarization, and language understanding, thanks to their ability to use self-attention mechanisms to capture long-range dependencies. Many modern NLP models, like BERT and GPT, are based on the transformer model [18].

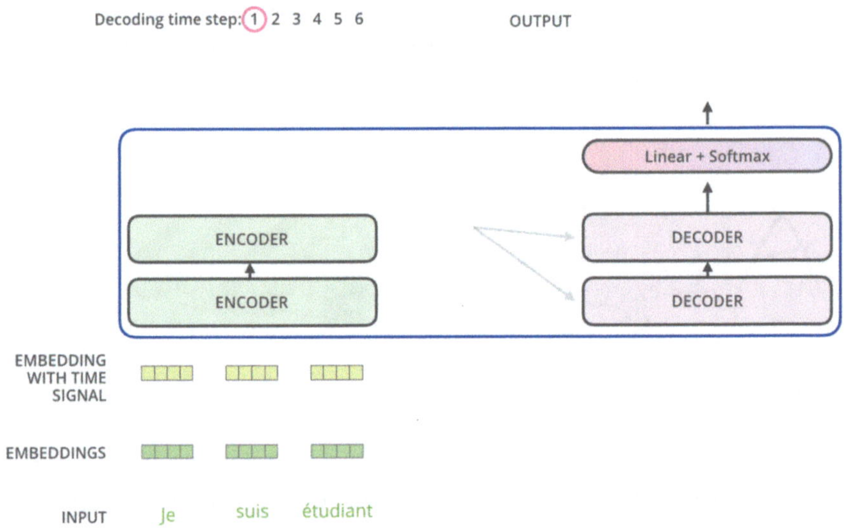

Transformers

2018: BERT was developed by Google AI researchers to get better at understanding natural language. With BERT, bidirectional training and transformer architecture capture context both left and right of a word and yield significant improvements in tasks like answering questions, sentiment analysis, and texting classification. With BERT's pretraining strategy and contextualized embeddings deterministically, it establishes a new baseline for language representation learning.

2021: OpenAI announces DALL-E, an AI model that generates images from text descriptions. DALL-E is capable of creating diverse and imaginative visual creations from textual descriptions by employing transform architecture with extensive image-text pairs.

2022: New model GPT-3.5 marks a milestone in LLMs. It displays the sophistication of natural language processing, generation, and conversation, showing the advancements of neural network-based language models and their use in chatbots, virtual personal assistants, and text-based AI applications.

2023: GPT-4 arrives, demonstrating further progress in GenAI. The new model has better language understanding, context retention, and text generation than previous models.

2024: If 2024 becomes known for anything, it will be the year GenAI came to the forefront—with advances like Stable Diffusion 3, Vlogger, Claude 3, Devin AI, and even ChatGPT-5 launching mid-year.

Continued Milestones in GenAI

GenAI, a branch of artificial intelligence responsible for producing original content, has experienced a series of significant breakthroughs that have advanced its case for the next chapter in its timeline.

Here are key moments that have transformed GenAI Vista:

- **WaveNet (2016)**: DeepMind's WaveNet marked a significant advancement in generative models for audio. WaveNet could generate realistic-sounding human speech, which opened doors for more human-like AI assistants and highly accurate text-to-speech synthesis.
- **Progressive GANs (2017)**: Progressive GANs, developed by NVIDIA, were a milestone in producing high-resolution, photo-realistic images. These GANs were able to generate images with unprecedented detail and clarity by progressively adding layers during the training process.
- **GPT-2 and GPT-3 (2019, 2020)**: OpenAI's generative pre-trained transformer (GPT) models were a dramatically new approach to GenAI in text generation. They were shown to produce coherent and contextually relevant sentences, allowing them to be used in various applications, from writing aids to the chatbot population.
- **DALL-E (2022)**: OpenAI released DALL-E to the public. DALL-E is a deep learning model recently developed that can create digital images from natural language prompts.
- **ChatGPT (2022)**: OpenAI launched ChatGPT, a conversational-based chatbot built upon GPT, and the platform hit one million users in five days.
- **GPT-4 (2023)**: The latest version is said to be more accurate and better at reasoning. Optional access to GPT-4 within the chatbot is now available for Premium ChatGPT users.
- **GPT-5 (2025)**: It will be a router model that intelligently delegates the task to the appropriate underlying model. about GPT-5, indicating it will not be a standalone model but a system integrating GPT-series and O-series models, such as O3.

GenAI was born at each of these milestones and continuously improved as challenges around computational power, quality of data, and training stability were overcome [19].

1.4 How Does Generative AI Work?

GenAI begins with prompts, entered as text, images, videos, designs, or musical notes, and employs diverse algorithms to produce essays, problem solutions, or realistic fakes in response. In the beginning, generative AI was complicated, requiring API submissions and specialized tools within languages like Python. However, nuclear user experience is changing; plain language requests are now accepted.

Generative AI pioneers are improving user interactions that adjust based on tweaks to style and tone. These models use different AI algorithms for content representation and process. Natural language processing in text generation converts the words and characters into sentences, parts of speech, entities, and actions and then uses different coding mechanisms to represent the vectors. Likewise, images become visual representations, likewise given by vectors. It is absolutely important to realize that these models may encode the biases (and other issues) present in the training data. After creating a world model, the developers select GANs and VAEs (Variational Autoencoders) to do specific tasks (i.e., generating realistic human faces or synthetic data for AI training). GANs and VAEs are the models behind generative AI that generate new data. These datasets learn from a large amount of input data and generate output that resembles the input either in style or structure. Generative AI models are trained on a large dataset, analyzing the data and patterns within it, and using that information to generate new content. You can divide the process into a few steps [20–22] (Fig. 3).

1. **Training Data**

The first step is to initially collect and prepare a huge dataset that acts as the base for the GenAI model. This data is vital because the model uses it to learn from patterns, structures, relationships, and so on. The nature and quality of the training data have a huge impact on the performance of the model and the variety of outputs it can generate.

2. **Data Preprocessing: Neural Networks and Training Process**

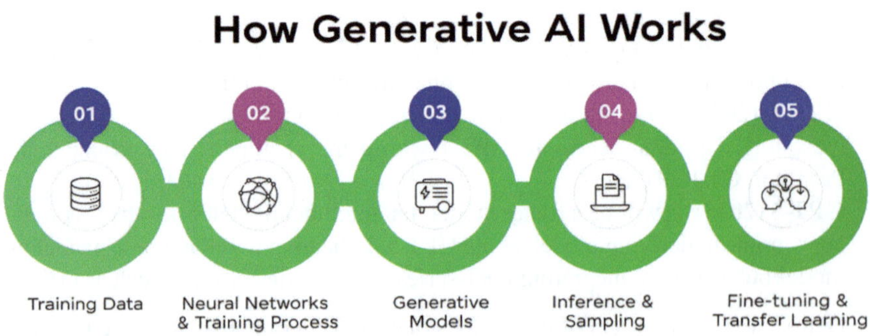

Fig. 3 Generative AI work flow

It uses some ML algorithms, primarily neural net models (deep learning). These networks are composed of layers of interconnected nodes (neurons) that function similar to the human brain. During training, it is fed data and learns to recognize patterns via backpropagation, effectively adjusting the weights of these connections in such a way that its predictions are as close as possible to the actual data.

3. **Generative Models**

The model can produce new content once it is trained. Examples of generative models are GANs, which utilize a generator and discriminator in a two-part structure to synthesize realistic data. VAEs—which encode something and decode it to build new, similar things—and Transformer-based models such as GPT that shine at understanding sequential data for text generation.

4. **Inference and Sampling**

Once the model is done training, it can begin the inference stage, where it uses the learned patterns and relationships to create new content. This usually requires the model to be given a prompt and generates new data. Generative processes frequently involve probabilistic sampling, and for instance, a scalar "temperature" parameter can be used to modulate the creativity and variability of the outputs.

5. **Fine-Tuning and Transfer Learning**

Finally, most GenAI systems are fine-tuned, which means they are initially trained on large, general datasets and then refined against specific datasets more suited to the desired applications. This approach, referred to as transfer learning, allows the model to generalize well to specific tasks or domains. With all this said, these four steps are the pillars of most GenAI systems—they usually follow these steps to create something new based on whatever data they have been trained on [23].

2 Core Concepts and Principles

GenAI is a game-changing technology that generates new content in multiple domains, including the use of text, images, and music, supported by user inputs and advanced algorithms. GenAI models are different from traditional AI models because they learn to synthesize new data that resembles the training data while avoiding duplication. There are some quintessential concepts behind this capability and principles that are needed to get acquainted with in order to better understand this potential and application. The next sections discuss these foundational aspects of generative AI.

Core Concepts

- **Generative Models**: Important models consist of GANs, Variational Autoencoders (VAEs), and autoregressive models. They learn the data distribution and generate samples from this data distribution.

- **Probability Distributions and Sampling**: GenAI builds on the basis of knowing and sampling from probability distributions to generate new data points that mimic the statistical properties of the training data.
- **Neural Networks and Machine Learning**: The underlying architecture of GenAI relies on neural networks and ML models that are crucial for processing and generating complex data structures.

Principles and Methodologies

- **Reinforcement Learning from Human Feedback (RLHF)**: By combining human feedback with generative models, RLHF tailors them to individual users while adding abilities needed for advanced tasks.
- **Natural Language Processing (NLP)**: NLP is the main driving force for GenAI, specifically in any application for generating text and understanding the language spoken by humans.

Applications and Impact

- **Wide Range of Use Cases**: GenAI is used in a broad spectrum of domains, from content generation to healthcare, gaming, and beyond, providing improved creativity and personalization.
- **Business Integration**: Businesses are advised to integrate GenAI into their business strategies to increase productivity and obtain a competitive edge.

Despite the promise of GenAI, there are also hurdles like data integrity, computational overhead, and ethical issues. They need to be considered if we are to realize the potential of GenAI and use their power responsibly across sectors [19].

2.1 Machine Learning Foundations

GenAI is based on top of ML, utilizing the same concepts and algorithms to adapt models to learn patterns, structures, and relationships in the data. This section of the book provides the fundamental ideas, methods, and parts that are the foundation of GenAI systems.

Types of Machine Learning: Supervised, Unsupervised, and Reinforcement

Generally, from a ML perspective, learning problems fall into three categories:

Supervised Learning

Think of a teacher in a classroom, showing her students images of cats and dogs, labeling each one. Supervised learning works in a similar way. It utilizes labeled data, with known and desired outputs associated with each input. The algorithm scans through this data to understand how input relates to output, enabling it to predict output for new, never-before-seen data. A few real-life examples are:

- Spam Filtering. Filtering spam emails using features learned through labeled examples.
- Image Recognition. Frame Classification: Images classified as containing some object (cats, dogs, etc.) as defined by labeled training data.
- Linear Regression. Predict house prices based on the size and area of the house.

Unsupervised Learning

So unsupervised learning is discovery. Like a child discovering the parts and pieces of a new toy box, but at their own pace. Unsupervised learning relates to the same idea, where data is unlabeled, and the algorithm finds patterns and relationships without labels and without having a deterministic target. This is helpful for tasks such as:

- Customer Segmentation. Segmentation groups customers with similar characteristics for targeted marketing campaigns.
- Anomaly Detection. Detecting anomalous data patterns, such as fraudulent payments.
- Dimensionality Reduction. In unsupervised learning, dimensionality reduction techniques can be used to compress high-dimensional data into a lower-dimensional space for efficient analysis.

Reinforcement Learning

These are agents that learn to determine the optimal action to take in each state by interacting with the environment in which it is based. GenAI Relevance: Used for fine-tuning generative models (e.g., via helping text generation with reinforcement learning).

Key Differences

1. **Data Availability**: Applying supervised learning (labeled data) Unsupervised: It works on data where labels are not present. Reinforcement: Learns by interacting with an environment.
2. **Learning Type**: Supervised: Where it learns from known labels and attempts to predict for future labels. Unsupervised: When it recognizes the patterns or structures in the data without any supervision. For your own sake to reinforce: It learns how to make a series of decisions to maximize the rewards.
3. **Examples**: Supervised: Image classification, spam testing, language translation. Unsupervised: Clustering customer preferences, anomaly detection. Examples Game playing (e.g., chess, Go) Autonomous driving.

2.2 Understanding Generative Versus Discriminative Models

Generative and discriminative models are two fundamental approaches in ML each with distinct objectives and methodologies. Generative models aim to learn the underlying distribution of data to generate new samples, while discriminative models focus on distinguishing between different classes of data. Understanding the differences and applications of these models is crucial for leveraging their strengths in various domains.

Informally:

- **Generative** models can generate new data instances.
- **Discriminative** models discriminate between different kinds of data instances.

A generative model could generate new photos of animals that look like real animals, while a discriminative model could tell a dog from a cat. GANs are just one kind of generative model.

More formally, given a set of data instances X and a set of labels Y:

- **Generative** models capture the joint probability p(X, Y), or just p(X) if there are no labels.
- **Discriminative** models capture the conditional probability p(Y | X).

A generative model includes the distribution of the data itself and tells you how likely a given example is. For example, models that predict the next word in a sequence are typically generative models (usually much simpler than GANs) because they can assign a probability to a sequence of words. A discriminative model ignores the question of whether a given instance is likely and just tells you how likely a label is to apply to the instance. Generative models tackle a more difficult task than analogous discriminative models. Generative models have to model *more*. A generative model for images might capture correlations like "things that look like boats are probably going to appear near things that look like water" and "eyes are unlikely to appear on foreheads." These are very complicated distributions. In contrast, a discriminative model might learn the difference between "sailboat" or "not sailboat" by just looking for a few telltale patterns. It could ignore many of the correlations that the generative model must get right. Discriminative models try to draw boundaries in the data space, while generative models try to model how data is placed throughout the space. For example, the following diagram shows discriminative and generative models of handwritten digits [20] (Fig. 4):

Generative Models

- **Purpose and Mechanism**: Generative models, including GANs and VAEs, learn the underlying joint probability of the input data to generate new data resembling the input. They are especially valuable in use cases where data generation is required (like image generation and the design of chemical compounds).
- **Design and Solutions**: One of the biggest challenges in generative models is to maintain the diversity of the outputs, often affected by mode. Recent developments have included novel loss functions and regularization techniques to produce more diverse and realistic outputs.

Discriminative Models

- **Objective and Functionality**: Discriminative models are trained between the classes, focusing on the boundary because they model the conditional probability

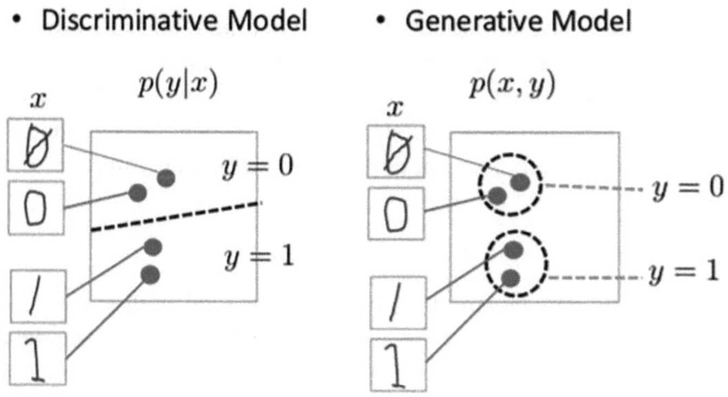

Fig. 4 Discriminative and generative models of handwritten digits

of the target given the application. They tend to be more efficient and simpler than other classification algorithms.

- **Advantages**: These modules are usually more accurate than generative models when it comes to classification, thanks to their simple architecture and relatively easy learning algorithms [24].

Hybrid Models

- **Integration of Generative and Discriminative Models**: Hybrid models, like adding a Gaussian-coupled SoftMax layer, leverage the dual prevention aspects of generative and discriminative model paradigms. They are capable of estimating both class posterior and class-conditional distributions, which improves a wide range of applications, including semi-supervised learning and confidence calibration.

Generative models are great for data synthesis and creative tasks, while discriminative models are great for smaller datasets and tasks that require specific classification. In this regard, hybrid models could be an exciting, if not ideal, research avenue to explore the merits of both paradigms [24].

2.3 Probabilistic Modeling and Latent Variables

Probabilistic modeling and latent variables are important components for understanding complex data structures and generating predictions across many fields. Latent variable models reveal latent structure and dependency in data, making the statistical models more interpretable and powerful in a sense. These papers shed light on various aspects of probabilistic modeling and latent variables, including their applications and methodologies.

- Conditional latent-variable models (e.g., Probabilistic U-Net) are employed to quantify annotation variability in segmentation tasks. These models shape the latent space such that the data is learned comprehensively under uncertainty. By adopting an information-theoretic style approach, one can enhance the mutual information between latent and output variables and thus improve segmentation performance.

High-Dimensional Multivariate Longitudinal Data

- Latent variable models are used for high-dimensional multivariate longitudinal data to model dependencies across variables and time. Such models allow us to infer and predict statistically given data that is of mixed type and partially observed. An information criterion is used to decide on the number of latent factors, demonstrated in a customer shopping behavior prediction analysis.

Bayesian Non-linear Latent Variable Modeling

- The Gaussian process latent variable model is extended to handle non-Gaussian observations via random Fourier features. This work enables MCMC inference with the characterization of posterior uncertainty. The generalized model works well across a range of applications, such as motion capture and text data.

Behavioral Latent Variable Models

- Interpretable latent variable models connect observations of behavior to latent cognitive processes, allowing for accurate predictions of behavior. These models can be applied to develop adaptive experiments (e.g., tasks involving delay discounting) to trigger specific behavioral responses and to confirm cognitive models.

Generative Probabilistic Matrix Models

- A generative probabilistic matrix factorization model with linear latent structures and statistical dependencies. This model can be used to generate scratch-structured training data for supervised learning, controlling both latent dimensionality and correlation structure to mirror the complexity found in real-world data.

Although these models are very powerful in modeling complex data structures, they still face challenges with regards to appropriate latent dimensionality and computational efficiency. Joining information-theoretic perspectives with engineered inference techniques could mitigate several of these issues, increasing the relevance of probabilistic models across domains [21].

3 Types of Generative AI Models

GenAI models are a class of artificial intelligence that generate new content, including text, images, or audio, by studying patterns and structures in existing data. These models are key to applications ranging from creating art to enabling interactive storytelling. GANs, VAEs, and Diffusion: Three main types of generative models with mechanisms and applications. Here is a rundown of those models and their features.

The field of GenAI models is proceeding rapidly, with new families of methods appearing as shown in Fig. 5. Deep learning techniques, namely, VAEs, GANs, DMs, flow models, and transformer models, have immensely improved image and text generation methods. VAEs, the heralds of generative models, employed an encoder–decoder architecture with probabilistic graphical models to learn latent representations for image synthesis. GANs were a milestone by using a generator and discriminator in an adversarial setting, which pushed the generator to produce images closer and closer to real data. Denoising models (DM) have recently risen as a phenomenon with remarkable performance on image generation quality. Flow models perform accurate probability estimation and fast sample generation using reversible transformations. Moreover, the self-attention mechanism of transformer models has greatly enhanced the performance on sequential data and is foundational in many generative deep learning models. In this article, we review the most common GANs currently applied to architecture and the most recent diffusion models.

Large language models (LLMs) such as the generative pre-trained transformer (GPT) series and bidirectional encoder representations from transformers (BERT), which are based on transformer architecture and learn on an incredible amount of text, have shown superior language understanding and generation abilities and can therefore be used in tasks such as text generation, question answering, and machine translation. Deep learning has effectively led to large vision models that exploit image data both for processing and generation, largely based on CNN and transformer architecture, opening state-of-the-art results in image classification and object detection as well as image generation. These large models have captured considerable interest both in academia and practical applications whilst breaking through the boundaries

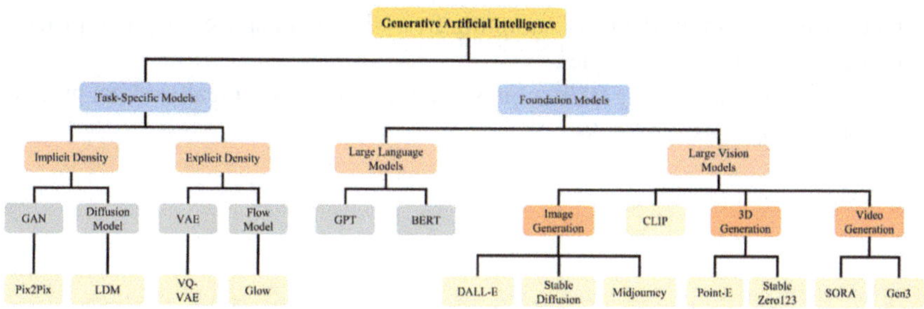

Fig. 5 Types of generative artificial intelligence models

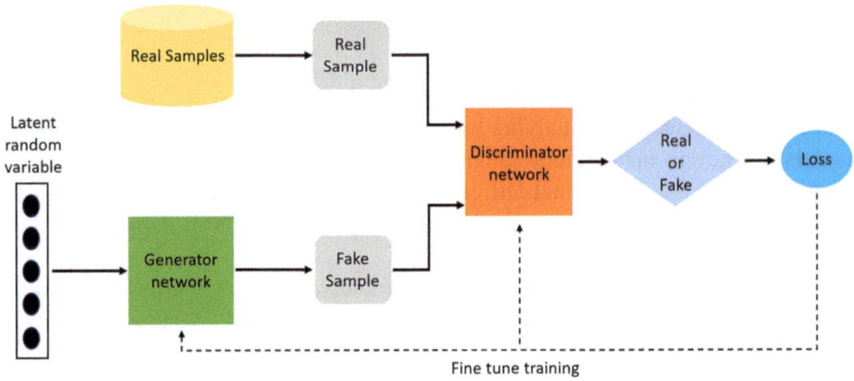

Fig. 6 Basic GAN architecture

of image and text generation technologies and showcasing the potential and wide spectrum of the state-of-the-art approaches in deep learning generative models [22].

3.1 Generative Adversarial Networks (GANs)

GANs are a class of deep learning model that includes two neural networks—a generator and a discriminator. The generator aims to transform random noise into realistic graph-structured matching samples to confuse the discriminator, and the discriminator aims to differentiate between real and generated samples. With ongoing competition and optimization, the quality of the generated samples improves until they are indistinguishable from each other by the discriminator [22, 23, 25] (Fig. 6).

3.2 Variational Autoencoders (VAEs)

- VAEs encode our input data into a latent space, which we then use in a decoder to reconstruct our input data. By sampling from the latent space, they can produce new data using this technique.
- VAEs are good at learning latent representations, but they typically do not produce as sharp images as GANs do.

3.3 *Other Models—Autoregressive Models (e.g., Transformers)*

- GenAI also involves transformers and autoregressive models, which enable text generation tasks.
- Hybrid models that integrate features from various generative models are also being experimented with to utilize the strengths of each approach.
- LLMs are deep learning models trained on massive amounts of text data to generate human language-like responses.
- Models are,
  - AI chatbots (ChatGPT)
  - Content generation and summarization (BERT, T5)
  - Code generation (GitHub Copilot).

3.4 *Diffusion Models*

- These models create data by adding noise to the data and then reversing the process to denoise it, which creates new data. They tend to produce good-quality images with good semantic coherence.
- Diffusion models (e.g., Stable Diffusion), while computationally expensive, produce high-resolution images when combined with segmentation methods.

Comparison of Generative Models

Model	Strengths	Weaknesses	Primary applications
GANs	High-quality outputs, realism	Training instability	Image and video generation
VAEs	Latent space representation, flexibility	Lower quality for complex outputs	Feature learning, anomaly detection
Autoregressive	Sequential data handling	Limited scalability for large outputs	Text, speech, time-series generation
Diffusion models	Detailed outputs, robust performance	Computationally intensive	Scientific simulations, image generation

Generative models have also demonstrated exceptional abilities, yet they continue to grapple with issues like interpretability, training stability, and bias within the generated output. Ensuring these challenges can be appropriately addressed requires collaboration between disciplines and further research to ensure models are robust and provide control over outputs [25].

3.5 High-Level Overview of Principal GenAI Techniques

In this section, we present some of the major GenAI methods that are relevant for this emerging field, along with their fundamental principles, development status, and challenges. GenAI is a class of AI models with the ability to produce new and original data by imitating the underlying patterns learned from the training datasets.

Figure 7 presents an overview of the GenAI models, which are categorized into five classes: autoencoders, autoregressive models, GANs, diffusion and flow-based models, and FMs. The FMs are further categorized based on either the generation tasks (text and images) or multimodality or training framework. FMs, including notable examples such as GPT and Stable Diffusion, are the driving force behind recent advancements in GenAI.

Figure 7 shows high-level overview of core GenAI techniques/models, e.g., autoencoders, GANs, diffusion and flow-based models, autoregressive models, and emerging FMs, along with basal manners and their architectures, given the wide range of FM.

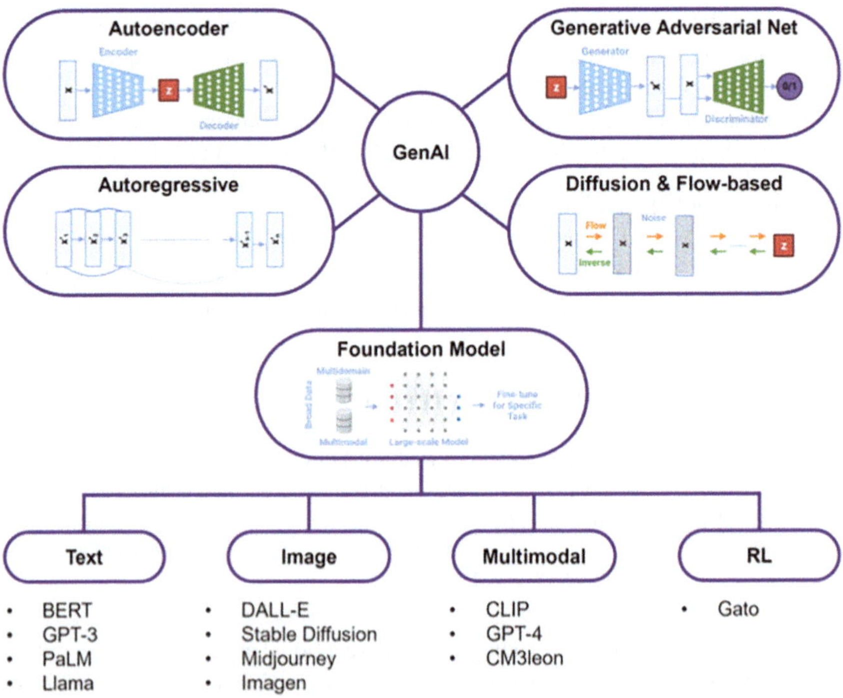

Fig. 7 Overview of the GenAI models

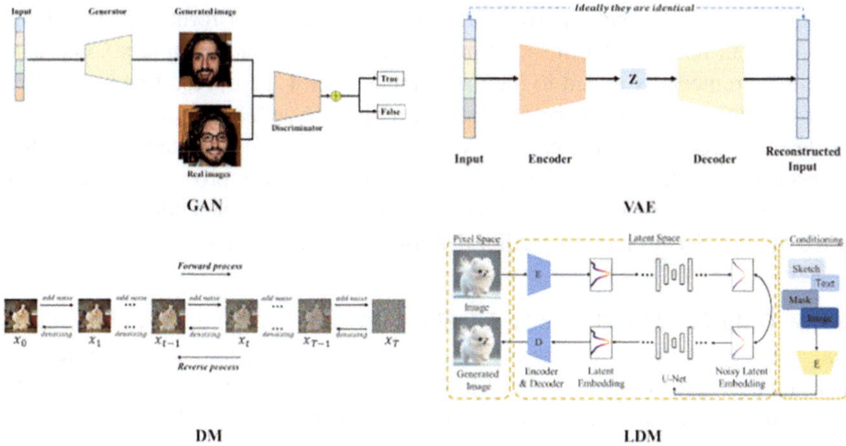

Fig. 8 The framework of GAN, VAE, DM, and LDM. *Z* is a compressed low-dimensional representation of the input

Generative Adversarial Networks

GANs consist of a generator and a discriminator, as shown in Fig. 8. More specifically, the generator generates synthetic samples from random noise, and the discriminator assesses if the transferred samples are like real samples. The goal is for the generator to produce samples that cannot be told apart from actual data by the discriminator. This two-way competitive architecture allows the model to strike a balance between the tasks of generative modeling and discrimination, acting as a catalyst for end-to-end training across the system. The many benefits of GANs also come with their own set of challenges, one of which is mode collapse during training. This happens when the generator overgeneralizes on a few patterns or samples, thus ignoring the wider range of characteristics of the dataset and generating output that would be narrow, repetitive, or low variant. This problem harms the reliability and generalization of the model.

Conditional GAN: The standard GAN models lack controllability, so conditional GAN (CGAN). CGAN allows conditional image generation through the integration of additional information, like text or handwritten sketches, to detail what features the resulting images should present. CGANs incorporate conditional information directly into this procedure with the generator taking in both noise and this conditional information. This method improves the accuracy of created outcomes. Also, adaptations like pix2pix and StyleGAN have been produced to enhance conditional image forgery and determine beyond.

Diffusion Models

Most DMs used today are based on the denoising diffusion probabilistic model (DDPM), which is a simplification of the DM under variational inference. As shown in Fig. 8, DMs involve forward diffusion and reverse denoising (inference) processes.

The forward process is similar to a Markov chain, where it transforms the input image into Gaussian noise. In the forward process, for a given data sample x0, Gaussian noise is incrementally added to the data sample over T steps, resulting in different noisy samples xt. As more timesteps are taken, the unique features of x0 become less and less apparent. As T increases towards infinity, xT is distributed similarly to a Gaussian with isotropic covariance. Moreover, the inference process can be seen as a series of denoising autoencoders with shared weights (autoencoder is commonly realized as U-Net) trained to predict denoised outcomes of their respective inputs, etc. [26].

Latent Diffusion Model: Unlike DDPM, LDM does not directly work with the images but in the latent space (known perceptual compression). LTMs project the data into an efficient latent space of lower dimensionality where the imperceptible high-frequency details are abstracted away. The structure of an LDM is shown in Fig. 8. The image is compressed by the encoder to a latent representation, and then the diffusion process is conducted in this latent representation space. End-to-end, LDM infers the data sample from the noise, the decoder gets the data back to the original pixel space, and obtains the generated images. To further boost the generation speed, the latent consistency model (LCM) was recently proposed to optimize the denoising inference step.

Foundation Models

In computer science, large-scale models, also known as foundation models, are deep learning models with many parameters and complex architecture, especially in natural language processing and computer vision tasks. These models require large compute resources to train but perform excellently across a wide range of tasks. Any DM can, in fact, be considered a hierarchy of n-in-1 neural networks as depicted by Fig. 9, demonstrating the ongoing search for stronger, flexible AI systems.

Large Language Model (LLM)

The transformer model has accomplished extraordinary results in the field of natural language processing (NLP) and includes four components: encoder, decoder, positional encoding, and the final linear and SoftMax layers. The encoder and the decoder are each made of multiple identical layers. Each layer includes several attention layer components and feed-forward network layer components. Positional encoding is also added to the text embeddings to provide position information to the words in the sequence. The most noted transformer has inspired two of the most popular transformer models: BERT and GPT. The only difference between these two transformer models is BERT uses a bidirectional pre-training language model and fine-tuning, but GPT uses an autoregressive pre-training language model and prompting.

GPT is designed to perform language processing tasks by leveraging large-scale unsupervised learning through a process called pre-training. The training process consists of two general stages a language model is first trained as an unsupervised model on a large-scale corpus without task-specific labels or annotation. Next, supervised fine-tuning follows in the second stage to accommodate task-specific

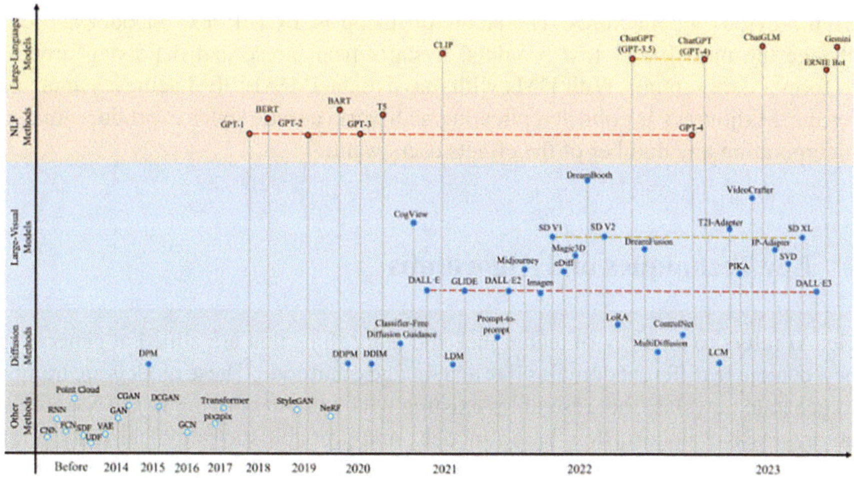

Fig. 9 The evolution of prominent large-scale models in computer science

applications or domains. Also, overall, the approach of BERT has turned into a breakthrough, getting the state of the art performance on different sorts of language problem. BERT was trained in two steps, a pretraining step, and then a fine-tuning step. For pre-training, the language model is trained upon massive text corpuses. Pre-training serves as the first step, the goal of which is to give the BERT model a good understanding of the language so that it can successfully deal with different ML tasks. Fine-tuning then takes the pre-trained BERT model and any smaller labeled datasets and fine-tunes the model's parameters. When retraining for individual tasks through this process, the model can be optimized for those specific tasks, allowing for increased suitability and performance of the model (e.g. through sentiment analysis and text classification).

The last few years have seen rapid, almost explosive, growth in the application of LLMs. We focus on the language models that are trained on such text data without any fine-tuning, known as the basic language models. These include Dialog-optimized language model (LaMDA) and OpenAI's GPT-3.

Large Vision Models

Since then, pretrained vision-language models, e.g., contrastive language-image pre-training (CLIP) in vision, have achieved strong performance on multiple downstream visual tasks by exploiting general zero-shot generalization ability. These models are often trained on hundreds of millions or billions of images–text pairs scraped from the web. Moreover, other lines of research address large-scale, visual input-prompted base models. The segment anything model (SAM) can segment from images and visual prompts (boxes, points, masks) without knowing the category a priori.

Diffusion Models (DMs) are the current foundation for any generative model, and they offer unique capabilities for both creativity and understanding that differs from previous architectures. The stable diffusion is a CLIP text encoder and can change the model with text prompts. It starts from noise and denoising process to form a data sample. With DMs with massive data, DALL·E 3 achieves amazing results. Midjourney is good at replicating real-world artistic styles to produce images incorporating any number of the effects users want.

4 Key Techniques and Algorithms

GenAI refers to an array of techniques and algorithms that are aimed at generating new content in different domains like text, images, and code. These models are trained to generate new data that resembles existing data without copying it. GANs, VAEs, autoregressive models, flow-based models, and probabilistic graphical models are key techniques that constitute the algorithms in GenAI. These processes are at the core of the construction and utilization of generative AI in various disciplines, such as healthcare, finance, and scientific research. Here is a comprehensive overview of these techniques and algorithms.

4.1 Foundational Concepts and Terminology

In the field of Generative AI, one commonly comes across the term 'foundation model (FM)'. Foundation models are large-scale neural network models trained using extensive and diverse datasets, contrasting with traditional ML models designed for specific tasks. Since FMs are trained on broad data, they can be adapted to a wide range of downstream tasks. These models utilize standard deep learning techniques (e.g., generative adversarial Networks, Transformers) and transfer learning methods. A ML model is a function (e.g., a polynomial function in case of a linear regression) that has unknown parameter values. The process of estimating these parameter values from existing data is known as 'pre-training.' Now, in the case of FMs, we are talking about very large and complex models; for example, OpenAI's GPT-3 has about 175B parameters. Therefore, to train such an FM model, we need 10,000s of GPUs over months of training time and a vast training dataset (trillions of words in the case of GPT-3). An objective of pre-training a large language model (LLM) is to enable it to learn facts and the relationships between words. This pre-training results in what is known as the Base Model (Fig. 10).

In language models, tokens represent sequences and do not convey meaning. Many existing tokenizers can be used to convert words into tokens. Embeddings, which are like vectors, associate meaning with these tokens; the direction of the vector signifies meaning, and the magnitude signifies certainty. There can be 1000s of dimensions for large language models (LLMs). We train the LLMs to predict the next embedding. However, the training tasks differ for other types of foundation models (FMs), such as image generation models or code generation models.

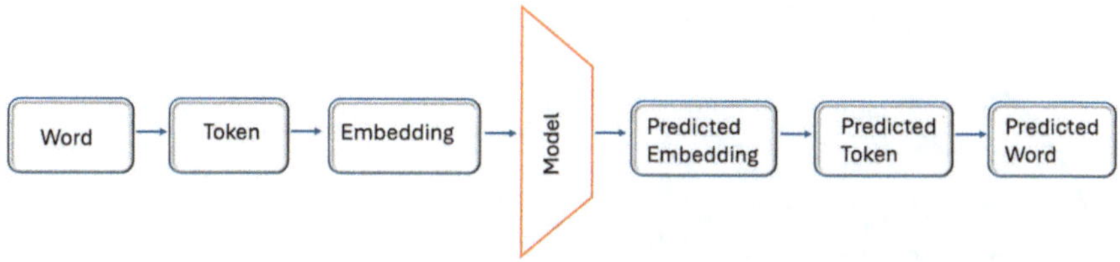

Fig. 10 High-level diagram of language model

Since the base model is trained on diverse data, we perform fine-tuning, which involves retraining the base model for tasks using high-quality data. For example, OpenAI's ChatGPT was fine-tuned on the GPT Base Model using carefully crafted question-response pairs. This fine-tuning process primarily requires a significant investment in creating high-quality data.

The next step in creating a GenAI model is to use reinforcement learning to evaluate the quality of responses and improve the performance of the model. In the case of LLMs, especially in the initial versions, humans evaluate the response by rewarding the model (you can consider this as a scale from 0 to 10). The training goal of the model is to maximize the reward. This process is known as Reinforcement Learning from Human Feedback (RLHF). RLHF not only enhances the model's performance but also generates high-quality data consisting of responses with reward scores. This data can be used to train a reward ML model, which can automate the evaluation process and reduce the need for human experts in future versions, thereby lowering costs. This technique is used by many GenAI models, including OpenAI's ChatGPT and Meta's LLaMA.

4.2 Large Language Models: Powering Generative AI

LLMs are built on the Transformer architecture introduced in Attention Is All You Need (AIAYN) paper in 2017.

In this section, we focus on LLMs and particularly the generative type of these models. LLMs are a significant development in artificial intelligence, especially when it comes to understanding and generating human-like text.

Figure 11 depict an overall view of LLMs, GenAI, and foundation models in the AI realm.

- A foundation model is a base model trained on a massive amount of data to handle a wide variety of tasks. The model can be modality-specific or multimodal. These are also referred to as pre-trained models. The data modalities include text, code, image, speech/audio, tabular data, and so on.

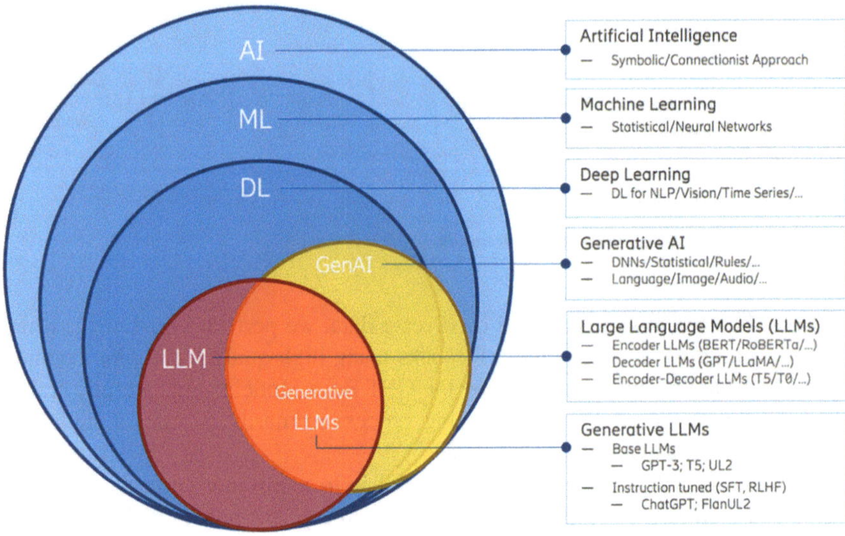

Fig. 11 Contextualizing generative AI and LLMs in the areas of AI

- An LLM is a foundation model trained on a large corpus of textual data with a self-supervised objective. The model's architecture can be encoder-based, decoder-based, or encoder-decoder-based.

GenAI is a type of artificial intelligence that creates new content based on what it has learned from existing data. GenAI, are models that are built to, by using ML algorithms, create new content that can be in the form of text, audio, video, images, or programming code.

LLMs are huge language models pre-trained on large amounts of text data, resulting in millions (or more) of model parameters without tuning on data for specific tasks. A fine-tuned language model starts with a pre-trained LLM, which is then further tuned on a smaller, task-specific dataset to optimize its performance on that specific task. LLMs are built using large amounts of pre-training data. The pre-training data, therefore, plays a pivotal role in their development.

To provide a clearer understanding of the LLM models, we categorize them into two types:

(1) Non-generative LLMs—i.e., encoder-only LLMs such as BERT/XLNet/RoBERTa, used mainly for natural language understanding tasks.
(2) Generative LLMs—i.e., decoder-only, e.g., GPT/LLaMA/MPT, and encoder-decoder, e.g., T5/BART/T0/UL2—are used mainly for natural language generation tasks. Generative Pre-trained Transformer (GPT) refers to the underlying architecture of most decoder-only generative LLMs [27].

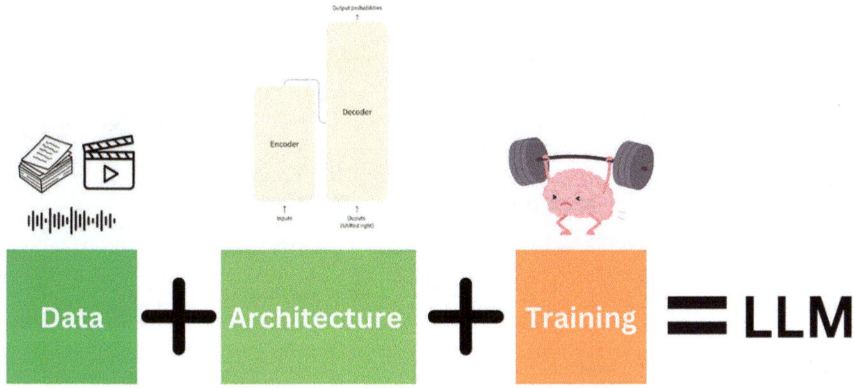

Fig. 12 Components of an LLM

Components of an LLM

LLMs are built upon three foundational pillars: data, architecture, and training (Fig. 12).

- **Data**. This is the backbone of any LLM. The capabilities and limitations of a model are determined by the quality, diversity, and size of the dataset it is trained on. The data typically being examples of informational text from books, web pages, news articles, and other written media. This allows the model to learn different linguistic patterns, styles, and breadth of human knowledge.
- **Architecture**. This is the underlying model structure, typically based on a Transformer architecture. Transformers are neural networks that are particularly adept at processing sequential data such as text—using mechanisms such as attention, they weigh the importance of various parts of the input data. Such architecture allows the model to comprehend and produce coherent text relevant to its given context.
- **Training**. The last step is the training, in which the model is trained on the data that was collected. The model then learns by updating its internal parameters iteratively to minimize prediction error. This is typically done by using supervised learning, where the model receives input data and the correct output and learns to create the expected output on its own. Training algorithms is not just about baking the CRUDer, but also about understanding the subtleties associated with the corpus of language, like context and the creative capacity (or lack of) to respond, vowels, alliteration, etc.

How Do LLMs Learn?

Understanding how these LLMs are trained is extremely important, so let us dive into the training process (Fig. 13):

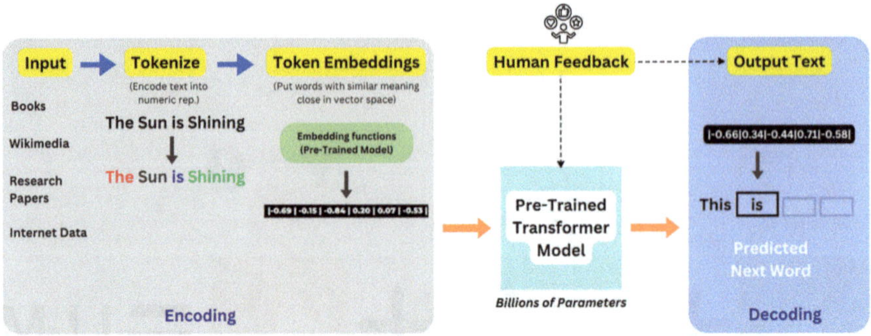

Fig. 13 LLMs training process

- **Input**. Initially, the model is trained on a massive and complex raw corpus of text, which includes everything from books, Wikimedia, research papers, and all sorts of places found across the internet. This is the raw material that the model uses to learn language patterns.
- **Tokenize**. In this step, the text is tokenized; basically, it means that we break it into small pieces (usually words and subwords). The input text is broken down into smaller pieces that can be properly processed by the model.
- **Token Embeddings**. Each token gets mapped to a numerical representation called a token embedding. These embeddings are vector representations capturing the semantics of the tokens such that semantically similar tokens appear close to each other in the vector space.
- **Encoding**. This is passed into the encoding layers of a transformer model. These "layers" will be processing the embeddings to understand the context of each word in a sentence. In particular, the model accomplishes this through the self-attention mechanism that allows it to incorporate information from surrounding words with embeddings that are dynamically adjusted.
- **Pre-trained Transformer Model**: The model's core architecture is a Transformer that has been pre-trained on the input data. It has trained itself to predict sections of text based on other sections by tweaking its internal weights to reduce the predictive error. This model consists of many layers of attention, feedforward neural networks, and a huge number of parameters to capture the complexities of the language.
- **Human Feedback**. This step, which is optional but also vital, involves humans giving feedback on the outputs of the model, with the goal of tuning the model even further. This feedback loop process can contribute to the convergence of output generation from the model with human standards in what might be termed as "humanity outputs."
- **Output Text**. The model makes predictions for the next word in a sequence based on the parameters laid out in pre-training and any fine-tuning. And that's the decoding process.

- **Decoding**. It's from the decoding step where the model turns back the processed embeddings into human-readable text. The model decodes this prediction from its numerical representation to a word after predicting the next word. When decoding, the common practice is to choose the most probable word from the output distribution. This prediction may be the final output, or it may be an additional input token for the model to predict the next words (G, H, I, etc.); this way, the model can feign generating longer stretches of text. This process of encoding and decoding is guided both by the training data it was given and by human feedback that continues to be provided.

4.3 Transformer Model

A Transformer model is a kind of deep learning model mainly used in natural language or sequential data processing. It was first proposed in the 2017 paper Attention is All You Need by Vaswani et al. The Transformer model introduced a new approach that was different from anything before that time, which relied on recurrence and convolutions for sequential data processing. Instead, it employs a mechanism known as "attention" to weigh the impact of different parts of the input during the processing of each part of the output. For example, this way the Transformer model learns the context and semantics of a sentence by understanding the relationship and dependency of the words, regardless of how far apart they are from each other. This has enabled transformer-based AI systems to achieve a level of understanding of human language that is unprecedented, with some tasks reaching or even exceeding human level. The state of the art in many natural language processing (NLP) tasks, most prominently with the advent of LLMs, is transformer models.

"The dominant sequence transduction models are based on complex recurrent or convolutional neural networks that include an encoder and a decoder. The best-performing models also connect the encoder and decoder through an attention mechanism. We propose a new simple network architecture, the Transformer, based solely on attention mechanisms, dispensing with recurrence and convolutions entirely."—"Attention Is All You Need," Google.

The Transformer model was developed as a reaction to a number of shortcomings in previous deep learning models, particularly their application to Natural Language Processing (NLP) [24] (Figs. 14 and 15).

Here's a quick overview of the main parts:

1. **Encoder**

It takes the input sequence and encodes it into several series of vector representations. In the case of encoders, for each layer, it has mechanisms for self-attention to attend to relevant segments of the input sequence. It enables the model to learn long-distance dependencies and understand the context of the input very well.

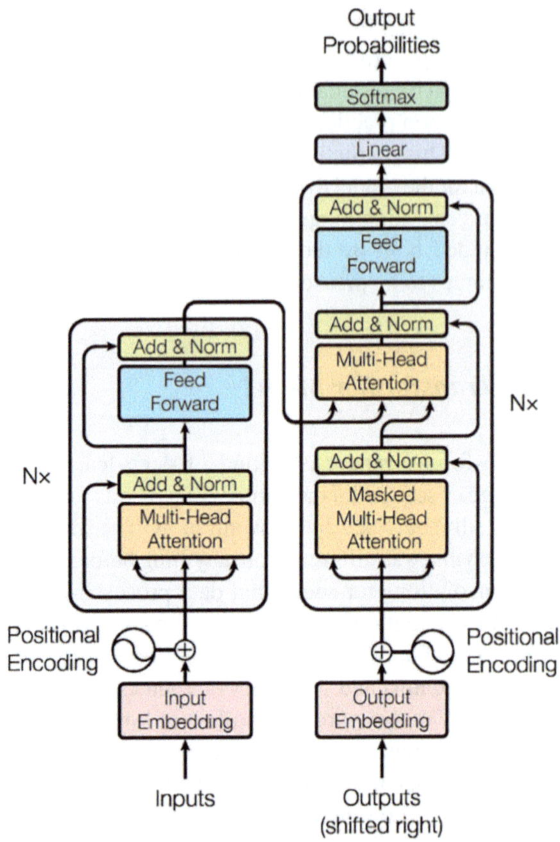

Fig. 14 Transformer model architecture (OpenAI, 2023)

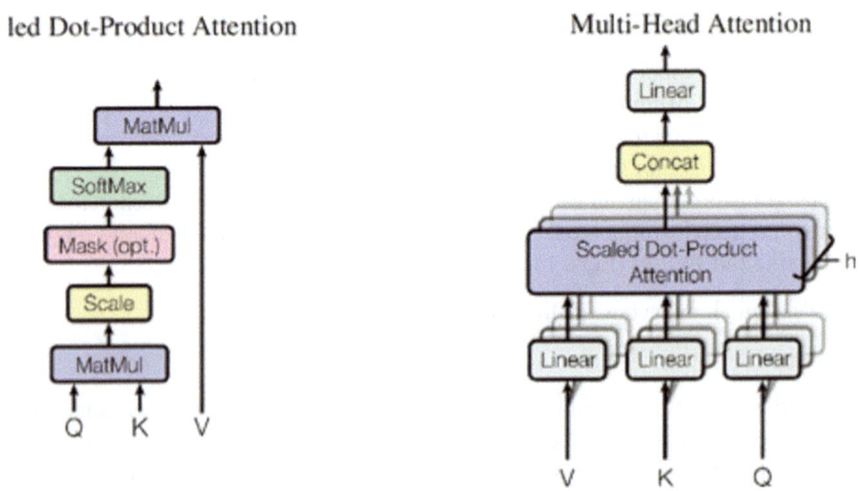

Fig. 15 *Left* Attention to scale product point. *Right* Multi-head attention consists of several layers of attention running in parallel (OpenAI, 2023)

2. **Decoder**

Using the representation from the encoder, this generates the output sequence. The self-attention component enables each layer in the decoder to attend to relevant parts of the encoded representation, while encoder-decoder attention allows the decoder to access relevant parts of its own output. This means the model can produce coherent outputs that are related to the input.

3. **Multi-head Attention**

This enables the model to pay attention to various features of the input data concurrently. This allows the model to pay attention to different features or patterns in the data, enabling it to build a more nuanced representation of the input and produce more accurate outputs.

4. **Positional Encoding**

The Transformer model does not operate in sequence, so it must be supplied with information about the orders of elements. This is also where positional encoding comes into play; it gives the model information about the relative position of an element in the input sequence.

5. **Residual Connections and Layer Normalization**

These techniques are used to stabilize and speed up the training of the model. There are some benefits of the Transformer model's architecture:

- **Parallel Processing**: Because it is designed for fast training and inference on GPUs.
- **Long-Range Dependencies**. It captures relationships between elements far apart in the input sequence.
- **Context Awareness**. Further contextualizes the input data.

4.4 LLM Hallucination Mitigation Strategies

To ensure the response generated by LLMs is accurate, researchers are developing multiple approaches. Some rely heavily on human intervention, such as reinforcement learning using human feedback (RLHF), while others require fresh datasets to train a model; this step is called fine-tuning. Retrieval-augmented generation (RAG)—where an external knowledge source is passed to the LLM.

To address these limitations, several techniques have been developed:

1. **Fine-Tuning**

Fine-tuning is the process of training an LLM on a specific task-related dataset. Trained on data from a specific domain, a model can learn the unique vocabulary and subtleties relevant to that domain, producing better and domain-sensitive output (Fig. 16).

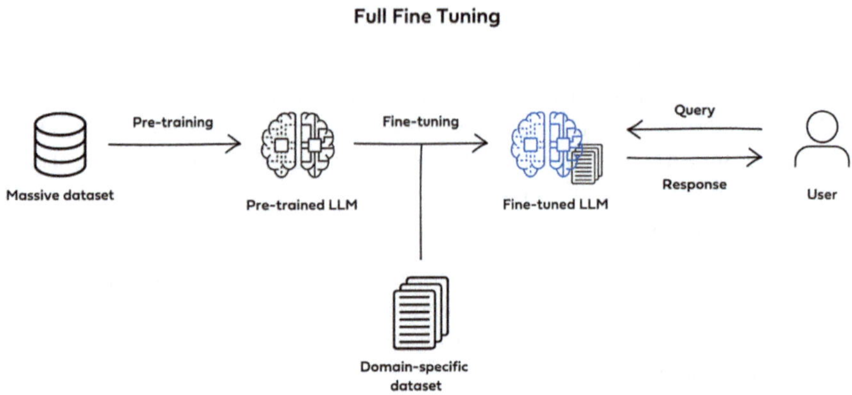

Fig. 16 Fine-tuning. *Image credits* Deci AI

Example: For instance, a legally structured LLM trained on legal documents can solve legal questions better.

2. **Prompt Engineering**

This specific task of designing the inputs for the LLM is called prompt engineering. We empower users with tools that help them direct the LLM usefully by giving it specific instructions and context in the prompt (Fig. 17).

For instance, a prompt that asks the LLM to "summarize the following news article in 50 words" will generate a more precise and informative answer than giving it the article directly.

3. **Retrieval Augmented Generation (RAG)**

RAG is a hybrid approach that integrates retrieval- and generation-based methods. It retrieves relevant documents from a large corpus due to the use of retrieval. Next,

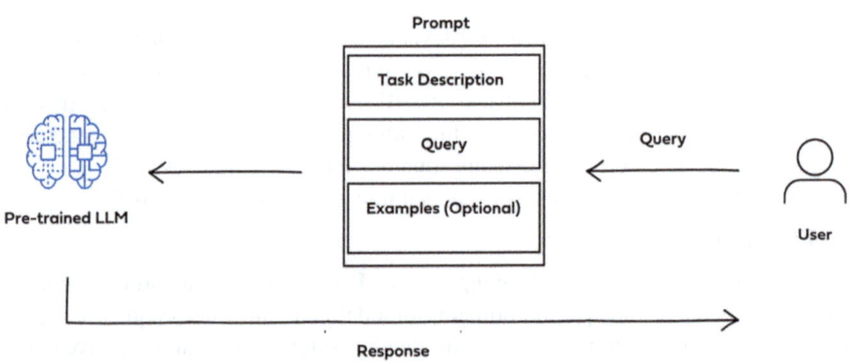

Fig. 17 Prompt engineering. *Image credits* Deci AI

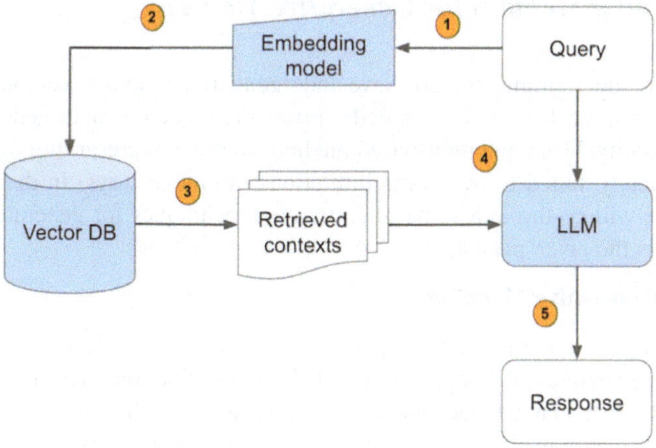

Fig. 18 RAG

it takes the retrieved documents as input and generates a new, customized answer (Fig. 18).

Let's understand the RAG workflow.

Query. It begins with a prompt, which is the input or question you are asking the model.

- **Embedding Model**. This request is then passed to an embedding model, which turns the request into a vector. It takes the raw text and converts it into a numerical representation, retaining the semantic meaning in a high-dimensional real vector between 0 and 1.
- **Vector Database**. This query vector is then used to look for relevant contexts in a massive document pool. These documents are further transformed into their corresponding vectors in a similar high-dimensional space.
- **Retrieved Contexts**. It does that by looking at how similar vectors are between the query and all the contexts (documents) and returning the most relevant ones. This is typically done through nearest neighbor search techniques.
- **LLM**. These retrieved documents are then given to the LLM alongside the (original) query.
- **Response**. The LLM then produces a response from both the initial query as well as from the new context provided by the retrieved documents.

Example: A researcher's implementing a RAG system can browse relevant scientific papers based on a researcher's query and draft a new and informative overview of these papers.

4.5 *Training Methods for Generative Models*

Generative model training methods have undergone tremendous development, using a variety of approaches to solve specific problems related to data generation and modeling quality. From quantum-based methods to privacy-preserving frameworks, these are ideally suited to overcome limitations in unique ways. In the upcoming sections, we will go through some main training techniques for generative models mentioned in the given papers.

Quantum Generative Modeling

- Quantum generative modeling (QGM) utilizes differentiable quantum circuits (DQC) to generate and sample from probability distributions. This method uses a phase feature map to encode data into a latent space, which is then processed with a trainable quantum circuit. A quantum Fourier transform is leveraged to perform fast sampling from parameterized distributions after mapping the model to the bit basis.
- The method is automatically differentiable, can be used to solve stochastic differential equations, and thus has applications in multidimensional generative modeling with qubit registers.

Unified Generative and Discriminative Training

- A unified approach where generative and discriminative training are unified to form multimodal large language models (MLLMs). By applying the semantic relationship between the input sample and the hidden state of the model, this method improves the capability of the model to get both global and fine-grained semantics.
- The method establishes dynamic sequence alignment and a new kernel to meet semantic distinctions, prevailing in the state-of-the-art results on generative tasks that demand cognitive and discrimination capabilities [24–27].

Diffusion-Based Image Synthesis

- In this approach, diffusion-based models are employed to synthesize images of multi-component assets that can be utilized to train defect inspection models. This approach provides the flexibility to control where and in which format the synthesized elements are placed, thus achieving a scene with a diverse set of objects, balanced defects across multiple components, and defect classes.
- Experiments on real-world asset images validate the approach and show that it is helpful for improving model accuracy.

Discriminator-Free Implicit Models

- An alternative discriminator-free approach for training implicit generative models employs an invariant statistical loss, which quantifies the difference between the samples of the model and those uniformly drawn from some distribution.

This prevents typical problems, such as unstable training and mode-dropping, introduced by adversarial discriminators.

- It is effective in both univariate and multivariate cases and successfully learns true distributions in many cases.

Privacy-Preserving GAN Training

- The training of GANs on differentially privatized data involves the use of entropic regularization of the Wasserstein distance to denoise the data distribution. This impacts the model's efficacy by tackling the regularization bias and privatization noise.
- The framework establishes sample complexity outcomes and provides experimental evidence of its advantages for privacy-preserving data collection.

These methods provide novel approaches, but ensuring model robustness, maintaining data privacy, and achieving computational efficiency are still critical focus areas for future work. Varying methods offer insights into how such obstacles may be surmounted to advance generative model training.

4.6 Regularization and Stability Techniques

This will be used to apply stability techniques, also known as regularization techniques, that are critical in the training of GANs, as they faced many issues in mode collapse in photonics as well as convergence issues during training. These strategies all serve to make GANs more robust, effective, and able to generate data that is closer to the real data distribution while also maintaining stable training dynamics. The following subsections cover different techniques and methods mentioned in the above articles.

Regularization Techniques

- One Class Classifier Regularization: Align GAN utilizes a One Class Classifier to regularize the mapping between input data and latent representation, helping the generator yield samples within the support of real data.
- Langevin Stein Variational Gradient Descent (LSVGD): This approach applies implicit regularization during the update dynamics by adding noise, which improves the diversity/spread of samples generated.
- Differentiable Regularization of Condition Number: A new regularizer that encourages 'well-behaved' (in the sense of low conditioning number) matrices, thereby enhancing the numerical stability of neural nets.

Stability Techniques

- Spectral Normalization and Adaptive Bound Control: By constraining geometric similarity to a certain value within a per-training threshold by a discriminator, the spectral normalization allows the layered structure matrix into a certain range characterized in terms of geometry, while ABCAS adapts a Lipschitz parameter by effectively taking the distance between real and fake distributions into consideration, thus stabilizing various GAN trainings.

- Pre-training for Support Alignment: AlignGAN pre-trains the generator of AlignGAN such that the supports of real and generated data align, keeping the alignment maintained with low-bandwidth noise convolutions.

Challenges and Considerations

Even with these enhancements, the GAN training still meets problems like mode collapse and the vanishing/exploding gradient. Ongoing advancements in regularization and stability methods are essential to adequately confront these challenges. Regularization can help with stability but can also reduce convergence/theoretical soundness, so it's a stubborn balance.

5 Applications of Generative AI

To clarify this field difference study, we presented various models in an application-type-wise manner in this paper. Well, there are so many areas that GenAI can be applied to. Figure 19 give an overview of the general class of applications. Some of these areas include:

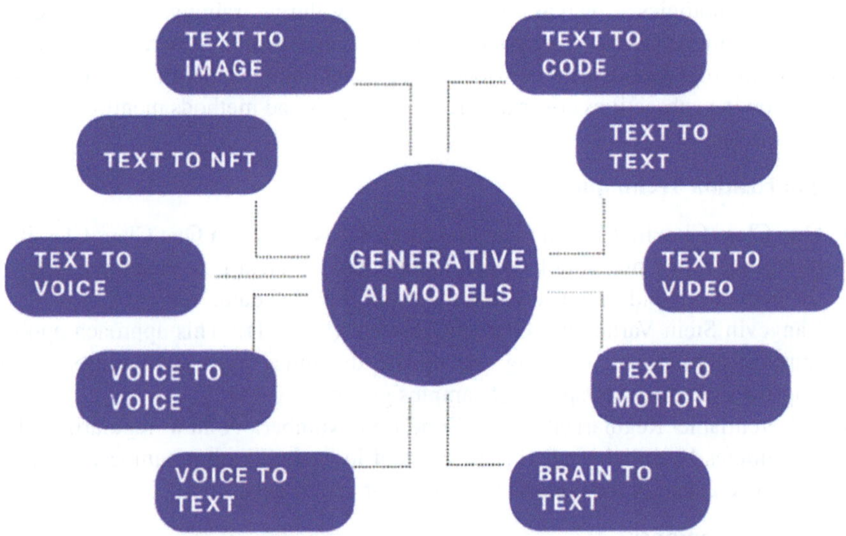

Fig. 19 Types of applications for generative AI

1. **Text Generation**: Using text-based data, GenAI can generate text, articles, stories, poems, etc. It mainly works on the language models like GPT-3, GPT-4, LaMDA, and LLaMA which are trained on words or tokens. This facilitates natural language processing, machine translation, text synthesis, and so on to deliver content types on the users mentioned above.
2. **Image and Visual Content Generation**: Realistic works are obtained with the help of models such as GANs, diffusion, transformers, and VAE. Usually, these systems are based on some neural systems, which are usually trained with text, image, or model-dependent sound inputs. In this context, tools such as DALL-E, Stable Diffusion, and Midjourney are often used.
3. **Music and Sound Production**: It can generate original music notes or sounds, so it includes generating new compositions, writing automatic music, mixing various music styles, etc. Prominent examples in this space are MusicLM and MusicGen, which both learn on sound waveforms of recorded music.
4. **Video and Animation Production**: There are many GenAI models focused on video for digital publishing, animated videos, commercials, games, etc., with a tremendous impact on this sector. Advanced models of these kinds have enormous implications in film, animation, and game character production that can redefine industry dynamics. An example is the models of the Runway ML category.

5.1 Various Industries Using Generative AI

1. **Medical and Biomedical Research**: GenAI can be used for data analysis and experimental result generation. It is involved in the design of new drugs and generative modeling of molecules such as SMILES, DNA, and proteins. What distinguishes biomedical datasets is that biological data is used during the training. The biomedical research field is exploring new frontiers with the help of generative models in healthcare services. Exploring high-dimensional genetic and biological data analysis could facilitate disease analysis, discovery of new drugs, drug design, and personalized medicine. This can involve the design and optimization of genetic sequences, the production of organisms with desired characteristics, and the development of sustainable biomaterials. There could be several purposes to generate or simulate new organisms.
2. **Generative AI in Robotics**: You can train a system on the actuation of a robot, for example, for motion planning, navigation, etc., to generate new trajectories. In the robotics domain, work has been done to create a robot that can learn the human brain architecture to achieve human-level thinking.
3. **Manufacturing**: In manufacturing, professionals can apply GenAI to find ways to operate more efficiently, predict when maintenance will be needed before it disrupts production, assist engineers in creating better designs faster, and develop a more resilient supply chain. Let's take a closer look at these possible manufacturing solutions.

4. **Software Development**: For example, for a software development team, GenAI will provide tools to generate and enhance code faster and with less expertise in using a programming language.

6 Challenges in Generative AI

Limitations and risks associated with GenAI.

GenAI is very new, and when using GenAI, we need to address several risks from data privacy, data security, workforce, and environmental ethics perspectives.

6.1 Limitations

In the world of business potential, GenAI faces limitations. Rely for the most part on legacy training data, which limits the accuracy and relevance of responses. Many models are not internet-connectable, preventing real-time updates.

Another challenge is the collection of high-quality training data. While the technology requires vast resources and struggles with contextual awareness, creativity, and abstraction. GenAI models are considered complex and confusing by users, which triggers fears of liability.

While insightful, GenAI fails to replicate the complex nuances of human cognition.

6.2 Concerns

The fast pace of GenAI development comes with significant challenges and moral issues. This is an important topic because there are data inaccuracies, biases, data privacy, and copyright issues involved. Because generative models do not confirm that what they are producing is factually correct, they may propagate lies, inaccuracies, and errors from training data. Stereotypes and discrimination can be rooted in biases derived from underrepresented data. To prevent biases, one needs huge amounts of diverse and representative training data. Data privacy and security are grave concerns, as inputs into open-source AI models may reveal sensitive data. AI-generated content that incorporates copyrighted material without authorization creates legal exposure and copyright infringement. The ownership of AI-generated content is a question with complex implications for both copyright eligibility and the ethics of its use. At the same time, as regulations are evolving and being implemented at various speeds across our MAs and MSN, MSN embraces the Ericsson guardrails, which scale ethical AI policies that ensure our training data is representative, unbiased, and respects user privacy.

6.3 Hallucinations

For instance, in text and image generation, GenAI models may also hallucinate (produce content unsupported by fact). LLMs also produce false or nonsensical text, and image models might generate scenes that never occurred. There are hallucinations because LLMs are trained on wildly inaccurate datasets, maximizing likelihood without giving a hoot about whether what they produce is accurate or realistic [24].

6.4 Legal

The GenAI capabilities pose security and legal threats primarily linked to data privacy, copyright compliance, identity theft, cybercriminality in general, and warfare and thus expand the overall business risk trajectory. For instance, one risk is the utilization of foundation models, which can lead to system problems such as algorithmic bias as well as copyright questions relating to ownership of the AI-generated material. With respect to identification and reputational risk, there is the genesis of deep fakes, which highlight the incredible disruption capability of AI by the target of such manipulations.

Countries are attempting to legislate and regulate the development of AI technologies. If you want to ensure usage exists on a sound basis, you want to regulate so that the many scenarios in which it can be harmful, proactively or not, do not lead to abuse. This is a collaborative process with a certain degree of hallmark balancing act of innovation and risk in progressive AI legislation. MSN will always comply with local laws and regulations.

6.5 Mitigating Ethical Risks

Detection and Authentication: To mitigate the adverse impact of deep fakes, sensitive detection tools need to be developed to recognize them. Researchers are testing technologies like digital watermarking, cryptographic signatures, and forensic analysis to verify the source of media content and identify manipulated or synthetic content.

Regulation and Policy: Policymakers and regulatory agencies are important players in addressing the ethical challenges of GenAI. As well as censorship, we must also encourage regulation to prevent the misuse and abuse of AI-generated content.

Education and Awareness: In addition to increasing public awareness around the potential threats from deep fakes (as well as from other artificial intelligence-generated content), it is important to build media literacy and critical thinking skills.

Teaching people what GenAI models can and cannot do will empower them to distinguish between manipulated and original content and ultimately prevent the spread of misinformation [25–27].

7 Future Directions

As GenAI continues to evolve, several promising research directions and opportunities emerge. This section explores potential avenues for innovation, interdisciplinary collaboration, and real-world deployment of GenAI technologies. Figure 20 illustrates a taxonomy of future directions in AI-driven UIs, categorizing potential advancements such as dynamic, context-aware UIs and emotionally adaptive interfaces. Detailed discussion on these future directions is provided.

(1) **Dynamic Contextual UIs and AI-Automated Interfaces**: The next evolution of the UIs entails complete automation, wherein UIs automatically mold themselves in accordance with user behavior and environmental context without manual customization. Though AI-automated interfaces use real-time data on user preferences, actions, and situational factors—including location, activity, or emotional state—to modify factors such as layout, functionality, and interaction modes accordingly. These UIs can learn and adapt over time, providing continuous improvement in accessibility and usability while reducing the cognitive burden on users. Dynamic, context-aware UIs are a giant leap in creating personalized and responsive ones. They can adapt interface elements in real-time based on live feedback, adjusting themselves to changes in lighting, user movement, or even user mental states. Such adaptive interfaces can be used extensively only if lightweight frameworks and technologies such as edge computing and federated learning allow them to run seamlessly on devices like mobile phones, which overall have less processing power. Since these frameworks do not share sensitive data with external servers, user privacy is guaranteed; thus, dynamic and contextually aware UIs can reach real-time personalization without sacrificing security or performance.

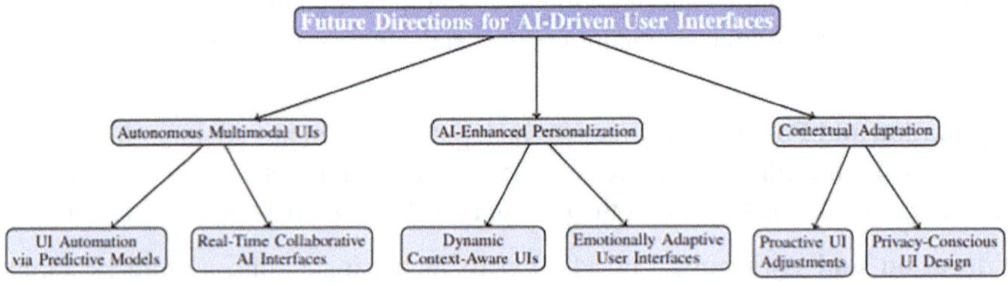

Fig. 20 Taxonomy of future directions in AI-driven UIs

(2) **Hybrid and New Modalities**: These include new interaction modalities that are paving the way for new and advanced experiences, like brain-computer interfaces (BCI), gesture recognition, and haptic feedback. The neural link provides a new method of interaction, as a user can gain control of a digital system through their neural pathways, freeing themselves from manual manipulation. Still under development, BCIs are showing high potential for accessibility aids for users with physical disabilities and virtual environments in gaming and VR. Haptic feedback and gesture recognition are also becoming more prevalent, especially in AR and VR-based applications. Through the interpretation of physical gestures, gesture recognition enables intuitive, hands-free user interaction, whereas haptic feedback provides tactile feedback and adds a new layer of immersion to the user experience. The incorporation of these modalities provides a richer and more intuitive interface, accommodating diverse user preferences and requirements. Hybrid interfaces that merge traditional graphical user interfaces (GUIs) with multimodal inputs present a comprehensive solution toward enabling context maintenance and smooth transitions between input modalities. This hybrid UI makes it possible for the users to toggle between text, voice, and visual interactions depending on their preferences, without being limited to a specific modality, which is a blessing when interfacing on mobile devices where the real estate on screen is limited. Using multimodal fusion techniques, such interfaces can correlate and align various types of inputs, such as text, voice, and images, seamlessly. Such integration would not only provide a more interactive experience but also enable the system to manage complex tasks effectively by adapting to the user preferences and the dynamic contextual clues. In general, these emerging modalities and hybrid approaches exemplify a trend towards more adaptive, intuitive, and accessible UIs fueled by advancements in AI that are more real-time and allow for a seamless user experience across platforms.

Proposed Solutions and Innovations

(1) **AI-Generated UIs and Proactive Adjustments**: GenAI has the transformative potential to enable dynamic UIs that adapt themselves according to user behavior and real-time observational contextualization, offering suitable interactive environments. GenAI models can analyze user behaviors and environmental variables to personalize UIs so as to emphasize important content, make navigational paths easier to navigate, and reduce cognitive load. This allows systems to give user-specific experiences that develop over time according to the manner in which they are utilized and make them more open and usable. Another major innovation responsible for unauthorized UI changes is predictive modeling. Utilizing historical user interactions, these models allow the system to infer user requirements and adapt interface elements (shortcuts, layout, and displayed objects) before users start an action. The added layer offers users a familiar foundation, allowing for faster navigation and more intuitive interactions, which is particularly beneficial in professional settings such as design, healthcare, and engineering, where efficiency is essential.

(2) **Emotionally Adaptive UIs and Autonomous Collaboration**: Emotionally adaptive UIs are yet the next frontier of responsive design, where interfaces cater to real-time assessments of user emotional states. These UIs can leverage multimodal inputs (like facial expressions, tone, or physiological indicators) to adjust design elements (color scheme, layout, etc.) and functionality to support or make more comfortable the experience of an individual. In applications such as mental health and customer service support, being able to respond to user emotions is especially valuable to such users' engagement and satisfaction. Autonomous multimodal collaboration is a critical feature of complex, collaborative environments, like healthcare operating rooms. Such AI driven UIs can fluidly blend multiple types of input (text, voice, gesture, and visual cues), enabling hands-free control and intuitive interaction in complex, high-risk environments. Autonomous multimodal collaboration enables efficient information exchange, enhances coordination of workflows, and lessens cognitive load on users, resulting in seamless and effective teamwork.

(3) **Ethical, Privacy-Conscious Design and Cross-Platform AR UIs**: Therefore, as AI-powered user interfaces become prevalent across all industries, ethical UI design is necessary. Design frameworks that respect privacy aim to enhance the freedom of users by increasing their transparency and control in the use of data. AIs built on top of these UIs can enforce data anonymization, build privacy-preserving interaction logs, and improve agency over first-party data, improving trust and transparency. Creating a uniform experience over multiple devices, cross-platform AR UIs is another novel solution that helps merge different device environments to synchronize digital and physical dimensions. These AR interfaces allow users to utilize virtual elements that are visible in their immediate surroundings, allowing for continuity of experience regardless of whether the experience is delivered via mobile devices, AR glasses, or a standard desktop system. This makes AR tools more flexible and usable for daily use, as it will improve accessibility and user experience.

(4) **Collaborative AI for Real-Time Co-creation**: Collaborative AI opens the door to real-time co-creation with a high level of interactivity in a highly immersive interface by users with AI. They are highly effective for creative domains like design, video editing, and educational content-hub creation, where collaborative user interfaces let users put down ideas in multiple ways (such as text, speech, and sketching) and allow AI to automatically contribute and build upon the human input. This allows the AI to instantly offer recommendations, improve designs, or create elements that work well with what the user has provided, creating a process of true interactive co-creation. The real-time collaboration with AI extends the user's capabilities and reduces the design cycle, thus providing a more engaging, productive, and immersive experience. Collaborative UIs can be smart partners, augmenting our creativity, helping people to create great content whenever they want, quickly, and providing personalization to their style and preferences. This opens up the possibility for AI to give users the ability to generate complex and innovative solutions, to facilitate creativity, and to provide dynamic educational tools that enhance learning experiences.

GenAI has been a disruptive force that is reshaping industries and pushing the frontiers of human creativity. The future of this technology is very exciting as it continues to develop. In this article, we explore the trends and predictions we expect to drive the future of GenAI. Covering everything from developments in deep learning to the integration of AI ethics, we identify the transformative pathways for GenAI, shine a light on the benefits and present use cases, and highlight strategic opportunities.

7.1 Emerging Trends in Generative AI

The evolution of GenAI Considering the recent advances in technology and the increasing demands of the industries, GenAI continues to evolve by the day with extensive research in the same field. Here are the trends that will be instrumental in framing the GenAI side of things for the upcoming times.

1. Advanced Deep Learning Techniques: The future of GenAI is in the continuous advancement in deep learning techniques. As neural networks grow to be more complicated, fashions will be able to create more sophisticated, correct, and sensible content. Both techniques, such as VAEs and GANs, will continue to improve and produce increasingly diverse and evocative output.
2. Cross-Domain Applications: GenAI will work across various domains. And models that can produce content in multiple formats—including, but not limited to, text, images, music (and soon video, etc.)—will become the default and will spur innovation across numerous industries at the same time.
3. Ethical Considerations and Bias Mitigation: Ethical issues, like bias present in generated content, will become a central theme to contend with. GenAI's future will be focused on developing methods in bias reduction to make content fair, diverse, and representative.
4. Interactive and Co-creative Systems: GenAI models are turning from passive content generators to interactive and co-creative systems with generative AI models. They will have meaningful interactions with users, helping create content, generate ideas, and even collaborate creatively [23, 25–27].

7.2 Predictions for the Future of Generative AI

1. Personalized Content Creation: GenAI will change the face of personalization by developing hyper-personalized content based on individual preferences. News articles and entertainment content will be operated dynamically for individual preferences.
2. Transforming Art and Creativity: AI-generated art and creativity will further redefine the landscape, creating blurred boundaries between human and machine-generated content. Artists will partner with AI, discovering new mediums and possibilities that were not possible before.

3. Redefining Content Generation: The time will come when AI creates content that is impossible to differentiate from human-generated content. It opens up new directions in content creation and questioning of our perception of an author and creativity.
4. Innovative Product Design: Clients, customers, and users stand to benefit from the integration of GenAI which shall develop prototypes that comply with user demands in mind. This will speed up the design process and create more user-centered products.
5. AI-Augmented Creativity: AI will enhance the human creative process by providing suggestions, improving drafts, and generating new concepts. AI-human co-creation will change creative work across all industries.

7.3 Advantages of the Future of Generative AI

- **Efficiency and Creativity Fusion**: The combination of all human creativity powered by AI-driven efficiency will unleash productivity and innovation like never seen before.
- **Unleashing Human Potential**: With AI-augmented creativity, humans will not only be viewers of art but also its pioneering creators, surpassing the limits of imagination in visual art, literature, and communication.
- **Accelerated Innovation**: Ideas and concepts will be generated at unprecedented rates and innovation cycles will be accelerated.

Potential Research Directions

Improving Model Robustness: Future research efforts may focus on enhancing the robustness of GenAI models, particularly related to addressing frequent challenges such as mode collapse, training dynamics issues, and vulnerability to input setting perturbations. Design novel training algorithms, regularization methods, and architectural alterations to advance the stability and robustness of generative models.

Incorporating Contextual Information: Content generated by the GenAI models can become more coherent and contextually relevant if we keep integrating contextual data and prior knowledge in their models. By utilizing context through techniques such as conditional generation, attention mechanisms, and reinforcement learning, models can generate outputs that are far more personalized and adaptive.

Exploring Hybrid Approaches: Generative models can be further developed by employing different forms of GenAI techniques as multi-generative approaches combining GANs and VAEs or transformer-based models combined with convolutional neural networks (CNN). This may be done by integrating techniques of hybrid approaches that exploit the strengths of each technique, alleviating their shortcomings, resulting in more performant and all-purpose generative artificial intelligence systems.

Semantic Understanding and Control: Giving GenAI models semantic control and comprehension capabilities enables them to assist in the process of controlling and guiding the generation process in a more intuitive way. Potentially, in this field, researchers will be able to create interpretable and controllable generative models, allowing content to be generated that reflects what the user types in as the attributes, style, or semantics.

Opportunities for Interdisciplinary Collaboration

Human–Computer Interaction (HCI): Collaborate with human–computer interaction experts to create interfaces, interactive tools, and environments that facilitate research in artificial intelligence and creative content generation. For interaction with generative AI systems to prove effective, HCI principles could be applied to create the controls, feedback, and workflows needed for collaboration.

Cognitive Science: Insights from cognitive science can help in designing GenAI models in a way that emulates the creative and cognitive processes of humans. This leaves us in a place where we may see generative models that are capable of creating content with some degree of human-like reasoning, imagination, and problem-solving in the long run as AI researchers work with cognitive scientists.

Arts and Humanities: The arts and humanities may offer alternative approaches to GenAI by including artists, designers, and scholars and nurturing interdisciplinary creativity within it. AI researchers and practitioners in creative disciplines should collaborate to widen the GenAI application results in exploratory art creation as well as cultural expression and aesthetic innovation.

Real-World Deployment and Applications

Creative Industries: There is a great possibility for GenAI technologies to bring a change to creative industries, such as entertainment, advertising, design, fashion, and more. The use of GenAI in these fields can aid creative professionals in maximizing their efficiency and creativity through minimized content creation loops, personalized features, and innovative story formats.

Healthcare and Education: These models can be used for medical image synthesis, drug discovery, and patient diagnostics like personalized treatment planning in the medical field. In education, GenAI can enable interactive learning, personalized tutoring systems, and adaptive content generation, helping to deliver education tailored to particular learning preferences.

Environmental Sustainability: GenAI techniques can also be used in the field of climate modeling and ecological conservation to optimize resource utilization, design sustainable products, and simulate environmental scenarios. GenAI remains an ever-present tool; therefore, by us generating solutions, our opinions and passions contribute to the movement towards better sustainable development and environmental stewardship [24, 26, 27].

8 Conclusion

Generative AI (GenAI) is among the most potent technologies, enabling machines to produce content like text, images, music, and videos. It uses sophisticated models such as GANs, VAEs, and Transformers, which detect patterns in data and generate new, unique outputs. It is one of those technologies that has immense potential for good—for making life-like images, for stimulating creativity, for helping with such chores as writing or design. Well, there are some challenges with GenAI like data bias, ethical issues, high computing power required, etc. The new tech is anticipated to have its interesting applications for the future as well—creative tools and problem-solving, for instance. GenAI can leverage the benefits of society, finding new ways of thinking and creating through both development and usage as long as we do it in a public, safe way.

Acknowledgement I would like to express my sincere gratitude to Mr. Nishant Saxena for his invaluable guidance, constructive suggestions, and insightful views during the preparation of this book chapter. His thoughtful inputs greatly enriched the quality and depth of the write-up, and I am truly thankful for his support.

References

1. A. Dash, J. Ye, G. Wang, A review of generative adversarial networks (GANs) and its applications in a wide variety of disciplines: from medical to remote sensing. IEEE Access **12**, 18330–18357 (2024)
2. https://www.sciencedirect.com/science/article/pii/S209526352400147X#bbib41
3. https://medium.com/@nnabwani/full-fine-tuning-peft-prompt-engineering-or-rag-69689ba58a52
4. P. Yadav, G. Rathwad, J. Jain, Generative AI: shaping the future while disrupting the present. Int. J. Multidiscip. Res. (2024). https://doi.org/10.36948/ijfmr.2024.v06i05.28085
5. H.K. Kılınç, Ö.F. Keçecioğlu, Generative artificial intelligence: a historical and future perspective. Acad. Platform J. Eng. Smart Syst. (2024). https://doi.org/10.21541/apjess.1398155
6. https://arxiv.org/html/2405.20024v1
7. https://www.cmu.edu/intelligentbusiness/expertise/genai-principles.pdf
8. https://www.linkedin.com/pulse/demystifying-machine-learning-supervised-unsupervised-bushra-akram/
9. https://www.researchgate.net/publication/328299387_If_software_is_narrative_Joseph_Weizenbaum_artificial_intelligence_and_the_biographies_of_ELIZA/citation/download
10. https://www.guvi.com/blog/what-is-generative-ai/
11. https://opencv.org/blog/what-is-generative-ai/
12. https://developers.google.com/machine-learning/gan/generative
13. https://www.sciencedirect.com/science/article/pii/S209526352400147X
14. https://arxiv.org/html/2411.04137v1
15. J. Correia, F. Baeta, T. Martins, Evolutionary generative models (2023). https://doi.org/10.1007/978-981-99-3814-8_10
16. A. Kaushik, Generative AI adoption in automotive vehicle technology: case study of custom GPT (2024). https://doi.org/10.47363/jaicc/2024(3)400

17. K. Curran, E. Curran, J. Killen, C. Duffy, The role of generative AI in cyber security (2024). https://doi.org/10.54517/m2796
18. M.B. Oluwagbenro, Generative AI: definition, concepts, applications, and future prospects (2024). https://doi.org/10.36227/techrxiv.171746875.59016695/v1
19. H.J. Jeon, B.V. Roy, Information-theoretic foundations for machine learning (2024). https://doi.org/10.48550/arxiv.2407.12288
20. L. Younes, *Introduction to Machine Learning* (2024). https://doi.org/10.48550/arxiv.2409.02668
21. V. Goar, N.S. Yadav, Foundations of machine learning (2024). https://doi.org/10.4018/979-8-3693-1598-9.ch002
22. I.A. Scott, G. Zuccon, The new paradigm in machine learning—foundation models, large language models and beyond: a primer for physicians (2024). https://doi.org/10.1111/imj.16393
23. M. Maskey, R. Ramachandran, T. Lee, K.J. Murphy, S. Roy, M. Ramasubramanian, I. Gurung, R. Ganti, Foundation models for science: potential, challenges, and the path forward (2024). https://doi.org/10.5194/egusphere-egu24-3202
24. H. Hayashi, A hybrid of generative and discriminative models based on the Gaussian-coupled softmax layer (2023). https://doi.org/10.48550/arXiv.2305.05912
25. D. Singh, M.A. Ali, S. Singh, P. Maurya, S. Suman, R. Rai, N. Tiwar, A. Rahman, Understanding and enhancing diversity in generative models. Int. J. Res. Publ. Rev. (2024). https://doi.org/10.55248/gengpi.5.0524.1231
26. C. He, Y. Jiang, Investigating the advancements in generative models (2023). https://doi.org/10.1109/aiac61660.2023.00037
27. D.M. Anstine, O. Isayev, Generative models as an emerging paradigm in the chemical sciences. J. Am. Chem. Soc. (2023). https://doi.org/10.1021/jacs.2c13467

AI-Assisted Fiber Optic Sensors for Simultaneous Measurement

Koustav Dey, V. Nikhil, and Sourabh Roy

Abstract In the last few decades, sensing mechanisms by employing the fiber optics has achieved huge attention owing to their unique characteristics. The machine learning (ML) approach has brought a thoroughgoing rehabilitation in the field of fiber optics-based sensing mechanisms due to its capabilities of extracting a huge chunk of information from a huge datasets that enhance the degree of performance. Various sensing structures including fiber Bragg grating (FBG), multi-single-multi mode (MSM), single-multi-single (SMS) mode have proved their efficacy in these aspects. The main bottleneck of measuring two parameters simultaneously is the crosstalk issue when utilizing the popular transfer matrix method. ML, which is trained with a large dataset collected from the experiment, is a viable solution to overcome this traditional issue. ML has demonstrated its effectiveness by mitigating the crosstalk issue to a higher degree and thereby enhancing the sensing performance. This unique technology has affirmed its potential in several domains to provide an enhanced, reliable sensing performance.

Keywords Fiber optic sensors · Artificial intelligence · Fiber Bragg grating · Cascaded fiber structures · Temperature measurement · Strain measurement

---

K. Dey (✉)
Lightwave Technology Laboratory, Department of ECE, Missouri University of Science and Technology, Rolla, MO, USA
e-mail: kdzmw@mst.edu

K. Dey · V. Nikhil · S. Roy
Department of Physics, National Institute of Technology, Warangal, Warangal, Telangana, India
e-mail: sroy@nitw.ac.in

Y. Vasimalla and S. Kumar (eds.), *Generative AI for Photonic Sensing*, Progress in Optical Science and Photonics 36, https://doi.org/10.1007/978-981-95-1561-5_3

1 Fiber Optic Sensors and Artificial Intelligence (AI): Brief History and Overviews

1.1 Fiber Optic Sensors

After the Laser invention in 1960, a new horizon opened and gave a boost to the research field, which is known as the 'fiber optic sensors (FOSs)'. The origins of fiber optic sensors can be found in the early light-directing concepts, such as the 1840s demonstration of total internal reflection by John Tyndall. With notable developments such as Charles K. Kao's work on pure glass fibers in the 1960s, modern fiber optics first appeared in the middle of the twentieth century. Utilizing optical interference and fiber deformation, the first fiber optic sensors for temperature, strain, and pressure were developed in the 1970s. Distributed sensing was made possible in the 1980s by technologies such as Optical Time Domain Reflectometry (OTDR), and fiber optic gyroscopes proved to be reliable in defence applications. Multiplexed sensing and increased industrial and environmental monitoring applications were made possible by the introduction of Fiber Bragg Gratings (FBGs) in the 1990s. Their application in industries including seismic monitoring, oil exploration, and healthcare has expanded since the 2000s thanks to developments like Distributed Acoustic Sensing (DAS) and integration with IoT and AI. Fiber optic sensors are already essential in many industries due to their high sensitivity and resilience to electromagnetic interference. Future research will concentrate on increasing sensitivity and broadening applications in space, energy, and autonomous technologies.

In the FOSs, an optical fiber is the heart of the whole system. FOS consists of an optical source (Light emitting diode (LED), Super-luminescent diode (SLD), a Laser diode (LD) etc.), an optical fiber, a transducer and a detector with the processing unit (Optical spectrum analyser (OSA), Multiwavelength meter etc.). FOSs have emerged as an unparalleled technology in the last few decades due to their numerous exceptional characteristics including high sensitivity, immunity to electromagnetic interference (EMI), ability to operate in harsh environments etc. The miniaturization and versatility made them ideal candidates for a wide range of applications. Due to the demand towards smarter and more efficient systems in the industry, the necessity for advancing sensing technologies like fiber optic-based sensing systems is of utmost importance along with driving innovation in the precision, reliability and integration into the monitoring solution for the next generation.

Among various fibre-based sensing configurations, sensors based on fiber Bragg gratings (FBGs) [1–5], single mode-multimode-single mode (SMS) and multimode-single mode-multimode (MSM) [6–13] are the most popular. A brief description of these popular sensing configurations has mentioned below.

1.1.1 Fiber Bragg Gratings (FBGs)

Bragg gratings are distributed Bragg reflector (DBR), which forms in a small section of the fiber by periodic variation of the RI along the core of the fiber [14–20]. It was first demonstrated by Ken Hill in 1978. This index modulation is permanent and called grating. When light from a broadband source is coupled to one side of the fiber, some part of the light will be reflected from all the periods of the grating. Each reflected light will combine coherently and will form a large, reflected signal at a particular wavelength depending on the periodicity of the grating. The reflected wavelength is called the Bragg wavelength. The remaining light will be transmitted down through the fiber.

In an optical fiber, the phase-matching condition is given by

$$\alpha_1 - \alpha_2 = \Delta\alpha = \frac{2\pi}{\Lambda} \tag{1}$$

where α is the propagation constant and Λ is the periodicity of the grating (Fig. 1).

In the case of FBG,

$$\alpha_2 = -\alpha_1 = \alpha \tag{2}$$

As forward (LP_{01}) and reverse propagating core mode couples together.

Hence, the phase-matching condition will become

$$\alpha - (-\alpha) = \frac{2\pi}{\Lambda} \tag{3}$$

$$2\alpha = \Delta\alpha = \frac{2\pi}{\Lambda} \tag{4}$$

Grating periodicity would be less than 1 μm as $\Delta\alpha$ is larger.

But

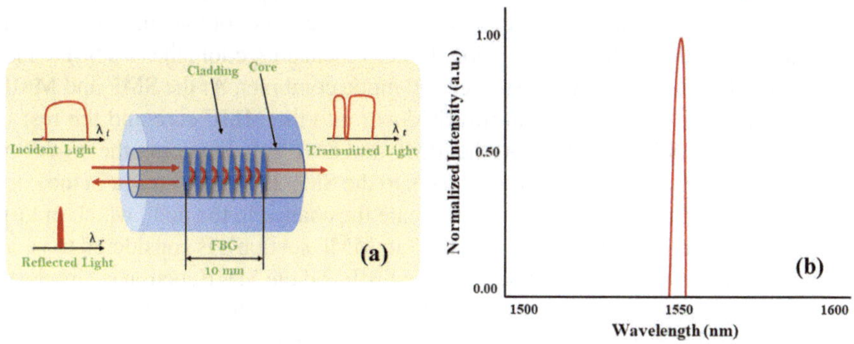

Fig. 1 **a** Schematic representation (FBGs) and **b** reflected spectrum of the FBG

$$\alpha = \frac{2\pi}{\lambda} n_{eff} \quad (5)$$

Hence,

$$2\left(\frac{2\pi}{\lambda} n_{eff}\right) = \frac{2\pi}{\Lambda} \quad (6)$$

Therefore, λ_B can be written as

$$\lambda_B = 2 \cdot n_{eff} \cdot \Lambda \quad (7)$$

From the above equation, it can be concluded that any changes in the effective refractive index and the periodicity will cause a change in the Bragg wavelength.

Furthermore, Bragg's condition needs to satisfy the conservation law of energy and momentum.

Energy conservation law:

$$hw_i = hw_r \quad (8)$$

Momentum conservation law:

$$k_i + K = k_r \quad (9)$$

w_i and w_f are the frequency of the incident and the reflected radiation. k_i and k_r are the wave vectors for the incident and reflected radiation. K is the grating wave vector.

1.1.2 Single-Multi-single (SMS) Fiber Structure

To prepare the SMS based cascaded structure, an optimized section of the MMF is spliced in between two SMFs. The schematic of the fabricated structure is illustrated in Fig. 2a. This cascaded fiber structure relies on the basic principle of the multimode interference (MMI). This incident is owing to the encounter of the core modes with the cladding modes in the MMF section. In this cascaded section, the two SMFs i.e., SMF1 and SMF2 work as mode splitter and mode combiner. At the SMF and MMF splicing point, some of the signal travels down into the MMF core and the rest is coupled to the MMF's cladding. At the MMF and SMF splicing point, the modes of the MMF from cladding will be linked back to the SMF2. The difference in the core and the cladding modes refractive index create the change in the path, which in turn causes the interference. The mode approach in MMF section's is considered to be as a statistically isotropic. It is assumed that the SMF and the MMF section is circularly symmetric and perfectly aligned with each other's. This will ensure the stimulation of the linearly polarized (LP) modes in the MMF section.

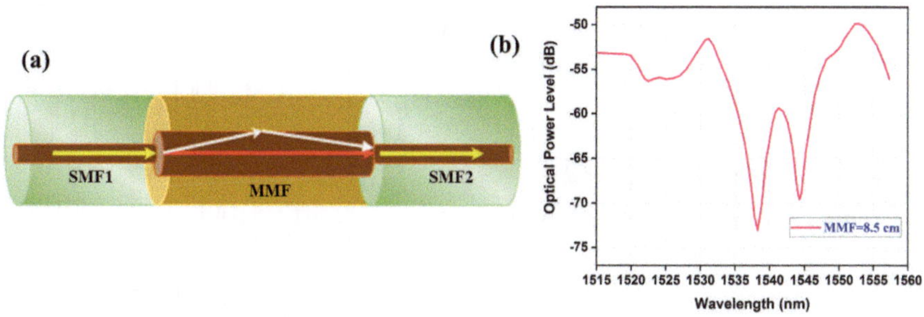

Fig. 2 **a** Schematic representation and **b** transmission output (MMF = 8.5 cm) of the SMS fiber structure

The stimulated modes in the MMF will flow down to it and then encounter at the splicing of the MMF and the SMF2. The distribution of the electric field at a distance say, L could be broken down as follows [12]:

$$E_s(r, L) = \sum_{m=1}^{s} \alpha_m \varphi_m(r) \exp(-i\beta_m L) \tag{10}$$

where α_m is the coefficient of coupling (CoC) between the LP_{01} mode and the LP_{0m} in the SMF and MMF fiber, respectively, $\varphi_m(r)$ are designated as eigenmodes in the core of the MMF and β_m denotes as the constant of propagation of every eigenmode inside the MM fiber section. The CoC can be intimated as

$$\alpha_m = \frac{\int_0^\infty E_s(r)\varphi_m(r) r dr}{\left(\int_0^\infty |E_s(r)|^2 r dr \times \int_0^\infty |\varphi_m(r)|^2 r dr\right)^{1/2}} \tag{11}$$

The power that transmitted through the cascaded SMS could be evaluated by the convolution integral of the $E_s(r, L)$ with the primary mode of the SMF2 $E_s(r, 0)$ as

$$P(L) = 10 \log_{10} \left(\frac{\left| \int_0^\infty E_s(r, L) E_s(r, 0) r dr \right|^2}{\int_0^\infty |E_s(r, L)|^2 |E_s(r, 0)|^2 r dr} \right) \tag{12}$$

Under external perturbation, the effective refractive index of core and cladding, along with the L might change which may changes the eigenmodes in the core i.e., $\varphi_m(r)$. As a result, there will be a change in CoC.

Therefore, it can conclude that, by changing the MMF length, the CoC between the LP_{01} and LP_{0m} could be varied which creates the change in the interference pattern of the SMS.

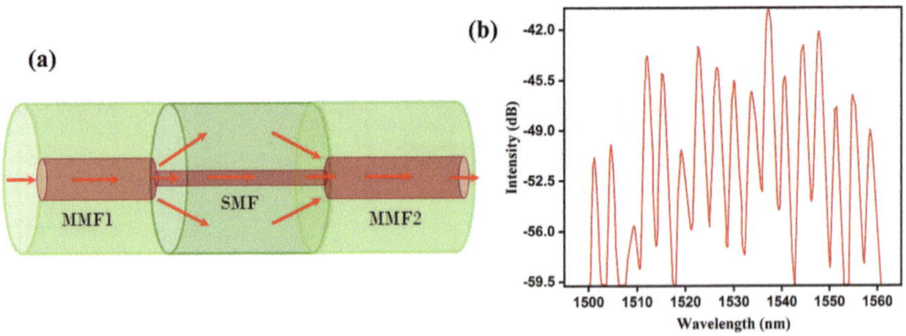

Fig. 3 **a** Schematic representation and **b** transmission spectrum of the MSM fiber structure

1.1.3 Multimode-Singlemode-Multimode (MSM) Fiber Structure

The multi-single-multi (MSM) mode fiber structure is based on the idea that the core and cladding modes interfere with one another in the SMF section. MMF can excite a large number of cladding modes into the SMF to mismatch in the mode field because of its large core diameter. Light that travels down to the MMF1 and is coupled to the SMF fiber segment excites higher-order modes [21–23]. While some of the field travels through the SMF's cladding, other parts couple to the SMF's core. At the interface of the SMF-MMF2, Cladding modes superposed with the core mode which led to the observation of an interference phenomena and designated as the transmission spectrum of the MSM fiber structure (Fig. 3).

2 Artificial Intelligence (AI)

Artificial Intelligence (AI) is the simulation of human intelligence processes by machines, particularly computer systems. The term was first coined by John McCarthy in 1956 at an academic conference. In FOSs, AI has been employed rapidly over the years to enhance data analysis capabilities, improving sensing accuracy and decision-making.

2.1 Necessity of the Integration of AI to FOSs

AI techniques such as machine learning (ML), and deep learning (DL) have been employed to analyse a huge chunk of data generated by the FOSs. It has introduced a comprehensive in-depth understanding and analysis in the field of sensing systems with higher accuracy.

- The integration is particularly important where the complex patterns and the relationship in the dataset must be demodulated and identified in order to enhance the performance of the sensing system. Often, the data collected using the FOS systems contains noise and various complex patterns that need to be demodulated to extract meaningful information. AI works effectively to achieve this.
- AI has the capability to anticipate the possible malfunctions or anomalies in the structures being monitored. This in turn enables proactive maintenance and reduces downtime.
- A properly and thoroughly trained AI model can adjust various sensing parameters in the fiber optic sensing systems that would enable the optimization for varying environmental conditions.

2.2 Importance of AI Assistance in Simultaneous Measurement of Two Parameters

In the field of fiber optics sensing technology, simultaneous measurements of two different parameters are essential for various applications especially where the different environmental or structural aspects must be monitored concurrently in order to achieve higher precision, reliability and a comprehensive monitoring. Among several other reasons, the most important aspect of simultaneous measurement is to mitigate the cross-sensitivity issue. This effect could lead to the inaccurate measurement if only one parameter is being monitored. In general, a transfer matrix method is employed to retrieve the temperature and strain data simultaneously. A and B, the two optical parameters are nearly linear functions of the external perturbations, respectively, then it may be expressed as

$$\begin{bmatrix} A \\ B \end{bmatrix} = \begin{bmatrix} a & b \\ c & d \end{bmatrix} \begin{bmatrix} \Delta T \\ \Delta \varepsilon \end{bmatrix}$$

where a, b, c and d denote the coefficient values corresponds to the absolute parameters. Nevertheless, it has been noticed that the above coefficients values become a non-linear upon applications of the changes of the either temperature or strain or both that leads to a high cross sensitivity issue. Furthermore, the accuracy of the detection of the popular matrix is poor [24]. This problem could easily be mitigated by employing the machine learning approach with higher accuracy [25–34].

3 Operational Principle

3.1 FBG Interrogation with the SMS

FBG have a particular wavelength depending on the periodicity and effective refractive index of the grating. Whereas the SMS spectrum will have a transmission spectrum over a certain window of the wavelength range. It is having several rising and the falling slopes which could be efficiently used for the interrogating the FBG(s) that falls in the slope regions. This technique depends on the amplitude conversion from frequency. Under the applications of the external perturbations, the peak of the FBG's will be shifted and will follow the slope regions as the intensity will be decreased with increasing the external perturbations. The scheme is shown in Fig. 4.

Let assume the output power through the SMS is P_{Out}, wavelength of the incident light is $\lambda_{inc.}$, falling edge slope is E_1, effective power of the incident light P_{eff} and the power spectrum is E_{eff}, then the change of the power captured by the detector could be estimated as follows:

$$P_{Out} = (E_1 \lambda_{inc.})\big(1 + E_{eff}\big)P_{eff} \tag{13}$$

From the above equation, it can be deducted that the output power varies in linear fashion with the wavelength of the FBG, which represents a conventional interrogation method.

Nonetheless, the power of the reflected light varies linearly with the external stimulus such as temperature and strain as follows:

$$\Delta P = P_{input} \times \big[(E_1 \times F \times \Delta\varepsilon) + \big(E_1 \times F' \times \Delta T\big)\big] \tag{14}$$

F and F' are related to the fiber parameters.

Based on the data above, it can be said that the wavelength changes of the Bragg peak (in case of FBG) or the dip in the interference spectrum (for MSM/SMS) due to the applied temperature and strain along with the intensity change could be

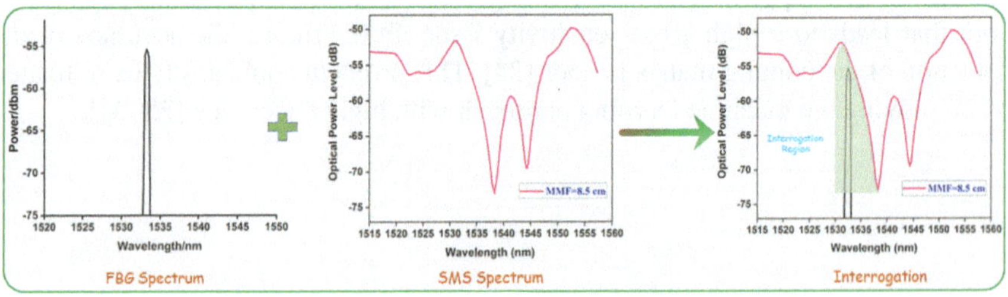

Fig. 4 The detailed schematic of the interrogation technique between the SMS and the FBG. The FBG has a particular wavelength, that satisfy the Bragg condition, will follow the falling slope region of the SMS over the wavelength of 1530–1538 nm [31]

monitored. The temperature and strain variation could be used as an input dataset for the modelling using ML.

The operating principle for the MSM based sensing structure has been outlined in the previous section.

4 Modelling Using ANN

Artificial Neural Network (ANN) is a structure of connecting neurons that convey data to the other neurons in the model. It was motivated by biological neurons and functions like a connective tissue in the living brain. To carry out the learning process, the nodes and links each have a specific weight that can be changed.

4.1 Modelling for SMS-FBG Based Sensing Scheme

Figure 5a displays the suggested ANN model that was used to the study for SMS-FBG based sensing structure. In this part, the modelling parameters have been introduced. Determining the input and output parameters is the initial stage of modelling. A limited and suitable labelled dataset is needed for the ANN model's training procedure. We used a MATLAB simulation program to create the dataset for this purpose. In this part, the multi-layer perceptron (MLP) design is used to construct the dataset.

The initial datasets are essential to the development of the ANN model. The accuracy of the model is solely reliant on the way the dataset is matched to the particular problem. Normalizing the input variables and shuffling them to prevent bias towards a certain dataset were part of the pre-processed of the gathered dataset. Figure 5b depicts the structural outline for the full employment of our new ANN model.

A dataset consisting of ten thousand data have been separated into training and testing of around 70% and 30%, respectively and were used to create our proposed ML model for the further analysis. The proposed model is trained with several prospects using the train data set. Lastly, several attributes are analysed using the test data set. The data points include the following: output parameters including power change (ΔP) and wavelength change ($\Delta\lambda$), and input parameters as change in temperature (ΔT) and strain ($\Delta\varepsilon$). There are 500 nodes or neurons in each of the nine hidden layers that make up the ANN model. In order to quickly get a stable MSE, it is determined that 500 neurons in each of the nine hidden layers are suitable. It is best to avoid adding more nodes and layers because this would increase the computational load. With a learning rate of 0.0001, the well-known "Rectifier linear unit" (ReLU) function was utilized as an activation function and "ADAM" optimizer for "gradient descent back-propagation" optimization.

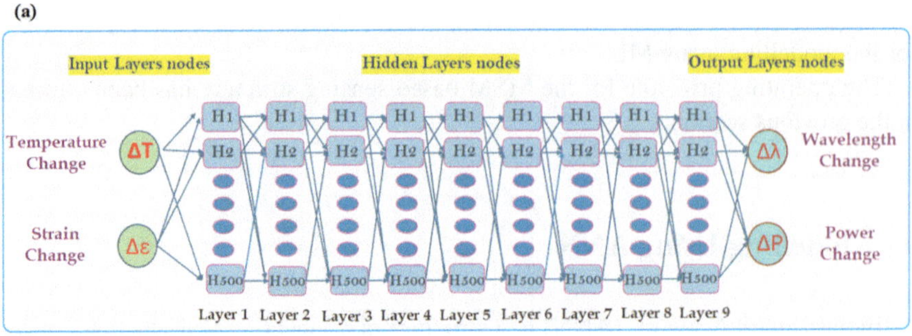

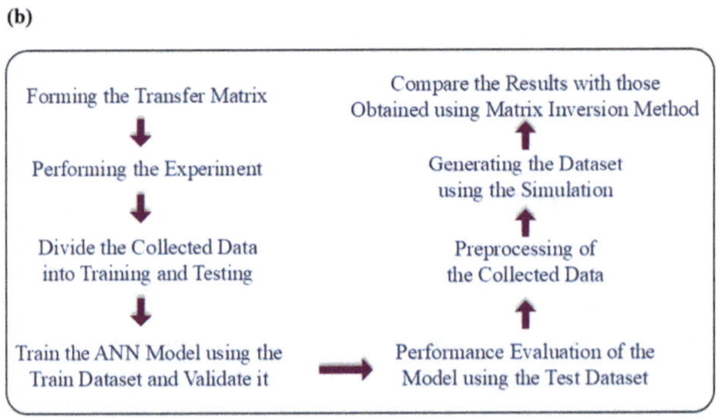

Fig. 5 **a** Illustration of the ANN structure for the modelling of the SMS-FBG based sensing structure and **b** general process chart for the implementation of the ML for the measurement simultaneously

Depending on the optimum values of the MSE the number of layers that need to be used for training the model is decided. From Table 1, we can see how the various layers affect the MSE values in detail.

Table 1 Tuning of the number of the layers [31]

Serial number	Number of layers	Value of the MSE
1	2	0.4667
2	3	0.0830
3	4	0.0831
4	5	0.0837
5	6	0.0235
6	7	0.0016
7	8	1.23×10^{-4}
8	9	**1.84×10^{-6}**
9	10	0.0821

As the 9 layers (highlighted in "bold") for the model ensure the lowest MSE, therefore it has been chosen for the modelling purpose.

4.2 *Modelling for MSM Based Sensing Scheme*

The suggested ANN model to carry out the MSM structure based simultaneous measurement is depicted in Fig. 6.

10,000 data sets were employed in our suggested model. These data sets were divided into 90% (or 9000) training data points for training the model and 10% (or 1000) points designated as 'test data' to examine various phenomena. The proposed model's input parameters are temperature (ΔT) and strain ($\Delta \varepsilon$), while the shift in wavelength of the two separate dips ($\Delta\lambda_1$ and $\Delta\lambda_2$) are the output parameters. The proposed ANN model with various layers and nodes have been depicted in Fig. 6. To avoid unwanted biases for a particular dataset fed to the input, those variables are appropriately randomized as part of the pre-process in the dataset.

In both aforementioned cases, it is necessary to pre-process the parameters brought into consideration in this work in order to input them into the suggested ANN model considering that they are not evenly distributed in the values. One kind of normalization technique uses minmax scaling to rescale the raw data in the range 0–1 by transforming it linearly. Controlling bias towards a specific dataset and the negative impact of feature magnitude differences is the key objective of this. The precision and speed of the convergence are also enhanced by this procedure.

The linear function can be explained as follows

$$X_{nom} = \frac{X - X_{\min}}{X_{\max} - X_{\min}} \tag{15}$$

X represents the characteristics of the sensing probe.

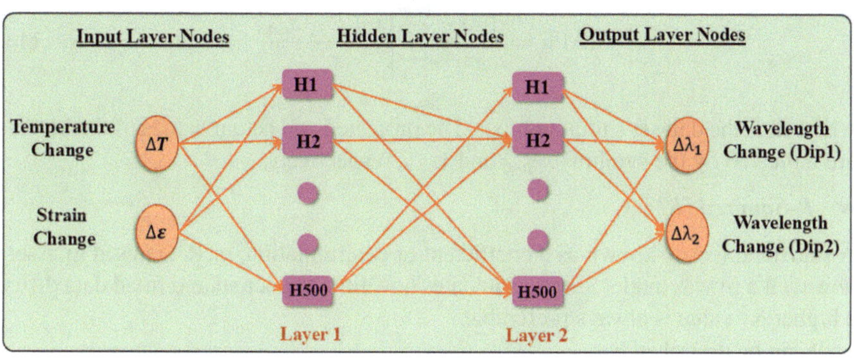

Fig. 6 Diagram of the ANN structure used to model the sensing structure based on MSM

A number of metrics, including 'mean squared error (MSE)', 'root mean squared error (RMSE)', 'mean absolute percentage error (MAPE)', and 'R-squared (R^2)', are used to evaluate the model's performance [26, 35–41].

i. ***Mean Squared Error (MSE)***

The average squared of mistakes is measured by the MSE loss function. The deviation of the estimated value with the actual values' average squared errors was included in this error of the average squared, and it may be explained as follows:

$$\text{MSE} = \frac{1}{m}\sum_{k=1}^{m} x_{i,est} - x_{i,act} \tag{16}$$

where, $x_{i,est}$ and $x_{i,act}$ represents the estimator's real and estimated values.

A lower MSE value is desirable because it shows a proper agreement of the estimated values with the actual values. The epochs have the smallest MSE always favoured due to superior execution.

ii. ***Root Mean Squared Error (RMSE)***

When assessing prediction quality, the most used metric is RMSE.

$$\text{RMSE} = \sqrt{\frac{1}{m}\sum_{k=1}^{m} x_{i,est} - x_{i,act}} \tag{17}$$

iii. ***Mean Absolute Percentage Error (MAPE)***

A statistical metric called MAPE is used to characterize how accurate a model is with respect to a specific dataset. A lower MAPE score indicates that the model fits the data better.

The following is a description of the MAPE:

$$\text{MAPE} = \frac{100\%}{m}\sum_{i=1}^{i=m}\left|\frac{x_{i,pre.} - x_{i,act}}{x_{i,act}}\right| \tag{18}$$

where m is the dataset's total number. The projected and actual values of the ith data are indicated by the symbols $x_{i,pre.}$ and $x_{i,act}$, respectively.

iv. ***R-Squared (R^2)***

A statistical metric known as a coefficient of determination, or R^2, is used to assess how well a fit was made. Since it indicates how little the actual and fitted data differ, a higher R^2 value is always preferable.

It can be described as

$$R^2 = 1 - \frac{\sum_i \left(x_{i,pre} - x_{i,act}\right)^2}{\sum_i \left(x_{i,avg} - x_{i,act}\right)^2} \tag{19}$$

Table 2 Optimization for multiple optimizers [29]

S. No.	Optimizers	RMSE
1	**ADAM**	**4.55×10^{-8}**
2	SGD	1.27×10^{-4}
3	RMS Prop	3.19×10^{-6}
4	Nadam	7.43×10^{-7}
5	Adadelta	0.1674
6	Adagrad	1.73×10^{-4}
7	Adamax	3.108×10^{-7}
8	Nadam	7.43×10^{-7}
9	Ftrl	0.2518

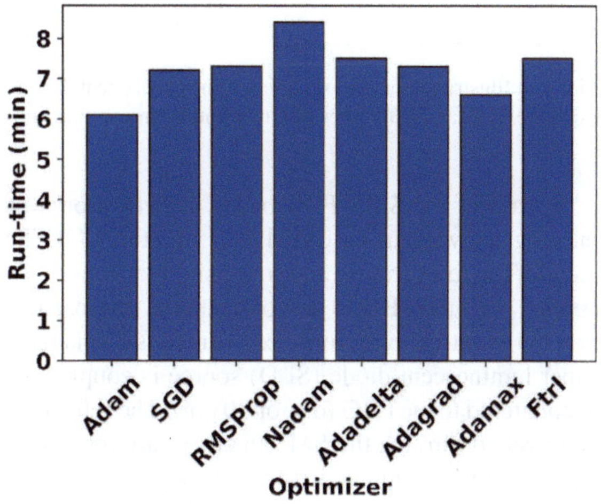

Fig. 7 Various optimizers and their corresponding run-time [29]

where $x_{i,avg}$ denotes the average value of the ith data.

We have used a number of built-in optimizers provided by the Python program's Keras module for this. For model training, the optimizer with the computational runtime and MSE with the lowest values are selected.

Owing to the smallest run-time and lowest MSE among all the considered optimizers as mentioned in Table 2, ADAM (highlighted in "bold") is utilized for the further analysis throughout the work as depicted in Fig. 7.

5 Details of the Experimental Set-Up

Following the model's training experiments, we will describe how to gather the inverse matrix's input data set in the next part.

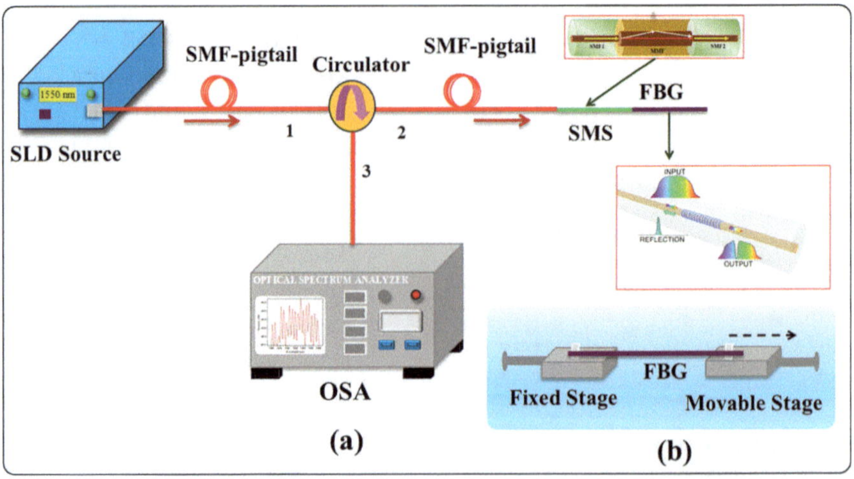

Fig. 8 **a** Illustration of the execution of the experimental for the proposed interrogation technique using the SMS and FBG fiber structure and **b** temperature and strain characterization

In case of the SMS-FBG based interrogation system, the FBG was etched and interrogated with the cascaded fiber structure of the SMS for simultaneous measurement of the temperature and train, simultaneously. The temperature and strain were applied onto the FBG to demonstrate the proposed scheme. The schematic illustration of the set-up to perform experiment is displayed in Fig. 8. Source light from the super luminescent diode (SLD) source is coupled to the 3-port circulator and then encountered to the FBG (dia. of 10 μm). The reflected light from the etched FBG will be travelled through the SMS fiber and arrives at the detector i.e., optical spectrum analyzer (OSA, Agilent 86142B) through another port of the circulator.

The shift in the Bragg wavelength along with the change in the intensity level could be monitored using the detector i.e., OSA. Figure 8b shows the scheme for the characterization of temperature and strain. Applying perturbations such as strain and temperature will alter the FBG's core Bragg wavelength. Moreover, the dropping edge part of the SMS will cause the FBG's power level to change. Plotting the change in power and the change in wavelength with the external perturbations yielded all coefficients in the matrix above.

The grating part has been etched in order to further boost the sensitivity. The sensitivity is increased by the grating's thinning, which produces a greater evanescent field wave (EFW). Furthermore, it was discovered that the etched fiber's young's modulus dropped, raising its thermal expansion coefficient that have a major impact on the thermal sensitivity and consequently the EFW. Consequently, the thermo-optic co-efficient of the fiber decreases. The co-efficient could be find out from the plot. To determine the coefficients, a temperature sensing experiment was conducted over the range of 0–2100 με and strain sensing in the range of 25–250 °C. Setting the temperature and strain readings over the previously designated range and figuring

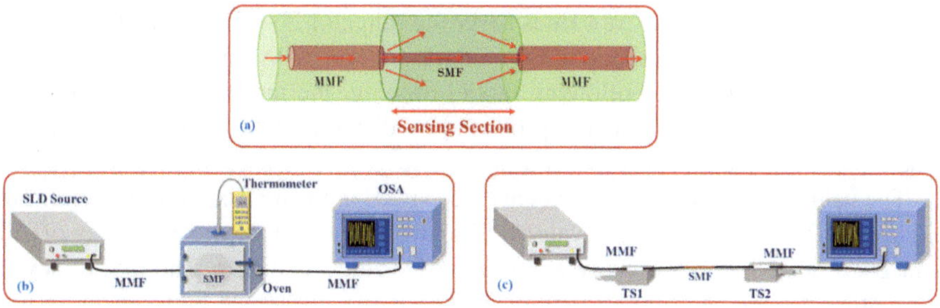

Fig. 9 **a** MSM fiber structural diagram and **b, c** experimental setup for characterizing temperature and strain

out the matrix's inverse will yield the relevant input datasets that will be utilized for training the model.

Two different dips of the MSM cascaded fiber structure have been taken into consideration for the simultaneous measurement of the temperature and strain. The schematic of the experimental set-up has been depicted in Fig. 9. The signal from the SLD is coupled to the MSM. The OSA will receive the light transmitted by the MSM and utilize it to monitor spectrum dips brought on by different disturbances.

The two dips behave differentially to strain and temperature. Plotting the graphs of the wavelength changes of two dips against the specified parameters yields the matrix coefficients. The anticipated dataset may be attained and prepared for model training once the inverse matrix has been acquired.

6 Results and Discussion

6.1 *SMS-FBG Based Sensing Configuration*

The Bragg wavelength shift and power level change of the FBG are simultaneously detected at the moment of temperature and strain perturbation. The coefficient can only have constant values if the nonlinear reactions of strain and temperature are disregarded. However, because the amplitude fluctuates non-linearly with the perturbation, the matrix technique produces a considerable percentage error between the estimated and observed values that cannot be disregarded. To overcome this bottleneck, the ANN approach could be an easy and straightforward solution. The impact of the proposed model is to mitigate the issue has been demonstrated and evaluated.

Plotting the graphs yields the coefficient values from the matrix, which are then used in the popular matrix. The following is the characteristics equation:

$$\begin{pmatrix} \Delta T \\ \Delta \varepsilon \end{pmatrix} = \frac{1}{|H|} \begin{pmatrix} 21.50\,\text{pm}/^{\circ}\text{C} & 2.79\,\text{pm}/\mu\varepsilon \\ -0.86\,\text{dB}/^{\circ}\text{C} & -0.079\,\text{dB}/\mu\varepsilon \end{pmatrix} \begin{pmatrix} \Delta\lambda \\ \Delta P \end{pmatrix} \tag{20}$$

A duo of 4000 ΔT and Δε values (over the range of 20–255 °C and 0–2150 με) make up the input data set, and 4000 pairs of Δλ and ΔP values are considered in order to train the model. Following model training, the model was tested using a fresh set of Δλ and ΔP, which is produced by different 2000 pairs of ΔT and Δε data.

The values of ΔT and Δε which have been used in the matrix above to determine Δλ and ΔP are called applied values. Similarly, to gather predicted values of ΔT and Δε, the ANN model receives Δλ and ΔP values as inputs. The inaccuracy in estimating the temperature and strain change is the difference between these two quantities. The coefficients of the matrix now need to be obtained using the similar datasets was used to train. This attests to enhanced sensitivity performance. Furthermore, one of the most crucial elements in claiming a better level of performance is the RMS inaccuracy.

First, utilizing the test data set, the FBG's power level and wavelength change have been plotted against variations in temperature and strain. The change of the power level along with the change in the wavelength of the etched FBG with change in the temperature and strain have been plotted in Fig. 10. The linearity of the graphs demonstrates the efficacy of the proposed ANN model. An inverse process simulation is performed for comparison. Each plot depicts the higher degree of linearity between the ANN model and the simulation. By employing the proposed ANN model, the new matrix equation could be derived as follows:

$$\begin{pmatrix} \Delta \mathrm{T} \\ \Delta \varepsilon \end{pmatrix} = \frac{1}{|H|} \begin{pmatrix} 33.18\,\mathrm{pm}/^{\circ}\mathrm{C} & 5.30\,\mathrm{pm}/\mu\varepsilon \\ -1.388\,\mathrm{dB}/^{\circ}\mathrm{C} & -0.114\,\mathrm{dB}/\mu\varepsilon \end{pmatrix} \begin{pmatrix} \Delta \lambda \\ \Delta P \end{pmatrix} \tag{21}$$

The sensitivity values for temperature and strain are of 33.18 pm/°C and 5.30 pm/με, respectively from the proposed ANN model. Because machine learning techniques increase the alignment of the input data, the ANN model's temperature and strain sensitivity is approximately 1.75 and 2.2 times greater, respectively, than the widely used traditional matrix method. Consequently, increased sensitivity was noted. Furthermore, the RMSE for strain and temperature measurements is 0.195 με and 0.025 °C, correspondingly. When comparing these statistics to those derived from the matrix method, the proposed modelling improves the temperature measurement by roughly 680 times and the strain measurement by 164 times.

The RMSE found by the matrix approach may be again decreased by employing the terms with higher order of the polynomial in the matrix. Nevertheless, it makes the equation more complicated, and in order to tackle this issue, a specific numerical model is required. Furthermore, this is not appropriate for more significant cross-sensitivity issues. On the other hand, the ANN technique may be used with any type of sensor, regardless of whether the input and output are linear or non-linear. Also, despite the errors in the dataset that fed to the input, the proposed model could be used effectively.

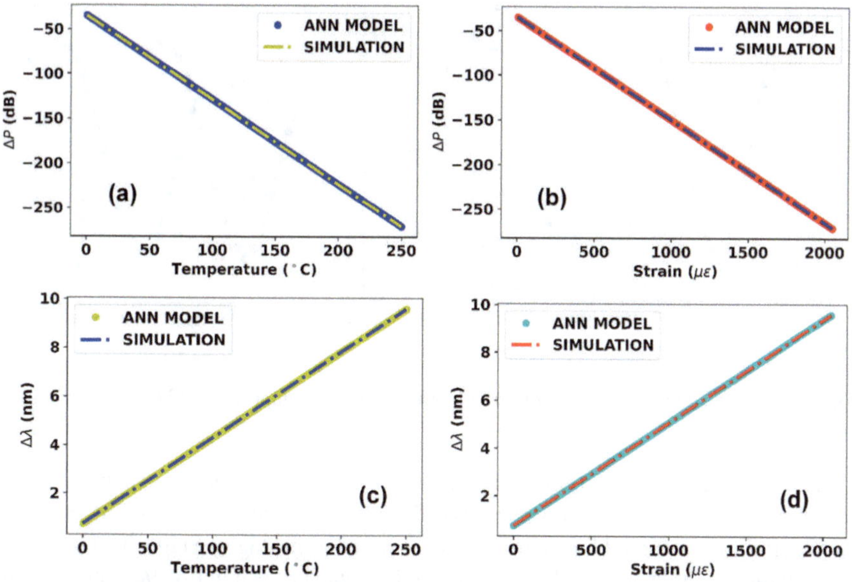

Fig. 10 The change in the power along with the wavelength due to the applied temperature and strain has been illustrated. **a, b** variation in the power and **c, d** change in the wavelength with temperature and strain, respectively for FBG [31]

6.2 MSM Based Sensing Configuration

Firstly, the change in the spectrum in the MSM at different temperatures and strains is studied in detail. The change in the wavelength for the two different dips of the MSM fiber structure have been demonstrated as a function of temperature and strain. The coefficients of the temperature and strain could be found from the plot and be added to the matrix equation.

To measure the wavelength shift, the fabricated MSM structure has been placed inside the oven to measure the wavelength shift with temperature as depicted in Fig. 9b. The temperature of the furnace was increased gradually to the target temperature and held for 2 min at each temperature at zero strain. The wavelengths for the two dips show a redshift as the temperature increases, as can be seen in the graphic. In Fig. 11, the wavelength shift for dip 1 is greater than that for dip 2.

According to Fig. 7, dip 1's wavelength is 1551.10 nm at 30 °C and about 1558.68 nm at 100 °C. As a result, dip 1's overall wavelength shift under the 30–100 °C temperature change is 7.58 nm. In contrast, dip 2's wavelength is 1543.84 nm at 30 °C and approximately 1547.79 nm at 100 °C. As a result, dip 2's overall wavelength shift under the 30–100 °C temperature change is 3.95 nm. As a result, during the 30–100 °C temperature change, dip 1's shift is nearly twice that of dip 2. Special precautions have been taken to prevent any unwelcome temperature fluctuations inside the oven because the MSM is greatly reactive to external disturbances.

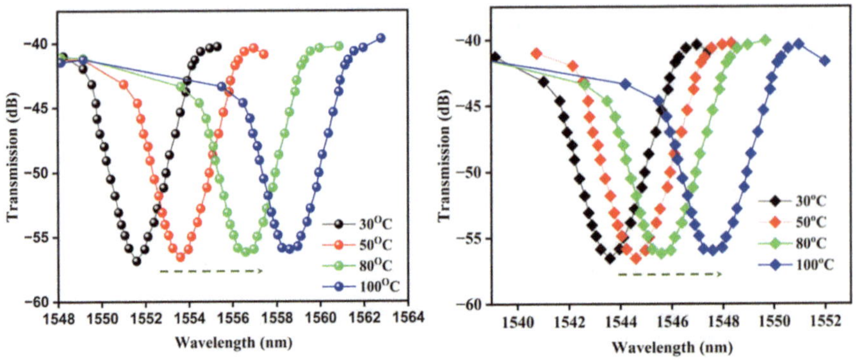

Fig. 11 Wavelength change for **a** dip 1 and **b** dip 2 at various temperatures [29]

The application axial strain has also a noticeable effect on the spectrum of the MSM structure. To analyze this impact on the proposed structure, the MSM cascaded configuration is attached between two micro translation stages. Stage, TS1 was kept fixed while the other stage (TS2) was movable as cab be seen in Fig. 9c. The conventional method is followed. TS2 is displaced to the right in order to induce strain on the cascaded fiber. The adapted strain is noted using the inscribed scale on the stage. Figure 12 shows that a shift towards lower wavelength is seen for the two dips when the strain on the manufactured MSM structure increases.

Dip 1 has a greater shift in Fig. 8a, but dip 2 exhibits a much smaller wavelength change in Fig. 8b.

To build the equation, the coefficient of the matrix must be found out. The matrix coefficient can be found by plotting the wavelength changes of the both spectra dips with variations in temperature and strain. As shown in Fig. 13, the co-efficient is calculated from the linear fitting.

Figure 13a, b show how strain and temperature affect both dips in the MSM spectra. The sensitivity for both dips show a monotonic change. Over the temperature

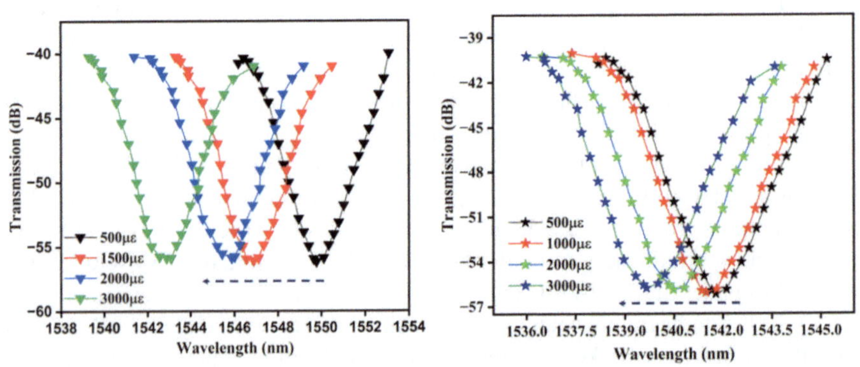

Fig. 12 Wavelength change for two dips: for **a** dip 1 and **b** dip 2 under various applied strains [29]

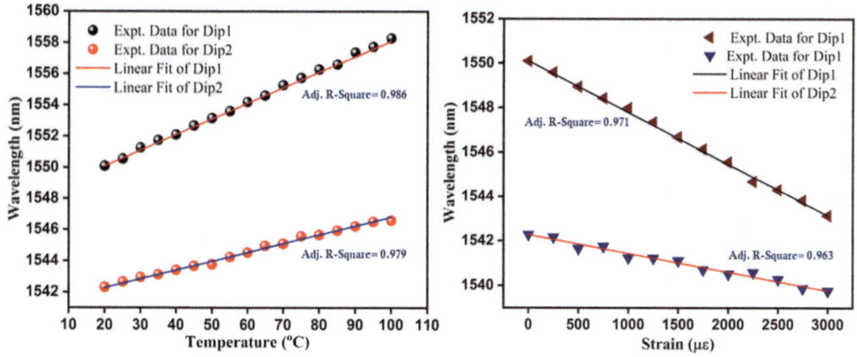

Fig. 13 Change of the wavelength with **a** temperature and **b** strain change, for the two captured dips [29]

range of 20–100 °C, the maximum shift for dip 1 and dip 2 is approximately 8.1 nm and 4.6 nm, respectively. For dip 1, the obtained sensitivity is around 105 pm/°C, while for dip 2, it is approximately 42 pm/°C.

With the variation in temperature in the SMF, the effective refractive index may also change in tandem with the dimension variations. As the temperature rises, so does the difference in effective refractive indices. The thermo-optic coefficient is also an essential parameter to measure the temperature details. Consequently, the selected dips will be moving towards the higher wavelength. Both dips show a blue shift for strain detection as the strain varies. The greatest wavelength shifts in dip 1 and dip 2 is roughly 7 nm and 2.5 nm, respectively, over the strain variation of 0–3000 με. For two dips, the obtained sensitivity is approximately 3.22 pm/με and 0.60 pm/με, respectively.

With the increase the axial strain on the SMF fiber section increases, the variations in the effective refractive index will diminish, which will cause the transmission dip to have a aptness to shift to a lower wavelength region.

After entering the parameter coefficients from the plots achieved from the experiment, the resultant transfer matrix equation looks like this:

$$\begin{pmatrix} \Delta T \\ \Delta \varepsilon \end{pmatrix} = \frac{1}{|D|} \begin{pmatrix} K_{\varepsilon 2} & -K_{\varepsilon 1} \\ -K_{T2} & K_{T1} \end{pmatrix} \begin{pmatrix} \Delta\lambda_1 \\ \Delta\lambda_2 \end{pmatrix} = \frac{1}{72.24} \begin{pmatrix} -0.60\,\text{pm}/\mu\varepsilon & 3.22\,\text{pm}/\mu\varepsilon \\ -42\,\text{pm}/^{\circ}\text{C} & 105\,\text{pm}/^{\circ}\text{C} \end{pmatrix} \begin{pmatrix} \Delta\lambda_1 \\ \Delta\lambda_2 \end{pmatrix} \tag{22}$$

For dip 1 and dip 2, the strain and temperature coefficients are denoted by the words $K_{\varepsilon 1}$, $K_{\varepsilon 2}$, K_{T1}, and K_{T2}, respectively. While $\Delta\lambda_1$ and $\Delta\lambda_2$ represent the wavelength shift for dip 1 and dip 2, respectively, ΔT and Δε represent the temperature and strain changes. D is the determinant of the matrix coefficient.

Lastly, a 10 min stability test is conducted at room temperature to investigate the impacts of wavelength drift and light source power fluctuation during a relatively long

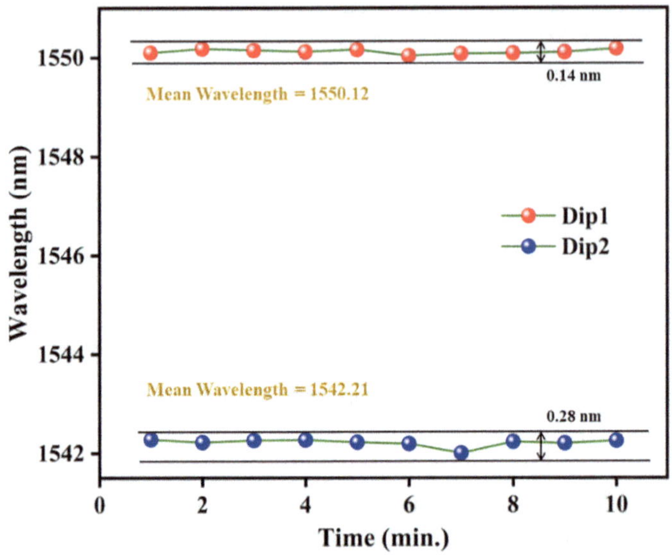

Fig. 14 Analysis the stability in terms of wavelength over 10 min [29]

working period. Figure 14 presents the numerical results. The wavelength fluctuations for dips 1 and 2 are estimated to be $\pm$ 0.14 and $\pm$ 0.28 nm, respectively. Dip 1 and Dip 2 have mean wavelengths of 1550.12 and 1542.21 nm, respectively.

Following experimentation, MATLAB is used to create a large dataset. The dataset is created by building a MATLAB matrix of the previously mentioned equation with the purportedly attained strain and temperature ranges at a tiny step size. 90% of the created dataset, or 10,000, is employed for training the model, and the rest 10%, or 1000, was utilizes for testing. The input set includes data sets of 9000 for ΔT and $\Delta\varepsilon$ for temperature and strain, respectively, throughout the ranges of 20–100 °C and 0–3000 $\mu\varepsilon$. Similarly, to effectively train the model, 10,000 data points were used for the two taken dips, $\Delta\lambda_1$ and $\Delta\lambda_2$.

Moreover, 1500 pairs of a new dataset of ΔT and $\Delta\varepsilon$ is employed for testing purposes, which was not used for model training. Temperature and strain measurements will be inaccurate due to the discrepancy the applied values with the estimated values.

The ANN is initially fed the aforementioned test datasets, and Fig. 15 shows the plotting of the wavelength changes of the two interference pattern dips with strain and temperature. The change in the wavelength with the temperature has been depicted in Fig. 15a while the change due to strain is depicted in Fig. 15b. For temperature there is a red shift while for strain a blue shift was noticed. The exceptionally high linearity of the resulting plots guaranteed that the suggested model would act as planned. For verification, a simulation is run using the inverse matrix technique. There is a great degree of coincidence between the graphs generated by the ANN model and the simulation. The new matrix that was produced by the model is:

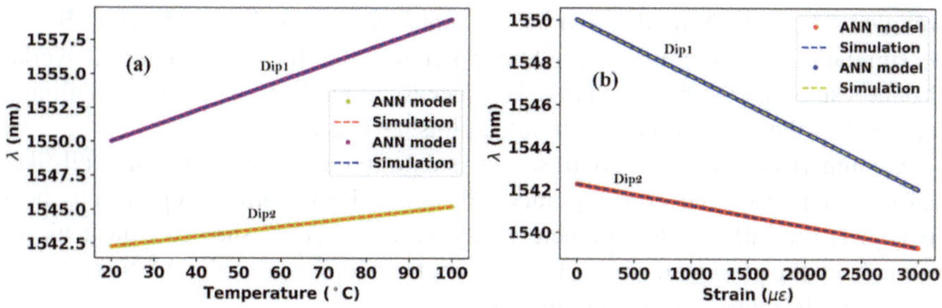

Fig. 15 Using the simulation data and the suggested machine learning model, ascertain how the wavelengths of the dips altered with **a** temperature and **b** strain [29]

$$\begin{pmatrix} \Delta T \\ \Delta \varepsilon \end{pmatrix} = \frac{1}{88.11} \begin{pmatrix} -0.69\,\text{pm}/\mu\varepsilon & 3.46\,\text{pm}/\mu\varepsilon \\ -48\,\text{pm}/^\circ\text{C} & 113\,\text{pm}/^\circ\text{C} \end{pmatrix} \begin{pmatrix} \Delta\lambda_1 \\ \Delta\lambda_2 \end{pmatrix} \tag{23}$$

Using the suggested model, the obtained sensitivity values for temperature and strain are 113 pm/°C and 3.46 pm/με, respectively. The MATLAB simulation is utilized to create the dataset used for the input of the ANN model rather than directly incorporating the experimental data. As a result, the provided dataset is more linear and coherent. We think that greater sensitivities are seen because the dataset is more consistent. Furthermore, the low cross-sensitivity necessary for a successful strategy is shown by the matrix's higher determinant value. Furthermore, the measured strain and temperature have RMSEs of 0.146 με and 0.017 °C, respectively. Notably, in compare with the popular matrix technique, the improvement from using the suggested ANN model is roughly 1295 times and 303 times more.

To execute the proposed method in practical situations, the model must be trained on unknown data chosen at random during the real measurement. This has been confirmed using an additional 1000 data that were not utilized for model testing or training. The findings are shown in Fig. 16.

The trained dataset's actual and projected values for temperature and strain are shown in a scattered plot in Fig. 16. The expected strain and temperature values are

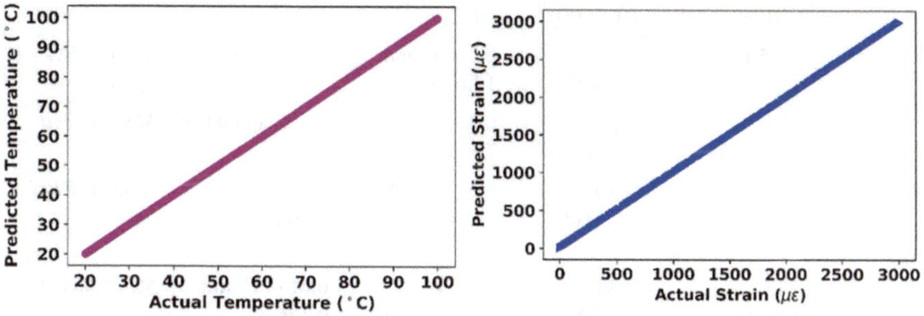

Fig. 16 A scatter plot of the datasets containing the trained one for **a** the temperature and **b** the strain to compare the real and projected values [29]

provided by our ANN model, and the real data are collected using MATLAB after the simulation. A continuous rising linear plot is visible for every scenario. Overlap of data results in continuous charts. The successful application of the modelling is validated by the appropriate match in the plots having a higher R^2.

A comparison table that contrasts our suggested approach with the well-liked matrix mechanism and contemporary ML-assisted comparable reports has been displayed to amplify it. The parameters highlighted in "bold" showcase the achieved results from the presented works. It is evident from Table 3 that the ANN model can enhance performance to a greater extent.

Employing higher-order polynomial with the matrix could decrease the RMSE that was experimentally achieved from the matrix which include a bit complexity to the matrix. For this aspect, a particular numerical model is also required. Furthermore, if there is a serious problem with the cross-sensitivity, this accession would not be

Table 3 A comparison of the transfer matrix and the ANN in many sensing metrics [29]

Methods	Temperature sensitivity (pm/°C)	Strain sensitivity (pm/με)	Resolution	RMSE	References
Transfer matrix method	12.60 (22–90)	1.14 (0–800)	0.26 °C and 9.21 με	–	[1]
	11.70 (40–300)	1.20 (0–500)	–	–	[2]
	47.00 (20–60)	2.20 (0–900)	–	–	[4]
	33.00 (0–50)	0.74 (0–708)	0.31 °C and 10.01 με	–	[4]
	11.65 (40–80)	–	0.3 °C and 21 με	–	[5]
	10.45 (100–450)	2.97 (0–1000)	1.91 °C and 6.7 με	–	[6]
	20.00 (− 100 to 300)	2.85 (0–1000)	–	–	[7]
	91.84 (30–90)	− 2.37 (0–8000)	0.41 °C and 0.07 με	–	[8]
	21.50 (20–60)	2.79 (0–900)	0.50 °C and 10.0 με	17 °C and 32 με	***SMS+FBG based***
	105 (20–100)	3.22 (0–3000)	0.33 °C and 7.6 με	22 °C and 44 με	***MSM based***
ANN	33.18 (25–250)	5.30 (0–2100)	6.6×10^{-4} °C and 4.0×10^{-3} με	0.025 °C and 0.190 με	***SMS + FBG based***
	113 (20–100)	**3.46 (0–3000)**	**8.85 × 10^{-5} °C and 2.89 × 10^{-3} με**	**0.017 °C and 0.146 με**	***MSM based***

recommended. Instead, even with non-linear and error-prone datasets of input, the ANN model might be applied to a broad range of sensing tactics.

7 Conclusion and Future Scope

In this chapter, a novel ANN-assisted fiber optic sensing system for simultaneous measurement of temperature and strain has been proposed and successfully demonstrated. Two different types of cascaded fiber optic sensing systems namely MSM and SMS-FBG based sensing approaches have been taken into account. The proposed edge-filtering based experiment has been continued in detail to gather input and output data for the SMS-FBG structure. The employment of the proposed ML-based edge-filtering methodology has significantly lessened the cross-sensitivity impact for measuring two parameters simultaneously in compare with the conventional matrix method. The ANN model produced strain and temperature sensitivities that were roughly 2.2 and 1.75 times greater, respectively, than the traditional matrix inversion method. Measurements of strain and temperature produced RMSEs of 0.195 $\mu\varepsilon$ and 0.025 °C, respectively. In comparison to the matrix method, the RMSE values for the temperature and strain readings decreased by 680 and 164 times, respectively. As a result, there was a notable improvement. In contrast, the measured RMSE for temperature and strain for the MSM-based structure is 0.017 °C and 0.146 $\mu\varepsilon$, respectively. In comparison to the transfer matrix approach, this guarantees a temperature increase of about 1294 times and a strain increase of about 301 times. Consequently, a noteworthy progress has been achieved. A straightforward and easily fabricated MSM fiber architecture, which has higher intrinsic sensitivity compared to the conventional FBG, takes its place. In comparison with the ML-assisted MSM technique ($D = 88.21$) to the ML-based FBG strategy ($D = 3.67$) yields a determinant value which is 25 times larger. This suggests that the effect of cross-sensitivity was significantly reduced [42]. The arguments undoubtedly support the assertions that the presented work is noticeably more innovative than the previously published FBG-based ML-assisted technique.

Thus, in this case, the suggested methodology validates a successful application of concurrently measuring two parameters. Consequently, the proposed ML-driven edge filtering technique may be used to various kinds of sensing configuration with a larger performance enhancement. For distributed sensing applications and for employing a large FBGs to follow health of the structural issues, this proposed technique could be further enhanced.

References

1. D.P. Zhou, L. Wei, W.K. Liu, Y. Liu, W. John, Y. Lit, Simultaneous measurement for strain and temperature using fiber Bragg gratings and multimode fibers. Appl. Opt. **47**, 1668–1672 (2008)
2. L. Xiong, D. Zhang, L. Li, Y. Guo, EFPI-FBG hybrid sensor for simultaneous measurement of high temperature and strain. Chin. Opt. Lett. **12**, 120605–120609 (2014)
3. A. Iadicicco, R. Ranjan, F. Esposito, S. Campopiano, Arc-induced long period gratings in polarization-maintaining panda fiber. IEEE Photonics Technol. Lett. **29**, 1533–1536 (2017)
4. Q. Liu, Z.L. Ran, Y.J. Rao, L. Shu-cheng, Y. Hui-Qin, H. Ya, Highly integrated FP/FBG sensor for simultaneous measurement of high temperature and strain. IEEE Photonics Technol. Lett. **26**, 1715–1717 (2014)
5. X. Yu, N. Song, J. Song, A novel method for simultaneous measurement of temperature and strain based on EFPI/FBG. Opt. Commun. **459**, 125020 (2020)
6. J. Tian, Y. Jiao, S. Ji, X. Dong, Y. Yao, Cascaded-cavity Fabry–Perot interferometer for simultaneous measurement of temperature and strain with cross sensitivity compensation. Opt. Commun. **412**, 121–126 (2018)
7. Z. Kang, X. We, C. Li, J. Sun, J. Wang, S. Jian, Up-taper-based Mach–Zehnder interferometer for temperature and strain simultaneous measurement. Appl. Opt. **53**, 2691–2695 (2014)
8. G. Zhang, Q. Ge, H. Wang, X. Wu, B. Yu, Stress applying TPMF based sensor for simultaneous temperature and strain measurement. Optik **254**, 168636 (2022)
9. K. Dey, R. Buddu, S. Roy, Performance of etched silica FBG for simultaneous strain temperature measurement. Silicon **14**, 4349–4356 (2022)
10. S. Sridevi, K.S. Vasu, S. Asokan, A.K. Sood, Enhanced strain and temperature sensing by reduced graphene oxide coated etched fiber Bragg gratings. Opt. Lett. **41**, 2604–2607 (2016)
11. K. Dey, S. Roy, P. Kishore, M. Sai Shankar, R.K. Buddu, R. Ranjan, Analysis and performance of edge filtering interrogation scheme for FBG sensors using SMS fiber and OTDR. Results Opt. **2**, 100039 (2021)
12. K. Dey, S. Roy, Detailed investigation of spectral properties of SMS fiber segment and its sensing performance under varying multimode fiber lengths. Infrared Phys. Technol. **127**, 104356 (2022)
13. K. Dey, S. Roy, Analyzing spectral properties and sensing performance of multi-single-multi mode fiber combination. Phys. Scr. **97**, 125507 (2022)
14. Z. Yang, H. Xu, K. Ni, X. Dong, Simultaneous measurement of force and temperature with a single FBG partially encapsulated with a metal canulla. Microw. Opt. Technol. Lett. **53**, 1656–1658 (2011)
15. W. Zhou, C.-L. Zhao, X. Dong, S. Zhang, C.C. Chan, P. Shum, Simultaneous measurement of force and temperature based on a half-corroded FBG. Microw. Opt. Technol. Lett. **52**, 2020–2023 (2010)
16. W.G. Zhang, X.Y. Dong, Q.D. Zhao, G.Y. Kai, S.Z. Yuan, FBG-type sensor for simultaneous measurement of force and temperature based on bilateral cantilever beam. IEEE Photonics Technol. Lett. **13**, 1340–1342 (2001)
17. K. Dey, V.D.R. Pavan, S. Roy, M. Sai Shankar, B. Ramesh, Interrogation of SMS for measuring of temperature and strain using half-etched FBG with enhanced sensitivity, in *Proceedings of SPIE 11355, Micro-Structured and Specialty Optical Fibres VI* (2020), p. 113550Z
18. X. Li, Y. Zhang, W. Zhang, T. Yan, Y. Zhang, High-sensitive optical force sensor based on enhanced effective index modulation. Optik **168**, 684–691 (2018)
19. K. Dey, V.D.R. Pavan, S. Roy, M. Sai Shankar, R. Buddu, Interrogation of SMS for measuring force and temperature using etched FBG with enhanced sensitivity, in *Proceedings of SPIE* (2020), pp. 113550Z1–6
20. J.K. Sahota, N. Gupta, D. Dhawan, Fiber Bragg grating sensors for monitoring of physical parameters: a comprehensive review. Opt. Eng. **59**, 060901 (2020)
21. Y. Ma et al., Torsion and temperature sensor based on polished MSM structure. IEEE Photonics Technol. Lett. **32**, 1117–1120 (2020)

22. B. Yin, Y. Li, Z. Liu, S. Feng, Y. Bai, Y. Xu, S. Jian, Investigation on a compact in-line multimode-single-mode-multimode fiber structure. Opt. Laser Technol. **80**, 16–21 (2016)
23. Z. Akbarpour, V. Ahmadi, and F. Arabpour Roghabadi, Enhanced Mach–Zehnder interferometer multimode–single-mode–multimode fiber optic refractive index sensor based on surface plasmon resonance. Preprints 2021110544 (2021). https://doi.org/10.20944/preprints202111.0544.v1
24. J. Sun, C.C. Chan, K.M. Tan, X.Y. Dong, P. Shum, Application of an artificial neural network for simultaneous measurement of bending curvature and temperature with long-period fiber gratings. Sens. Actuators A Phys. **137**, 262–267 (2007)
25. N. Vangety, K. Dey, S.K. Ghosh, S. Roy, Analyzing specklegrams of plastic optical fiber using convolutional neural network for weight recognition. Opt. Eng. **61**(12), 126112 (2022)
26. F.O. Barino, A.B. dos Santos, LPG interrogator based on FBG array and artificial neural network. IEEE Sens. J. **20**(23), 14187–14194 (2020)
27. K. Dey, V. Nikhil, S. Choudhary, S. Roy, Analysis of the optical properties of a fiber Bragg grating using machine learning approach, in *2021 IEEE Photonics Conference (IPC)* (2021), pp. 1–2. https://doi.org/10.1109/IPC48725.2021.9592988
28. A.C. Zimmermann, C.L.N. Veiga, L.S. Encinas, Unambiguous signal processing and measuring range extension for fiber Bragg gratings sensors using artificial neural networks—a temperature case. IEEE Sens. J. **8** (2008)
29. K. Dey, N. Vangety, S. Roy, Machine learning approach with higher accuracy for simultaneous measurement of the dual parameter by MSM fiber structure. Measurement **221**, 113426 (2023)
30. G.C. Kahandawa, J.A. Epaarachchi, H. Wang, D. Followell, P. Birt, Use of fixed wavelength fibre-Bragg grating (FBG) filters to capture time-domain data from the distorted spectrum of an embedded FBG sensor to estimate strain with an artificial neural network. Sens. Actuators A Phys. **194** (2013)
31. K. Dey, N. Vangety, S. Roy, Machine learning approach for simultaneous measurement of strain and temperature using FBG sensor. Sens. Actuators A Phys. **333**, 113254 (2021)
32. J. Sun, C. Chan, X. Dong, P. Shum, Application of an artificial neural network for simultaneous measurement of temperature and strain by using a photonic crystal fiber long-period grating. Meas. Sci. Technol. **18**(9), 2943 (2007)
33. N. Vangety, K. Dey, S. Roy, Analyzing specklegrams of multimode fiber using convolutional neural networks for stress recognition. Opt. Fiber Technol. **75**, 103166 (2023)
34. M. Basu, S. Ghorai, Sequential interrogation of multiple FBG sensors using LPG modulation and an artificial neural network. Meas. Sci. Technol. **26**(4), 045104 (2015)
35. N. Vangety, K. Dey, S.K. Ghosh, S. Roy, Quality assessment and statistical analysis of specklegrams in a weight induced plastic optical fibers. Opt. Eng. **61**, 126112 (2023)
36. L.V. Nguyen, C.C. Nguyen, G. Carneiro, H. Ebendorff-Heidepriem, S.C. Warren-Smith, Sensing in the presence of strong noise by deep learning of dynamic multimode fiber interference. Photon. Res. **9**, B109–B118 (2021)
37. Z. Ding, Z. Zhang, 2D tactile sensor based on multimode interference and deep learning. Opt. Laser Technol. **136**, 106760 (2021)
38. S. Lu, Z. Tan, G. Li, J. Yang, A sensitized plastic fiber sensor for multi-point bending measurement based on deep learning. IEEE Photonics J. **13**, 1–7 (2021)
39. D.L. Smith, L.V. Nguyen, D.J. Ottaway, T.D. Cabral, E. Fujiwara, C.M.B. Cordeiro, S.C. Warren-Smith, Machine learning for sensing with a multimode exposed core fiber specklegram sensor. Opt. Express **30**, 10443–10455 (2022)
40. H.S. Efendioglu, T. Yildirim, O. Toker, K. Fidanboylu, Design of intelligent fiber optic statistical mode sensors using novel features and artificial neural networks, in *2012 International Symposium on Innovations in Intelligent Systems and Applications* (2012), pp. 1–4
41. Y. Liu, G. Li, Q. Qin, Z. Tan, M. Wang, F. Yan, Bending recognition based on the analysis of fiber specklegrams using deep learning. Opt. Laser Technol. **131**, 106424 (2020)
42. F. Farahi, D.J. Webb, J.C.C. Jones, D.A. Jackson, Simultaneous measurement of temperature and strain: cross-sensitivity considerations. J. Lightwave Technol. **8**, 138–142 (1990)

Generative AI Models for Sensor Data Processing

R. Kanthavel, R. Adline Freeda, and R. Dhaya

Abstract It has revolutionized the production of realistic data in various disciplines and had increasingly important applications in sensor data processing. This work explores the Generative Artificial Intelligence (AI) models applied to sensor data, including diffusion models, variational auto encoders, and Generative Adversarial Networks. Sensor data often contains noise, missing values, and non-stationary behaviors that degrade the capabilities of traditional processing methods. Sensor data is widely used in Internet of Things (IoT) applications, healthcare data analysis, and environmental monitoring. Generative AI models will overcome such problems by generating realistic, high-quality sensor data and thereby improve sensor fusion, anomaly detection, and data augmentation. Hence, these are crucial models for real-time monitoring and decision-making in sensor-based systems by creating new sensor values while also allowing better accuracy of data points and even predicting missing values. This chapter discusses applications, difficulties, and future research prospects while giving a broad overview of cutting-edge generative AI algorithms and how they've been tailored to deal with sensor data.

Keywords Generative Artificial Intelligence (AI) models · Sensor data · Internet of Things (IoT) · Real-time monitoring · Decision-making

1 Introduction to Sensor Data Processing

Sensor fact processing may contain information from all varieties of sensors, including healthcare, production, transportation, environmental monitoring, and smart cities. Conversely, sensors interact with the physical or environmental environment, delivering real-time information about temperature, pressure, humidity, mild

R. Kanthavel · R. Dhaya (✉)
School of Electrical and Communications Engineering, PNG University of Technology, Lae, Papua New Guinea
e-mail: dhayavel2005@gmail.com

R. Adline Freeda
Department of Information Technology, KCG College of Technology, Chennai, India

Y. Vasimalla and S. Kumar (eds.), *Generative AI for Photonic Sensing*, Progress in Optical Science and Photonics 36, https://doi.org/10.1007/978-981-95-1561-5_4

motion, and biochemical reactions. Sensor information streams play a crucial role in decision-making, automation, and predictive insights for real-time structures [1].

- **Sensor Statistics Types**: Sensor statistics can be of many types, depending entirely on the type and utilization of sensors. The most common instructions are listed below:
  - **Environmental Facts**: Temperature, humidity, air excellent.
  - **Motion and Positioning Statistics**: Accelerometer, gyroscope, GPS Heart rate, blood pressure, and oxygen stages are biometric facts.
  - **Acoustic and Visible Statistics**: Sound waves, camera feeds, and video feeds.

1.1 Characteristics of Sensor Data

Large-scale sensor statistics typically exhibit the following characteristics:

- **High Volume**: The deployment of a wide range of sensors and the large volumes of raw data suggest that the flow of statistics is unending.
- **Noisy and Uncertain Information**: Information from sensors can be noisy or incomplete due to environmental interference or sensor malfunctions; therefore, preprocessing and cleaning techniques must be state-of-the-art.
- **Real-Time Constraints**: Many programs require real-time or near-real-time processing of sensor data to make quick decisions, such as in healthcare tracking or driving.
- **Time-Based Absolutely Nature**: Sensor readings are implicitly time-series records [2].

Hence, the evaluation includes temporal dependencies and styles. Problems with sensor data processing Issues with sensor statistics processing embody:

- **Inaccuracies and Noise**: Environmental noise or hardware issues can cause sensors to become noisy or provide inaccurate readings. We should strive to implement robust filtering and smoothing techniques.
- **Missing Values**: Real sensors occasionally malfunction and retrieve incomplete information, which complicates additional analyses.
- **Heterogeneity**: Multisensor streams can be heterogeneous; consequently, they're able to be specific with several codecs, gadgets, and scales; as a result, integration and an analytical approach end up very hard for the purpose that they come from precise origins.
- **High Dimensionality**: Data generated through the hundreds and lots of sensors in a multi-sensor community can from time to time be too excessive and for this reason require applications of dimensionality discount and data summarization.

Non-traditional sensor data processing methods Traditionally, we executed sensor data processing using statistical and signal processing techniques for noise filtering, characteristic detection, and feature extraction [3]. The following are some examples of the most commonly implemented methods:

- The Kalman filter estimates the tool's kingdom from noisy observations. The artwork takes the form of a recursive set of hints.
- Fourier Transformation is a methodology that converts time-domain information into frequency-domain information suitable for signal processing.
- **Time-Series Analysis**: Usually, we apply these techniques, which rely on ARIMA models, to predict future values using only statistics. While those traditional approaches are appropriate in their respective rights, several challenges relate to noisy, complex, high-dimensional data streams from modern sensors. The need for advanced AI techniques
- Therefore, due to the scaling of such sensor networks and their particularly complex nature, traditional record processing approaches are unable to meet the demanding situations of handling large volumes of data, locating lost styles, and making accurate predictions. The end-to-end outcome under consideration for change was Artificial Intelligence (AI), a subset of Generative AI, which models complex distributions to generate synthetic facts for augmentation or perhaps to reveal hidden styles in sensor networks [4].

Generative AI models address several challenges in sensor statistics processing, including fact augmentation, imputation, prediction, and filling in missing sensor values, thereby enhancing the quality and continuity of data streams. The path goes through an area where generative AI models, like GANs, VAEs, and diffusion models, are changing how sensor data is processed and giving us new ways to deal with noise, missing data, and large sensor networks.

2 Generative AI Overview

In other words, generative AI refers to a class of machine learning algorithms that possess the ability to produce new data that bears resemblance to their trained data. Unlike traditional AI models, generative models concentrate on tasks like classification, regression, and pattern recognition. On the contrary, they try to create new instances of data with features similar to the input data. These have been very successful models in applications across image generation, natural language processing, and time-series forecasting. They find their increased application in sensor data processing.

What Is Generative AI: Generative AI refers to a set of models that, based on a specific dataset, learn an underlying pattern and distribution in order to generate new samples from the learned distribution. This method produces new data points that are nearly indistinguishable from the provided training data sets, enabling its application in the following areas:

- **Data Augmentation**: The process of generating more diverse training data for a machine learning model.
- **Anomaly Detection** is the detection of anomalies, or data points that are different from the rest, by comparing them against what the model has learned to define as "normal" data.
- **Data Imputation**: This entails filling in missing data or data that may have become corrupt [5].

2.1 Types of Generative AI Models

The rise of generative models has had a significant impact on the development of a number of different architectures. Some of the most common types of generative AI models include (Fig. 1):

VAEs: Variational Autocoders (VAE) are potential models designed to learn the latent structure of data by encoding the entrance to a low-dimensional latent location and then organizing the data from this representation. This process allows the latent site to capture the most important properties of data, so that generation of new samples can only be activated by sampling from this continuous site. VAE-er is widely used

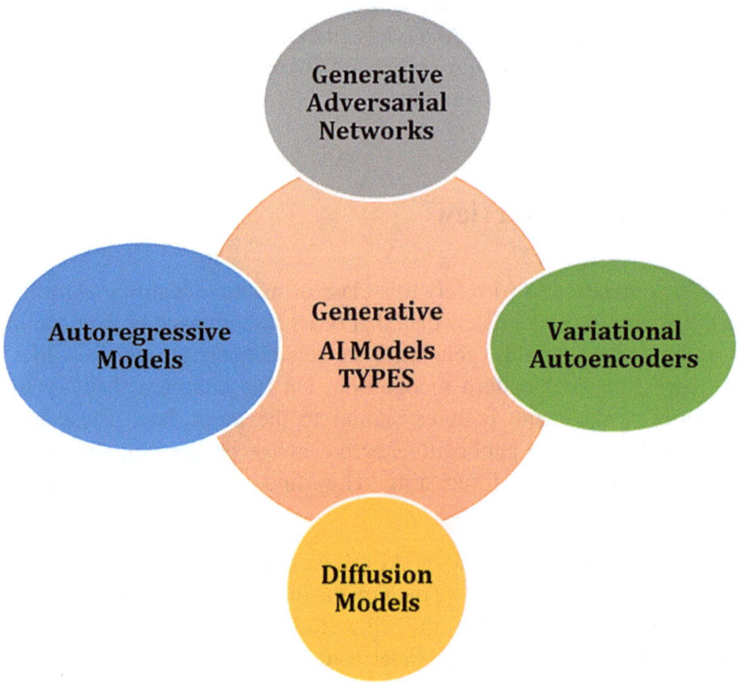

Fig. 1 Types of generative AI models

in applications such as restoring computer structure, deviations and generating new sensor data based on pattern in historical data sets. One of the biggest benefits of VAE-er is his ability to manipulate the latent space, which provides a high level of control over generated data. This feature makes them particularly effective for continuous data production, including features such as time series data synthesis.

Diffusion Models: Diffusion models are a type of generative model that generate data through a process of progressively adding noise to real data and then reversing this operation to remove the noise. This iterative process allows the model to learn how to generate complex, high-dimensional data effectively. Diffusion models have demonstrated exceptional performance in tasks requiring the generation of intricate data structures. Common use cases include time-series generation, sensor data prediction, and environmental simulations. One of the key strengths of diffusion models is their ability to handle data with deep temporal dependencies, making them especially well-suited for tasks like sensor data streams where maintaining temporal coherence is crucial.

Autoregressive Models: Authorized models predict each item depending on the items already generated, making them very effective for sequential data. Examples include GPT for text generation and sound synthesis. These models are widely used in time chain forecasts, generation of sensor data and future maintenance. Their most important strength lies in capturing dependence between past and future states, which leads them to be properly for applications where sequential relationships are important.

2.2 Generative AI Workflow

- The system to generate generative AI fashions has the following steps, which can be exceedingly big for the training and alertness of such fashions. Figure 2 illustrated the Generative AI Work flow.
- **Data Collection**: Sensors collect appropriate consultant statistics, with bureaucracy serving as the central component of the training process.
- **Preprocessing**: Sensor records normally result in quite a bit of noise and incompleteness; consequently, a few applied preprocessing techniques could consist of normalization, filtering, and handling missing values to clean the records.
- **Model Training**: We perform the training of the generative model, GAN-VAE, in conjunction with the preprocessing of sensor records, enabling it to learn the underlying statistical distribution.
- **Data Generation**: The model, once trained, can create new sensor records or decorate the prevailing one by simulating synthetic sensor readings with residences of the actual authentic information [6].
- It is crucial to smoothen or de-noise the generated information during the submit-processing step to ensure its suitability for downstream tasks, such as anomaly detection or predictive renovation, in real-time.

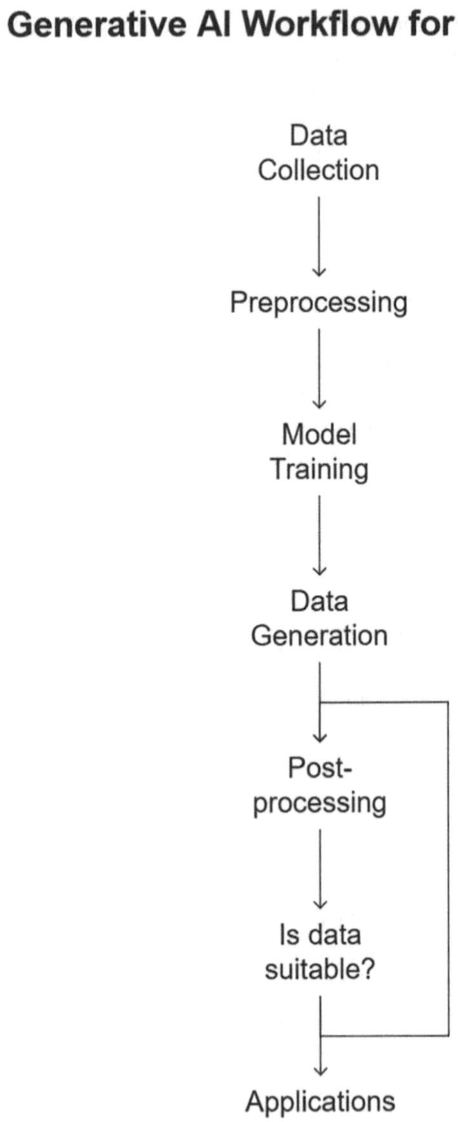

Fig. 2 Generative AI workflow

Generative AI models show promising outcomes inside several industries. Key programs incorporate data augmentation through the generation of enhanced sensor statistics to train more robust machine learning models, particularly when labelled data is scarce, difficult, or expensive to obtain. Other software areas include anomaly detection, where generative models investigate the typical behaviour of a sensor network to identify anomalies, faults, or outliers.

Predictive Maintenance involves predicting future sensor values or device disasters by generating plausible future states based on current and historical data. Simulation: This process generates synthetic sensor data for use in real-time simulations and testing environments, such as growing or simulating self-riding motors or intelligent cities. The advantages of generative AI in sensor data processing.

Generative AI models bring several benefits to the field of sensor record processing.

- **Improvement of Quality of Data**: Sensor data may be noisy or lacking in certain factors; generative AI models can fill in these gaps by generating new, high-fidelity readings from the sensors.
- **Smoothing Out Sparse Data**: When sensors are sparse or only occasionally sampled, generative models may provide synthetic data points to fill in missing points or time gaps.
- **Reducing Computation Overhead**: Through the use of generative artificial statistics, which presents variability in assessment to real-world records, generative AI eliminates the need for extensive sensor deployment, thereby saving resources and reducing computational overhead.

Generative AI models demonstrate their strength in handling challenging problems such as developing noisy, high-dimensional, real-time data from sensors. Some of the main architectures used for sensor data processing include GANs, VAEs, and diffusion models. These architectures equip data with attributes that facilitate aspects of augmentation, anomaly detection, and predictive modelling. The growing demand for more intelligent smart systems and real-time data analysis necessitates generative AI to refine the quality, reliability, and utility of sensor data in all applications [7].

3 Variational Auto Encoders (VAEs) in Sensor Data Processing

Variational auto encoders represent a powerful class of generative models that gain much attention nowadays in several fields, such as sensor data processing. This is probably because it holds an approach to learning latent representations of data in tasks such as data reconstruction, anomaly detection, and generation of synthetic sensor data. The following chapter will provide an overview of what VAEs are and how they relate to the processing of sensor data, including their applications within real-world domains. Variational Autoencoders (VAEs) are a class of generative models that combine the strengths of autoencoders with probabilistic modeling. Unlike traditional autoencoders, VAEs consist of two main parts: an encoder and a decoder.

- **Encoder** (also called the compressor): The encoder compresses input data into a lower-dimensional latent space, where more important features are stored. In contrast to traditional autoencoders, which map the input data to a single point in the latent space, VAEs map the data to a probability distribution, typically a Gaussian distribution. This probabilistic approach allows VAEs to sample from this distribution and generate new data, providing them with the ability to generate diverse and smooth variations of the trained data.

- **Decoder**: The decoder reconstructs the input data from the latent space representation. Since the data is represented as a distribution rather than a single point, the decoder can generate more diverse and varied outputs.

Training VAEs

VAEs are trained by minimizing a loss function that balances two key components:

1. **Reconstruction Loss**: Ensures the model can reproduce the input data from the latent space representation.
2. **Kullback–Leibler (KL) Divergence**: Encourages the latent distribution learned by the encoder to be close to a predefined prior distribution, usually a standard normal distribution.

The combination of these two objectives allows VAEs to generate smooth interpolations in the latent space and control the data generation process. This makes VAEs useful for tasks like data augmentation, denoising, and missing data imputation.

Advantages of VAEs for Sensor Data: VAEs are particularly effective in handling noisy, high-dimensional, and complex sensor data. Their probabilistic nature allows them to handle uncertainty and variability inherent in sensor networks. Key applications of VAEs include:

- **Anomaly Detection**: VAEs can detect anomalies by identifying data points that deviate significantly from the expected reconstruction, making them useful for detecting outliers in sensor data.
- **Filling in Missing Sensor Readings**: By predicting missing values from the learned latent distribution, VAEs can efficiently fill in gaps in sensor data.
- **Synthetic Data Generation**: VAEs can generate synthetic sensor data, enriching training datasets for machine learning models and improving their performance [8].

Use Cases and Applications: Despite some computational complexity, VAEs are a powerful tool in fields like healthcare, industrial monitoring, and environmental sensing. They help solve challenging data processing tasks, including:

- **Healthcare**: Improving diagnostic models by handling noisy or incomplete medical sensor data.
- **Industrial Monitoring**: Enhancing predictive maintenance models by filling in missing sensor data and detecting anomalies.
- **Environmental Sensing**: Generating synthetic data for environmental monitoring, such as predicting air quality or weather patterns.

VAEs are highly versatile and effective models for handling complex sensor data, offering solutions for anomaly detection, data augmentation, and missing data imputation.

GANs were primarily designed to generate images and videos, but their powerful ability to render more realistic data compelled a lot of attention. Subsequently, GANs have been applied widely to many forms of data, including sensor data. By addressing such critical concerns as augmentation of data, reduction of noise, and anomaly detection in sensor data, this section takes up how GANs work in dealing with sensor data, the application to sensor data processing, and real-world use cases.

3.1 How GANs Work?

GANs consist of two neural networks:

- **Generator**: The task of the generator is to generate synthetic data that closely approximates the real data.
- **Discriminator**: The discriminator, as the name implies, aims to distinguish between real data and the data generated by the generator.

The Two Networks Engage in a Competitive Process: The discriminator verifies whether the sample data are genuine and coming from the dataset or fake, generated by the generator. The generator is improved iteratively; at some point, it starts to generate more realistic data, whereas the discriminator learns to distinguish fake data correctly. This adversarial training continues until the generator delivers output that is almost indistinguishable from the real.

3.2 Why GANs for Sensor Data?

Sensor data pose particular challenges for which GANs are well suited:

- **Data Augmentation**: In many real-world applications, real sensor data is relatively scarce or expensive. So, GANs further enable the generation of synthetic sensor data retaining all the characteristics of real data to create bigger sets on which machine learning models can be trained [9].
- **Noise Reduction**: The sensor data itself also comprises much noise. Usually caused due to environmental reasons, hardware imperfections, or errors during the transmission process. A GAN can generate noise-free data because it can learn the underlying distribution of clean data and then create it in the absence of noise.
- **Anomaly Detection**: GANs can learn the "normal" distribution of sensor data; by recognising that any significant deviation from the learnt distribution could be marked as anomalous, it allows for timely detection of anomalies in sensor malfunctions or abnormal system behaviour.

3.3 Applications of GANs to Sensor Data Processing

GAN have emerged as a powerful tool in data processing, which enables the production, growth and growth of sensor-based information. Traditional sensor systems often suffer from problems such as noise, lack of values and insufficient labels, which can affect the accuracy and reliability of the machine learning model. The GANS sensor addresses these challenges by learning the underlying pattern in data and generating realistic data sets that have not yet been used for training strong models. The use of GANs extends into different domains, including non-conformity detection, future maintenance and data increase. GANs improves the decision in industries such as health care, production, autonomous systems and smart cities by mimicking realistic sensor reading. This introduction suggests how GAN sensors change data processing, improves data quality, more accurate and effective AI-powered solutions.

Data Augmentation for Machine Learning Models: Data growth is an important technique in machine learning that improves the model performance by artificially expanding the training dataset. This involves generating new data points from existing people through various changes, such as rotation, flipping, scaling and noise addition. This approach helps improve the model better, reduce the overfitting and improvement of the strength, especially when working with limited or unbalanced datasets. In deep education, data text is widely used in data vision (e.g., crop and color nervous image technology), natural language treatment (e.g., by imitating the diversity variations of data, enables the growth model to learn more representative patterns, which improves accuracy and adaptability in real-world applications.

- **Problem**: Sensor data creation is challenging and costly in many areas, such as in health care, environment monitoring, or industrial systems, particularly if large amounts of labelled sensor data need to be created.
- **Solution**: GANs synthesise real-looking statistical data from synthetic sensor readings so that a bigger dataset with more diversification can be constructed. The ability becomes very helpful, especially for imbalanced datasets, and some sensor events are rare, like rare machine failures.
- **Example**: GANs can be applied to generate synthetic accelerometer data that will be used for the prediction of the occurrence of failure in industrial equipment, and as more failure cases are incorporated in the training set, the model is better at predicting the early signs of equipment failure.

Noise Filtering and Denoising: This approach of using GANs for noise filtering significantly improves the quality of sensor data, enabling more reliable decision-making in real-time systems and applications.

- **Issue**: The sensors suffer from hardware and environmental conditions like temperature change fluctuations and interference.

- **Solution**: Noise is to be removed while important information needs to be preserved so that critical information will be retrieved with precise data analysis accuracy. A GAN can be used to map noisy sensor data to its noise-free version; thus, it learns the noise pattern and removes it in the output.
- **For instance**, in environmental monitoring, sensor data acquired might be noisy due to weather conditions. A GAN trained on clean historical data can create a noise-free version of the current sensor readings, thus enhancing the accuracy of real-time environmental predictions [10].

Sensor Network Analysis Detection: Sensor network analysis and detection is a critical aspect of managing and monitoring large-scale sensor systems. In various applications, such as environmental monitoring, healthcare, industrial IoT, and smart cities, sensor networks are designed to gather data from numerous devices spread across different locations. These systems generate vast amounts of data, which can be prone to errors, anomalies, and even security threats. Analyzing and detecting issues within these networks is essential for ensuring data accuracy, reliability, and security.

- **Problem**: Sensor data anomaly detection is a critical application for healthcare monitoring.
- **Solution**: A GAN can be trained to learn the normal behaviour of sensor networks by modelling the distribution of regular sensor data; once trained, such a GAN can be used to generate normal data. Discrepancies between real and generated data, those which could not be reasonably learnt by the model, might highlight anomalies.
- **Example**: GANs could be applied to traffic sensor data in smart cities. Due to ordinary traffic flows, anomalies like traffic congestion and accidents or unusual congestion can be identified and acted upon in real-time.

3.4 Data Missing Data Imputation

Missing data is a common issue encountered in sensor networks and other data-driven applications. Sensor data can be incomplete due to various reasons such as sensor malfunction, communication errors, environmental conditions, or power failures. Missing data can lead to biased or incorrect conclusions if not handled properly. Data imputation refers to the process of filling in missing values with estimated ones to ensure the integrity of the dataset and improve the accuracy of analyses.

- **Problem**: Sometimes sensors fail or send only partial data as communication gets lost, power is lost, etc., leaving holes in the stream of data.
- **Solution**: GANs would predict and fill in missing values; they learn from the general data distribution. The generator, when trained, fills in the blank sections with realistic sensor readings consistent with those of other data points in their vicinities.

- **Example**: In health IoT systems, GANs can predict and fill gaps in heart rate or oxygen saturation values from a wearable, thus eliminating loss of data and ensuring continuous monitoring [11].

Data Fusion of Multi Sensor: Data fusion refers to the process of integrating data from multiple sensors to produce more accurate, reliable, and comprehensive information than what could be achieved by using each sensor individually. In multi-sensor systems, data fusion combines data from different types of sensors or from sensors with varying characteristics, helping to overcome limitations like noise, sensor failures, or inconsistent measurements. This process plays a critical role in fields such as autonomous systems, robotics, environmental monitoring, and healthcare.

- **Problem**: In many applications, data is collected by several heterogeneous sensors that differ in format and characteristics. Combining those data for accurate analysis becomes a problem.
- **Solution**: GANs learn from the relation and then generate data to put together the various sensor streams coherently and in a fused manner. The generator creates united data that, in effect, infuses information from a variety of sources of sensors and improves the accuracy of the data.
- **Example**: This can be applied in autonomous vehicles; in such cases, GANs can prove useful to align camera, LIDAR, and radar data to create an accurate map of the environment when one cannot gather specific information or perhaps in cases of malfunctioning [12].

3.5 Architectural Variants of GANs for Sensor Data

The generative advertising network (GAN) has gained significant popularity in data generation work, and their applications for sensor data processing are especially remarkable. They provide a powerful way to generate, increase or restore sensor data, especially when there is noise, incomplete or inadequate. GAN-er has many architectural variants that can be adapted to sensor data, serve each specific measure or solve special challenges related to sensor data. Here are some of the most important architectural variants of GANs for sensor data: As powerful as the basic architecture of GANs is, some variants have been designed to especially handle specific issues better with sensor data. Figure 3 elaborates the Architectural Variants of GANs.

- **Conditional GANs (cGANs)**: These models condition the generation process on additional information, e.g., class labels or sensor metadata. For example, in a health care scenario, a cGAN may be conditioned on patient age or diagnosis to generate realistic sensor data tailored to certain populations.
- **Uses CycleGANs**: Especially for transforming sensor data lacking paired training data, for example, noisy and clean sensor data. They allow for unpaired learning, which makes them applicable in transforming one type of sensor data into another without direct one-to-one training examples.

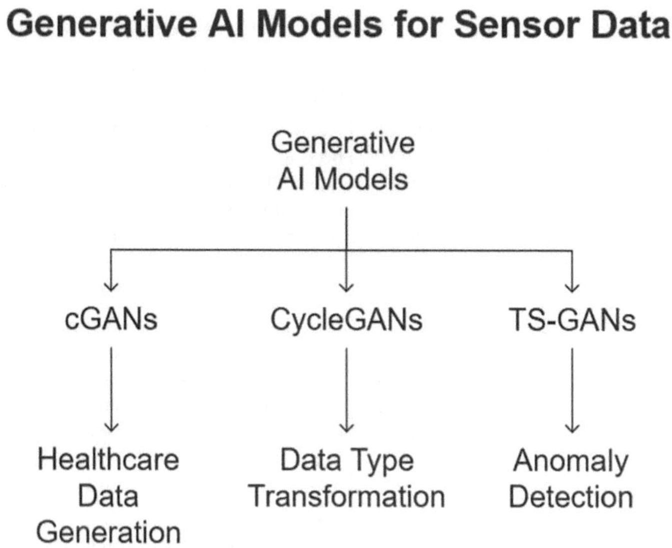

Fig. 3 Architectural variants of GANs

- **Time-Series GANs (TS-GANs)**: They are specifically designed to take into account the temporal dependencies inherent in time-series data, such as sensor data. TS-GANs can, for instance, produce moderately realistic time-series sequences that are potentially very helpful when used to, for example, make certain forecasts in the field of real-time sensor network anomaly detection [13].

Challenges and Considerations: Despite the remarkable advantages of Generative Adversarial Networks (GANs) in sensor data processing, several challenges and considerations must be addressed for optimal performance. These challenges affect various stages of GAN deployment, from training to real-world application. Below are some key challenges and considerations. Here's a Table 1 is summarizing the Challenges and Considerations in using Generative Adversarial Networks (GANs) for sensor data processing.

3.6 Training Instability

- **Problem**: GANs are known for their training instability, which arises due to the adversarial nature of the generator and discriminator. Both components are continuously competing against each other, and if the balance between them is not carefully managed, one can overpower the other, resulting in issues like mode collapse or failure to converge [14].
- **Impact**: Achieving convergence becomes even more problematic when dealing with high-dimensional sensor data, such as data from multiple sensor modalities

Table 1 Challenges and considerations in using Generative Adversarial Networks (GANs) for sensor data processing

Challenge	Problem	Impact	Solution
Training instability	GANs are prone to instability during training due to the competition between the generator and discriminator	Achieving convergence becomes difficult, especially with high-dimensional sensor data	Use techniques like feature matching, label smoothing, or Wasserstein loss. Fine-tune learning rates and regularization strategies
Quality of data	GANs rely on high-quality, well-labeled datasets, which can be sparse, noisy, or incomplete in sensor data	Poor data quality leads to unrealistic or low-quality outputs, undermining performance in tasks like anomaly detection or data imputation	Preprocess data through denoising, normalization, and augmentation. Use semi-supervised or transfer learning for sparse datasets
Computational resources	GANs require significant computational power for training, especially with large-scale sensor data	Long training times, high costs, and resource bottlenecks make real-time deployment challenging	Utilize model pruning, lightweight architectures (e.g., MobileNets), and hardware accelerators (e.g., GPUs/TPUs). Cloud solutions for scalability
Overfitting and generalization	GANs may overfit when trained on small or unrepresentative datasets, leading to poor generalization	Overfitting limits the ability of the GAN to generalize to real-world sensor data, reducing its practical utility	Use techniques like dropout, data augmentation, and cross-validation. Regularization methods can improve generalization
Mode collapse	The generator may produce a limited range of outputs, failing to capture the full diversity of sensor data	The generator may repeatedly produce similar outputs, reducing the utility of the generated data	Use techniques like minibatch discrimination, unrolled GANs, or multiple discriminators. Monitor output diversity during training
Data alignment and synchronization	Sensors collect data at different time intervals or frequencies, leading to misalignment	Temporal misalignment can make it challenging to fuse multi-sensor data, reducing the quality of generated data	Use techniques like interpolation, time-series alignment, or recurrent networks (e.g., LSTMs) for synchronization. Time-averaging or resampling can help

(continued)

Table 1 (continued)

Challenge	Problem	Impact	Solution
Real-time generation and application	Generating sensor data in real-time is challenging due to the time required for GANs to generate data	Real-time applications, such as autonomous vehicles or environmental monitoring, may be hindered by latency	Use lightweight GAN variants (e.g., FastGAN), smaller generator architectures, or edge computing to reduce latency and enable real-time applications
Ethical and privacy concerns	Generating synthetic sensor data, especially for sensitive applications, raises privacy and ethical issues	Privacy violations or misuse of data can hinder the adoption of GANs, particularly in regulated sectors	Anonymize generated data to ensure it doesn't contain identifiable information. Use techniques like differential privacy, federated learning, and secure multi-party computation

(e.g., temperature, pressure, and humidity sensors) that need to be synthesized or filtered. The complexity of sensor data further exacerbates training issues, making it harder to tune hyper parameters and stabilize training.
- **Solution**: Advanced techniques, such as feature matching, label smoothing, or Wasserstein loss, can be used to stabilize training. Additionally, careful tuning of the learning rate and regularization strategies may help mitigate instability.

Quality of Data

- **Problem**: GANs depend heavily on high-quality, well-labeled, and diverse datasets for training. In sensor data, this becomes particularly challenging because sensor readings are often noisy, sparse, or incomplete, especially in real-world environments where sensor malfunctions or external interference may occur [15].
- **Impact**: Poor or sparse data can lead to the generator producing unrealistic or low-quality outputs. If the training dataset doesn't sufficiently represent the true distribution of the sensor data, the generated sensor data may fail to capture essential patterns, leading to poor performance in downstream applications such as anomaly detection or data imputation.
- **Solution**: Data preprocessing and cleaning techniques, such as denoising, normalization, or data augmentation, can improve the quality of sensor data before feeding it into the GAN. Additionally, semi-supervised or transfer learning methods can help when labeled data is sparse, or when leveraging pre-trained models to improve results [16].

Computational Resources

- **Problem**: Training GANs, especially on large-scale or high-dimensional sensor data, can be computationally expensive. Sensor networks generate large volumes of continuous data, often requiring significant processing power and time to train models that can capture the complex relationships in the data.
- **Impact**: The need for high computational power can result in long training times, high costs, and resource bottlenecks, making it difficult to deploy GANs in real-time applications, particularly in industries that require rapid processing and decision-making (e.g., autonomous vehicles or real-time healthcare monitoring).
- **Solution**: To mitigate computational demands, techniques such as model pruning, using more efficient architectures (e.g., lightweight models like MobileNets), and leveraging hardware accelerators like GPUs or TPUs can be helpful. Cloud-based solutions may also offer scalable resources to handle large datasets [17].

Overfitting and Generalization

- **Problem**: Overfitting is a common issue in GANs, particularly when working with small or unrepresentative datasets. The generator may memorize the training data rather than generalizing to new data, resulting in poor performance when applied to real-world sensor data that differs slightly from the training set.
- **Impact**: The generated sensor data may closely resemble the training data but lack generalization to real-world conditions or new sensor configurations. This can undermine the practical use of GANs, especially in dynamic environments where sensor data constantly changes.
- **Solution**: Techniques such as dropout, data augmentation, or using more diverse training datasets can help reduce overfitting. Cross-validation and regularization methods are also essential for improving model generalization and ensuring robust performance in real-world applications [18].

Mode Collapse

- **Problem**: Mode collapse occurs when the generator learns to produce a limited range of outputs (modes), failing to capture the diversity of the real data distribution. This is a frequent issue in GANs and can lead to the generator repeatedly producing similar or even identical sensor data, reducing the utility of the generated data.
- **Impact**: In sensor data applications, mode collapse can result in the generator failing to produce data that reflects the full range of possible sensor states, which is crucial for applications such as anomaly detection or sensor data augmentation.
- **Solution**: Techniques like minibatch discrimination, unrolled GANs, or using multiple discriminators can help address mode collapse. Additionally, careful monitoring of the generator's output diversity during training can help ensure that the generator learns to produce a wide variety of outputs [19].

3.7 *Data Alignment and Synchronization*

- **Problem**: In many sensor networks, data from different sensors may be collected at different time intervals or frequencies. This temporal misalignment can make it challenging to integrate or fuse data from multiple sensors, particularly when generating multi-modal sensor data using GANs.
- **Impact**: Misalignment in sensor data can lead to poor data fusion results and affect the quality of the generated sensor data, as GANs may struggle to synthesize or generate data that aligns temporally or spatially.
- **Solution**: Techniques such as interpolation, time-series alignment, or using recurrent networks (e.g., LSTMs) can help synchronize sensor data before feeding it into the GAN model. Additionally, using techniques like time-averaging or resampling can improve the alignment of multi-modal data [20].

3.8 *Real-Time Generation and Application*

- **Problem**: Generating sensor data in real-time is often necessary for applications like autonomous vehicles, real-time environmental monitoring, or industrial IoT. GANs, especially complex variants like WGANs or CycleGANs, can require considerable time to generate data, which may not be feasible in real-time scenarios.
- **Impact**: The inability to generate data in real-time can limit the practical deployment of GANs in time-sensitive applications, as generated sensor data may not be usable for immediate decision-making or action.
- **Solution**: Lightweight GAN variants, such as FastGAN or using smaller generator architectures, can help speed up the generation process. Edge computing can also be leveraged to process sensor data locally, reducing latency and enabling real-time applications.

3.9 *Ethical and Privacy Concerns*

- **Problem**: When generating synthetic sensor data, especially for sensitive applications like healthcare, there are concerns about privacy and the potential for data misuse. For example, GANs used to generate medical sensor data (e.g., ECG or EEG) could inadvertently reveal sensitive patient information if not handled properly.
- **Impact**: Privacy violations and the ethical implications of generating synthetic data can hinder the widespread adoption of GANs, particularly in regulated sectors like healthcare, finance, and personal monitoring.

- **Solution**: Ensuring that generated data is anonymized and does not contain identifiable information is crucial. Techniques such as differential privacy, secure multi-party computation, or federated learning can be employed to safeguard sensitive data while enabling the use of GANs for sensor data generation.

While GANs offer transformative potential for sensor data processing, there are several challenges and considerations that must be addressed. Training instability, quality of data, computational resources, and issues like mode collapse and overfitting all pose significant barriers. However, by adopting advanced techniques, optimizing model architectures, and carefully addressing ethical concerns, these challenges can be mitigated, enabling GANs to be more effective in generating, enhancing, and analyzing sensor data. As research progresses and new methods emerge, GANs are expected to play an increasingly vital role in the evolution of sensor data applications.

4 Sensor Fusion and Multi-modal Data Generation

Sensor fusion and multimodal data generation are crucial standards in the fields of records science and artificial intelligence, as they significantly enhance the quality, accuracy, and richness of facts in analytical modelling and selection-making. These strategies offer a based tech. These strategies provide a foundational method for combining statistics from various sources, leading to a more profound, detailed, and nuanced understanding of intricate structures and environments. Term used to term "sensor fusion" refers to the process of gathering data from various sensors to create records that are more precise, dependable, and comprehensive than what I could likely determine using my own sensors. Opted adds richness to the understanding of the real physical world, with strengths taken from distinctive sensors that can complement the information provided.

4.1 Advantages of VAEs for Sensor Data

- **Sensor Fusion Has a Huge Advantage**: Data from two or more sensors may be accurate in relation to the measurement. For instance, place monitoring in a navigation gadget tends to be appreciably more accurate with the aid of records amassed from a GPS sensor blended with IMUs.
- **Noise Robustness**: The quality of output in sensors tends to be variable, and the amount of noise and blunders isn't uniform in all places. Sensor fusion reduces this noise level by performing either an average or intelligent integration of data, making the outputs far more trustworthy.

- **Multisensory Data** can therefore provide an adequate perception of the context in which it is used. For instance, cameras, LIDAR, and radar sensors fused in self-driving vehicles can create a more robust perception of the surroundings, which will improve the decision-making capabilities.
- **Redundancy and Fault Tolerance**: Sensor fusion maintains redundancy and will continue to provide better accuracy even when one or more sensors fail. It is critical for all safety-critical applications, ranging from aviation to medical monitoring [21].

Multi-modal Data Generation: Multi-modal data synthesis, also known as multi-modal data generation, involves creating synthetic data from multiple modality sources such as text, images, audio, and sensor data. This is an approach that is becoming increasingly important in training machine learning models, where it is necessary to analyze several kinds of data in combination.

4.2 Applications for Multi-modal Data Generation

Creating multi-modal datasets helps the machine learning model learn across the varied data types and improve the generalizing capability of models over various scenarios. For example, integrating patient records as text, medical imagery, and physiological data as sensor data in a healthcare framework contributes to developing a better diagnostic model.

- **Handling Data Imbalance**: In most applications, the data for a few modalities would be relatively sparse compared to others. Synthetic multimodal data generation may thus help balance datasets to better condition those sparsely represented modalities in training models.
- **Contextual Understanding**: Multi-modal data generation then, multi-modal data generation enables the learning of relationships between diverse kinds of data. Such an example would be generating pairs in the same dataset that consist of images and their respective descriptions, enabling a model to better understand—through context—the visual information while operating on natural language processing and computer vision tasks.
- **Applications in Robotics and Autonomous Systems**: Multi-modal data generation improves robot training by simulating environments that include visual, auditory, and tactile information. This enhances the robot's capability to respond effectively in real-world scenarios where it perceives diverse sensory input [22].

4.3 Techniques for Sensor Fusion and Multi-modal Data Generation

Kalman filtering is one of the most common applications of sensor fusion, namely, mathematical algorithms that estimate unknown variables by combining measurements over time; these are particularly useful in tracking and navigation applications.

- **Deep Learning Methods**: You can apply deep learning models to various multi-modal data generation models, such as CNNs and RNNs. For instance, GANs and VAEs can use the joint representations to generate synthetic data that combines multiple modalities.
- **Bayesian Methods**: Bayesian methods allow for the integration of a variety of sources of information into a model while at the same time quantifying uncertainty. Bayesian methods, ideally suited to updating beliefs based on evidence from multiple sensors, are a unique set of methods for sensor fusion.
- **Graphical Models**: Probabilistic graphical models, or neural networks, are very effective at representing and analysing multi-modal data regarding the relationship between different modalities. These models have the potential to generate and fuse data by establishing a connection between various types of data.

Sensor Fusion and Multi-Modal Data Generation are limited.

Sensor fusion and multi-modal data generation have many advantages, yet there are several challenges too:

- **Data Alignment**: The process of data validates alignment. Aligning data from various sensors or modalities can be challenging, especially when they operate at different frequencies, which can lead to various types of time delays, alignment techniques.
- **Dealing with Missing Data**: In real applications, several sensors either failed or indeed provided incomplete data; hence, it is very significant for the development of robust methods to deal with missing information, especially in the cases of sensor fusion efficiency and data generation [23].
- **Scalability**: As the number of sensors and modalities increases, the complexity of data integration also increases. Specifically, we need to develop large-scale data fusion and generate efficient algorithms without compromising performance.
- **Model Complexity**: We should design effective models that can integrate and generate multimodal data. The more complex this model gets, the further it drives away from being interpretable for applications that make critical decisions. Table 2 summarizes the Sensor Fusion and Multi-Modal Data Generation.

Sensor fusion and the generation of multi-modal data form the foundations for modern data analysis and machine learning. The resultant data are qualitative and improve model performance, allowing for deeper and more meaningful discovery of complex systems. Simultaneously, with advances in generative AI and maturity

Table 2 Sensor fusion and multi-modal data generation

Aspect	Description	Key applications
Definition	Sensor fusion integrates data from multiple sensors to create more precise and reliable datasets. Multi-modal data generation synthesizes data from various sources (text, images, audio, sensors) to enhance AI models	AI, robotics, healthcare, autonomous systems
• Advantages of sensor fusion	Enhances accuracy by combining multiple sensor inputs, reduces noise, ensures redundancy and fault tolerance	Navigation (GPS + IMUs), self-driving cars (LIDAR + radar + cameras)
• Multi-modal data generation	Creates synthetic multi-modal datasets to improve machine learning model training and generalization	Medical diagnostics (text + images + physiological data), NLP, computer vision
Techniques	**Kalman filtering**: Estimates unknown variables over time • **Deep learning (CNNs, RNNs, GANs, VAEs)**: Generates synthetic multi-modal data • **Bayesian methods**: Integrates information while quantifying uncertainty • **Graphical models**: Maps relationships between different data modalities	Tracking, AI-driven analytics, robotics, predictive maintenance
Challenges	• **Data alignment**: Synchronizing sensors with different sampling rates • **Handling missing data**: Dealing with incomplete or failed sensor inputs • **Scalability**: Managing increasing complexity as sensor networks grow • **Model complexity**: Ensuring interpretability in high-stakes applications	Large-scale IoT networks, industrial automation, real-time AI analytics
Future potential	Advances in generative AI and sensor technology will expand applications in healthcare, autonomous vehicles, and smart cities	Wearable health devices, automated decision-making, urban planning

of sensor systems, applications will grow and unleash innovation potential in areas ranging from health to automated systems to smart cities.

5 Diffusion Models for Time-Series Sensor Data

Diffusion models belong to a new class of generative models that are very effective for processing time series data, such as sensor data. Thanks to an original approach that converts noise into structured data through a sequence of iterative diffusion steps, they can capture strong dependencies that are inherent in time sequences, and the produced synthetic data are highly quality. Such a capability makes diffusion models suitable for a wide range of sensor data processing applications.

5.1 Overview of Diffusion Models

Deeply, the diffusion models primarily rely on the assumptions of a Markov chain, which are easily comprehensible through two main phases. It consists of a forward diffusion process and a reverse diffusion process.

Forward Diffusion Process: The process begins with a time-series data sample, such as a temperature or humidity sensor reading, and then adds Gaussian noise in steps across many time steps. This transformation essentially transforms the original data into a representation of noise, thereby masking the inherent structure beneath the surface. We can mathematically write it as a chain of transformations, adding noise at each step until we reach almost complete randomness.

The model then learns how to get back to the original data from this noisy state using the reverse diffusion process. This is an iterative improvement of the process of denoising itself, but this time the model is given the noise level as input. This lets the model create realistic data samples from random noise. It basically tries to minimize the gap, at each moment in time, between original data and output obtained after denoising. This means that the model learns the data's underlying distribution well [24].

Advantage Over Time-Series Sensor Data: Diffusion models seem to be really powerful for capturing the implicit temporal dependencies within time-series sensor data. Unlike the traditional generative models, which probably treat individual data points in isolation, diffusion models take into account the whole sequence of observations. As a result, they can learn the complex temporal relationships inherent to dynamic systems, which are critical for successful modelling. Another virtue that makes diffusion models promising is their robustness against noise. Sensor data quite often suffers from various sources of noise due to either environmental or hardware-related limitations. By design, diffusion models can add and then remove noise, making them very resilient to this variability and improving the quality of the generated outputs.

5.2 *Applications of Diffusion Models in Sensor Data Processing*

There are many practical applications of diffusion models in the world of time-series sensor data, diffusion models have many practical applications.

- **Anomaly Detection**: Diffusion models learn the normal distribution of sensor data, flagging as anomalies instances highly biased toward deviation from this normal distribution. This normal distribution is flagged as anomalies. Reconstruction errors, in the course of reverse diffusion, help effectively identify outliers and can give insights into the system's performance.
- **Data Imputation**: Missing values in time-series data severely restrict both sampling from the learned distribution; diffusion models can approximate and impute these missing values, ensuring the data set remains complete and useful for analysis. Obtaining labelled time-series data is a challenging task. Labelled time-series data is extremely hard to get. This may pose an inadequacy issue when it comes to training data for learning algorithms and machines. The high-quality synthetic time series data generated by the diffusion models bears a striking resemblance to sensor readings. These diffusion models enable the creation of data that can complement any previous set, thereby enhancing the model's performance [25].
- **Predictive Capability**: Predictive maintenance and resource management rely on the ability to infer future sensor readings with sufficient accuracy. The diffusion models use the temporal dynamism of the time series as their basic approach to infer future values based on historical data, thereby providing valuable information for proactive decision-making.
- **Time-Series Classification**: Sensor data classification from time series is rather challenging; it mainly arises out of inherent data complexity and variability. Diffusion models can act as feature extractors, which might transform low-level sensor readings into more informative representations to support classification. Issues and Considerations [26].

While some merits of diffusion models do exist, there are some key challenges facing them. First and foremost, such models are computationally expensive; for high-dimensional time-series data, training a model is usually computationally expensive. In fact, the diffusion process itself is in essence an iterative process that tends to be computationally demanding as well as time-consuming. Diffusion models may also be sensitive to hyper parameters, such as the number of diffusion steps and the level of noise considered in training. Obtaining a good setting requires extensive experimentation and tuning. The quality and quantity of the training data strongly influence the quality of the generated data; poor or noisy training data can be detrimental in this regard. In a word, the diffusion models show a very promising direction in the field of time-series sensor data processing; indeed, anomaly detection, data imputation, synthetic data generation, and forecasting tasks found an effective solution. The capability to model complex temporal dynamics makes them specifically suitable for applications in sensor networks where high-integrity and quality data come first. Considering the pace at which research and development are moving on in the domain of diffusion models, we expect applications based on sensor data processing to experience enormous growth; these will support effective tools in each and every area, from manufacturing to healthcare and environmental monitoring [27].

6 Future Directions and Emerging Trends

Increased technology Sensor fusion integration results in the creation of multi-modal data, which is influenced by advancements in AI, machine learning, and sensor technologies. Therefore, it is crucial for researchers, practitioners, and organizations to keep up-to-date with future directions and emerging trends in this area.

6.1 Smarter Integration of AI and Sensors

We will further integrate all these AI technologies with sensor technologies to enhance intelligence and create autonomous systems. There will be a greater use of machine learning algorithms, particularly deep learning, to process and analyse data from multiple sensors. This synergy would drive real-time decisions by adapting to the dynamics of a given environment and condition.

- Intelligent sensors will incorporate artificial intelligence directly, allowing for the analysis and processing of a vast amount of data at the point of generation. This reduces the amount of raw data flowing back to central systems and makes responses against detected anomalies effective [28].
- **Edge Computing**: Sensor fusion with edge computing will enable real-time processing near the source. The trend will include autonomous vehicles, industrial IoT, and cities with this target for the next wave of applications—smarter and responsive.

6.2 Advances in Generative Models

It is sufficient to say that generative models will continue to evolve at a pace that surpasses the current evolution in generative speed. As GANs, VAEs, and diffusion models continue to grow stronger and more efficient in data synthesis, they will synthesize highly realistic multimodal datasets.

- **Improved Realism and Robustness**: Future generative models will strive to produce data that not only mimics real-world distributions but also encompasses a wide range of scenarios, including event frequency. This will be especially useful in areas where robust training datasets are required. Generative pre-training on large-scale datasets will establish a common standard for fine-tuning on target tasks. This can lead to an improvement in the performance level and a reduction in labelled data required for training.

6.3 Cross-Modal Learning

Therefore, training the models to understand and integrate knowledge from multiple modalities will greatly benefit cross-modal learning. The models will support better contextual comprehension and integration, enabling holistic decision-making processes.

- **Multitask Learning**: Develop models that know how to learn a diversity of tasks in a cooperative, rather than competitive, way, and to share representations across modalities. Today, we could easily envision a model that processes video, audio, and sensor streams simultaneously for applications like object recognition and environmental monitoring.
- **Transfer Learning Across Modalities**: Developing techniques for knowledge transfer across modalities, enabling models trained on one kind of data.

6.4 Emphasis on Explainability and Interpretability

The deployment of AI systems in more high-stakes domains will increase the demand for explainability and interpretability in sensor fusion and multi-modal data creation. Future models will, by definition, be transparent enough for users to comprehend the decision-making process regarding the fused data from sensors [29]. Techniques to visualize which sensors and modalities have the most influence will be crucial.

- **Compliancy**: With the increasing scrutiny surrounding the use of AI applications in sensitive areas such as fitness care and finance, it is crucial that these fashions adhere to moral guidelines and legal requirements. This could imply understanding how the fashions work and whether there is any bias within the sensor information itself.

Practical Applications and Industry Adoption: This will wear down the proliferation of sensor fusion and multi-modal statistics generation into sensible packages across industries, influencing a wide range of business strategies from choice-making downwards. Health will integrate data from wearables, sensors, imaging devices, and electronic fitness data to provide more effective treatments or cures for individual patients, all while continuously monitoring them. We anticipate that multimodal methods will enhance the analysis process when utilizing telemedicine services. Fully autonomous development relies on sensor fusion and multimodal statistics technology to understand such a complex environment and ensure equally safe navigation. Therefore, it could enhance the ability to interpret a wide range of real-time sensory inputs with superior cognitive skills [30].

- **Smart Cities**: The latter will be leveraging the multi-modal information emanating from hosts of sensors, together with cameras over visitors and surroundings, to tinker with city planning, optimize site visitor control, and deliver tips on public protection [31].

Ethical Considerations and Data Privacy: With the introduction of multi-version facts, Sensor Fusion raises moral issues and data privacy as key drivers of future trends.

- **Data Governance**: This refers to the established hierarchy of appropriate data governance systems for the enterprise, which ideally manage the collection, storage, and sharing of sensor data while adhering to the evolving privacy regulations. Future Directions and Emerging Trends in Sensor Fusion and Multi-Modal Data is explained in Table 3.

Only impartial series and version training can ensure that AI structures aren't programmed to generate biased outcomes, by negating the biased standards within the selected records and in version training. Sensor fusion and multi-modal record technology are at the forefront of exciting trends and are enhancing the capabilities of intelligent devices [32]. Such developments will assist organizations that have embraced emerging technologies, prioritizing ethical concerns and focusing on the real world, in exploring new areas of innovation and improving decision-making. Over time, this will be necessary to realize the full benefits of sensor fusion and multimodal data creation.

7 Conclusion

Censorship reimbursement and multimodal data processing are transformative approaches that increase data analysis, machine learning and decision-making in industries significantly. By integrating data from many sensors and adding them into different types, these methods give more accurate and broad understanding of complex systems, which improves accurately and reliability. In addition, these techniques enable generic models, AI integration and data processing of adjacent time and intelligent decisions to take the front data processing. Organizations quickly use these techniques, and operate innovation in areas such as health care, autonomous systems and smart cities. However, this progress also increases important moral concerns, including the privacy of data, clarity in the model and prejudice. As the sensor fusion and multimodal data generation continue to develop, it is necessary to solve these challenges to ensure responsible and fair implementation. In the future, the focus will be expanded beyond technological progress to the development of

Table 3 Future directions and emerging trends in sensor fusion and multi-modal data

Trend	Description	Key applications
Smarter integration of AI and sensors	AI-driven sensors will process data at the point of generation, reducing reliance on centralized systems and improving real-time responses	Autonomous systems, IoT, anomaly detection
Edge computing	Real-time sensor data processing near the source for improved efficiency and faster decision-making	Smart cities, industrial IoT, autonomous vehicles
Advances in generative models	GANs, VAEs, and diffusion models will enhance data realism and robustness, improving synthetic multimodal datasets	AI model training, data augmentation, predictive maintenance
Cross-modal learning	Training models to integrate multiple data modalities for improved contextual understanding and decision-making	Multitask learning, transfer learning across sensor data types
Explainability and interpretability	Ensuring AI-driven sensor fusion models are transparent, interpretable, and compliant with ethical and legal standards	Healthcare AI, financial risk assessment, regulatory compliance
Industry adoption and practical applications	Sensor fusion will enhance real-world applications across industries, improving automation, efficiency, and decision-making	Healthcare (wearables, remote monitoring), smart cities, autonomous navigation
Ethical considerations and data privacy	Establishing governance frameworks to ensure responsible sensor data collection, storage, and usage	Data security, bias mitigation, regulatory compliance

moral and responsible structure, using the entire potential of these technologies to run positive changes in different domains.

References

1. C. Izuzu, The role of generative AI in interpreting internet of things at the edge. Int. J. Internet Things **12**(1), 1–7 (2024). https://doi.org/10.5923/j.ijit.20241201.01
2. R. Kanthavel, R. Dhaya, Proactive cooperative communication enabled fifth generation wireless sensor network using SCTP, in *2021 National Computing Colleges Conference (NCCC)*, Mar 2021 (IEEE), pp. 1–5
3. J. Wang, Y. Chen, Y. Gu, Y. Xiao, H. Pan, SensoryGANs: an effective generative adversarial framework for sensor-based human activity recognition, in *2018 International Joint Conference*

on Neural Networks (IJCNN), Rio de Janeiro, Brazil (2018), pp. 1–8. https://doi.org/10.1109/IJCNN.2018.8489106
4. S. De, M. Bermudez-Edo, H. Xu, Z. Cai, Deep generative models in the industrial internet of things: a survey. IEEE Trans. Ind. Inf. **18**(9), 5728–5737 (2022). https://doi.org/10.1109/TII.2022.3155656
5. R. Dhaya, R. Kanthavel, K. Venusamy, AI based learning model management framework for private cloud computing. J. Internet Technol. **23**(7), 1633–1642 (2022)
6. Y. Hu, BSDGAN: balancing sensor data generative adversarial networks for human activity recognition, in *2023 International Joint Conference on Neural Networks (IJCNN)*, Gold Coast, Australia (2023), pp. 1–8. https://doi.org/10.1109/IJCNN54540.2023.10191928
7. R. Dhaya, V. Sadasivam, R. Kanthavel, Reliable point to multipoint hierarchical routing in ScatterNet sensor network. Int. J. Adv. Netw. Appl. **2**, 738–744 (2011)
8. K. Radhakrishnan, D. Ramakrishnan, O.I. Khalaf, M. Uddin, C.L. Chen, C.M. Wu, A novel deep learning-based cooperative communication channel model for wireless underground sensor networks. Sensors **22**(12), 4475 (2022)
9. Y. Cao, S. Li, Y. Liu, Z. Yan, Y. Dai, P.S. Yu, L. Sun, A comprehensive survey of AI-generated content (AIGC): a history of generative AI from GAN to ChatGPT. arXiv preprint arXiv:2303.04226 (2023)
10. S.M. Yuan, Z.W. Hong, W.K. Cheng, Artificial intelligence and deep learning in sensors and applications. Sensors **24**(10), 3258 (2024). https://doi.org/10.3390/s24103258. PMID: 38794112; PMCID: PMC11125570
11. C. Zhang, C. Zhang, S. Zheng, Y. Qiao, C. Li, M. Zhang, C.S. Hong, A complete survey on generative AI (AIGC): Is ChatGPT from GPT-4 to GPT-5 all you need? arXiv preprint arXiv:2303.11717 (2023)
12. R. Dhaya, P. Indhuja, S. Sinduja, M. Swetha, Finest power efficient steering algorithm for wireless sensor networks for surveillance, in *2015 IEEE 9th International Conference on Intelligent Systems and Control (ISCO)*, Jan 2015 (IEEE), pp. 1–6
13. K. Wang, C. Gou, Y. Duan, Y. Lin, X. Zheng, F.Y. Wang, Generative adversarial networks: introduction and outlook. IEEE/CAA J. Autom. Sin. **4**(4), 588–598 (2017)
14. C. Doersch, Tutorial on variational autoencoders. arXiv preprint arXiv:1606.05908 (2016)
15. A.B.L. Larsen, H. Søndberg, O. Winther, Autoencoding variational inference. arXiv preprint arXiv:1511.06434 (2015)
16. M. Fraccaro, H. Søndberg, O. Winther, Disentangled sequential autoencoders. arXiv preprint arXiv:1605.08124 (2016)
17. J. Shang, X. Li, X. Wang, Variational autoencoders for anomaly detection in sensor networks. Sensors **19**(24), 5467 (2019)
18. Y. Bengio, A. Courville, P. Vincent, Learning deep architectures for AI. Found. Trends® Mach. Learn. **2**(1), 1–127 (2013)
19. K. Sohn, H. Lee, X. Yan, Learning structured output representation using deep conditional generative models, in *Neural Information Processing Systems (NeurIPS)* (2015)
20. W. Zhou, Z. Li, Variational autoencoders for sensor data imputation. IEEE Access **8**, 134312–134320 (2020)
21. S.K.B. Sangeetha, R. Dhaya, Deep learning era for future 6G wireless communications—theory, applications, and challenges, in *Artificial Intelligent Techniques for Wireless Communication and Networking* (2022), pp. 105–119
22. S. Mo, Y. Shi, Q. Yuan, M. Li, A survey of deep learning road extraction algorithms using high-resolution remote sensing images. Sensors **24**(5), 1708 (2024)
23. S.H. Rafique, A. Abdallah, N.S. Musa, T. Murugan, Machine learning and deep learning techniques for internet of things network anomaly detection—current research trends. Sensors **24**(6), 1968 (2024)
24. A. Jabbar, X. Li, B. Omar, A survey on generative adversarial networks: variants, applications, and training. ACM Comput. Surv. (CSUR) **54**(8), 1–49 (2021)
25. R. Kanthavel, S. Indra Priyadharshini, D. Sudha, K. Sundara Velrani, R. Dhaya, Multi-hoped cooperative communication-based wireless underground sensor network design. Int. J. Commun. Syst. **35**(10), e5174 (2022)

26. L. Jin, F. Tan, S. Jiang, Generative adversarial network technologies and applications in computer vision. Comput. Intell. Neurosci. **2020**(1), 1459107 (2020)
27. M. Andronie, G. Lăzăroiu, M. Iatagan, C. Uţă, R. Ştefănescu, M. Cocoşatu, Artificial intelligence-based decision-making algorithms, internet of things sensing networks, and deep learning-assisted smart process management in cyber-physical production systems. Electronics **10**(20), 2497 (2021)
28. R. Kanthavel, K. Anathajothi, S. Balamurugan, R.K. Ganesh (eds.), *Artificial Intelligent Techniques for Wireless Communication and Networking* (Wiley, 2022)
29. Z. Zhang, L. Wang, C. Lee, Recent advances in artificial intelligence sensors. Adv. Sens. Res. **2**, 2200072 (2023). https://doi.org/10.1002/adsr.202200072
30. D.P. Kingma, M. Welling, Auto-encoding variational Bayes. arXiv preprint arXiv:1312.6114 (2013)
31. A. Bandi, P.V.S.R. Adapa, Y.E.V.P.K. Kuchi, The power of generative AI: a review of requirements, models, input–output formats, evaluation metrics, and challenges. Future Internet **15**(8), 260 (2023). https://doi.org/10.3390/fi15080260
32. R. Dhaya, R. Kanthavel, F. Algarni, Research perspectives on applications of internet-of-things technology in healthcare WIBSN (wearable and implantable body sensor network), in *Principles of Internet of Things (IoT) Ecosystem: Insight Paradigm* (2020), pp. 279–304

AI in Optical Fiber Sensors and Sensing Network

Anamika Kumari and Vibha Vyas

Abstract In the buzz of artificial Intelligence (AI) in all domains, optical fiber sensors and sensing networks are no exception. When the whole world is exploring AI in their own specific applications, the integration of AI in optical fiber sensors also supports its evolution. The current article is an attempt to understand the process and applications. Millions of data are generated wherever there are sensors and so do these optical fiber sensors. AI uses these data as fuel and plays a transformative role in improving the effectiveness and efficiency of the sensors, making it smarter. Previously, sensors were expected to measure parameters only at the present moment; however, with advancements in technology and rapidly evolving sensing capabilities, superior sensors are those that can provide critical insights into future parameter measurements, even if not the precise values. Information to future measurements can only be provided if there are enough data of the past and the present. This can lead to the detection of anomalies and preventive maintenance. It can indicate the potential failure or malfunction in the system. It can also be used to predict failure in a system based on historical data. It can help automatic calibration of optical fiber sensors. Integration of AI with optical fiber sensors can enhance the functionality, data processing capability, and can also help in decision making, which is crucial while maintaining the efficiency of the system. This chapter covers the way AI has brought about change in the application of fiber optic sensors and also gives insight on its impact on the sensing network in industrial applications.

Keywords AI in optical fiber sensors · AI · Optical fiber sensors

A. Kumari (✉) · V. Vyas
Electronics and Telecommunication Engineering Department, COEP Technological University, Pune, Maharashtra, India
e-mail: kumaria20.extc@coeptech.ac.in

V. Vyas
e-mail: vsv.extc@coeptech.ac.in

Y. Vasimalla and S. Kumar (eds.), *Generative AI for Photonic Sensing*, Progress in Optical Science and Photonics 36, https://doi.org/10.1007/978-981-95-1561-5_5

1 Introduction

Sensing is the measurement of a physical, chemical or biological changes in the environment and converting them into data. In any analytical system, sensors are important part as it helps in producing the data which is later needed for decision making. Sensors are an integral part of many systems and industries so making the most out of these sensors are very important aspect. In earlier times when the sensors were limited and resources to save their data were limited most of the decisions and analysis was performed from the realtime data. This was not because they were not aware that the information of the past could be really meaningful for them but the reason was unavailability of past data. With time as science evolved the capacity and capability of the sensors have increased several times. But still the growth of data storing capacity was limited. With the limited data, statistics i.e. a branch of mathematics was able to take care of it with people or rather humans working around it. They were able to manage and draw conclusions with the data. In the last few decades when the data handling capacity and number of sensor have just exploded to a level where it is no more possible to handle data by humans. This brings the need of something that can help us automate and make machines do these tasks. At this juncture the need isn't just to handle the data and tasks but also to take decisions as good as humans or may be better. This is possible only when there is intelligence in the process. Humans take decisions based on its previous and past data. Similar to this, lies the concept of artificial intelligence which is definitely an artificial thing but trained in a way to take decisions probably better than humans. Training of the system is important which is done with the previous data. AI enhanced system has an added layer to any existing process which takes care of decision taking process. AI is getting integrated in all technical domains and it brings change in the sensing domain too. Optical fiber sensing is one of the many sensing techniques that can be used to measure a variety of different physical parameters like stress, strain, temperature and chemical composition. The information gathered can be utilised in various industries and systems.

Optical fiber sensing has an added edge over other methods in terms of its precision, durability, and resistance to harsh environments. The need of AI is obvious and here the discussion is focused on the optical fiber sensors and the impact of AI on optical fiber sensor domain. Optical fiber sensing is a technology that uses optical fibers not only to transmit data but also to measure various physical parameters such as temperature, pressure, strain, vibrations, and chemical composition in real-time. This technology has gained significant traction in industries like civil engineering, oil and gas, aerospace, and healthcare due to its precision, durability, and resistance to harsh environments [1]. They identify alterations in environmental parameters (temperature, pressure, strain, etc.) by examining fluctuations in light characteristics such as intensity, phase, wavelength, or polarization. In optical fiber sensor there are some broad categories and each one being utilised in their specific applications. Fiber Bragg Gratings (FBGs) that in general is used to quantify strain, temperature, and pressure, evanescent wave sensors that identify chemical and biological interactions, interferometric sensors is used mainly for measuring vibrations and displacements and LSPR-Based Sensors is another type which is used to detect molecular interactions by surface plasmon resonance. Figure 1 presents an overview of the integration of sensing technology with AI, illustrating the enhanced applicability resulting from this synergy. This figure depicts the framework of sensing technologies (left) and artificial intelligence technologies (right), alongside the evolution of varied applications in self-powered, implantable, and wearable sensors, as well as smart home systems, machine learning (ML), VOC sensing, and human-machine interfaces.

In the current chapter, Sect. 2 discusses the optical fiber sensors and sensing network and the technological gaps that paves the path of integration of AI with the sensing domain. Section 3 dives into the details of the role AI plays in an enhanced systems and how it can eventually uplift the overall experience of the user. Section 4

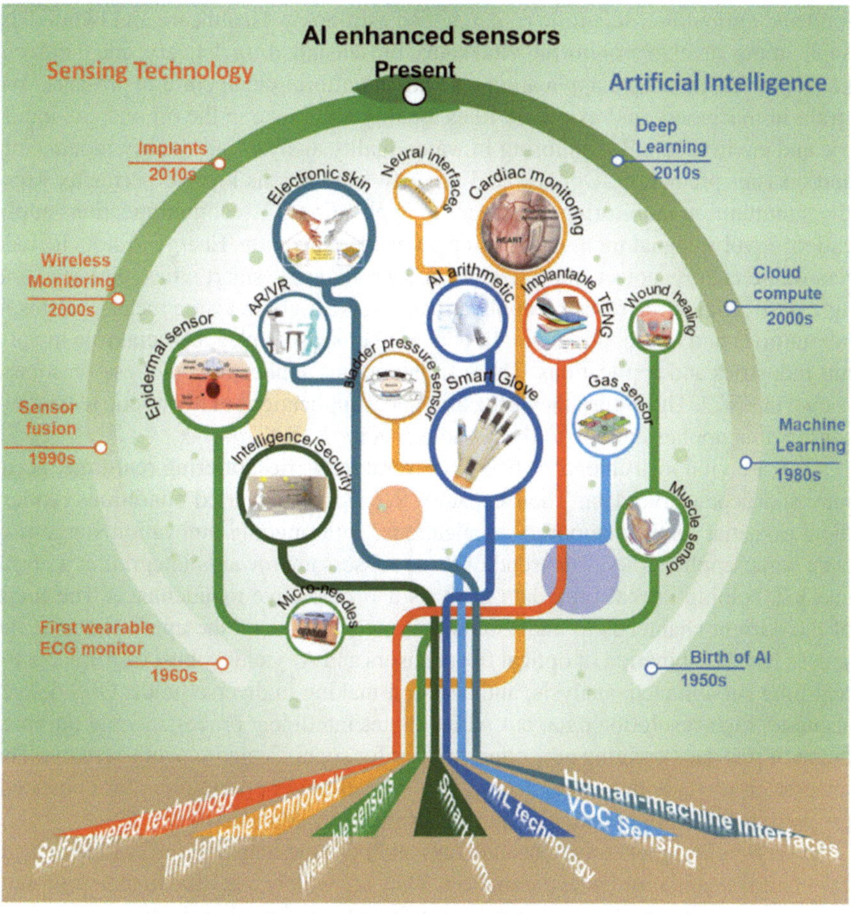

Fig. 1 The advancement of artificial sensors facilitated by intelligent technology [2]

gives the overview of the different AI algorithms which are being used with reference to different areas of its implementation. Section 5 elaborates on the applications for which the discussed AI algorithms are being implemented. Finally, the chapter ends with a conclusion where the current trend and the future scope have been discussed.

2 Optical Fiber Sensors and Sensing Network

Optical fiber sensors (OFS) are versatile devices used in various industries due to their high sensitivity, immunity to electromagnetic interference, and ability to operate in harsh environments. They are used in healthcare, biomedical applications, structural health monitoring, industrial applications, environmental monitoring, defense and security, telecommunications, energy sector, smart cities and IoT, food and agriculture, scientific research, and astronomy. Healthcare and biomedical applications involve monitoring vital signs, biosensing, drug delivery, and therapeutic responses. Industrial applications include monitoring temperature, pressure, and strain in machinery and assembly lines, quality assurance in the oil and gas industry, and environmental monitoring in water quality analysis, air quality monitoring, and disaster prediction. Defense and security applications involve perimeter security, ammunition monitoring, and security applications. Telecommunications applications involve signal monitoring and network optimization. Energy sectors include renewable energy, power transmission, nuclear industry, smart cities, and food and agriculture. Scientific research applications involve high-precision measurements in laboratories and telescopes. Overall, optical fiber sensors play a crucial role in various industries and applications. There are numerous applications of these sensors as shown in Fig. 2. This figure shows the biological applications of fiber optic plasmonic sensors that applies equally to the broad category of OFS.

OFS are vital instruments utilized in several industries, offering real-time, accurate, and dependable data. Their capacity to function in varied conditions renders them essential for contemporary applications, with ongoing innovation suggesting even wider applicability. The trend of these sensors are towards integration with AI and IoT helping in real time data analysis and predictive maintenance. The focus of the current chapter is to understand the impact of AI on the applications of the sensor. The combination of optical fiber sensors and AI yields robust instruments for real-time surveillance, analysis, and decision-making in diverse fields. OFS deliver accurate, high-resolution data, but AI facilitates intelligent processing and interpretation of this data, creating new opportunities for sophisticated sensing systems. The integration of AI with sensing technology creates new opportunities for enhanced, precise, and real-time sensing solutions across multiple industries.

In OFS light signals travel through optical fibers and help measure various physical, chemical, and biological parameters. They are widely used due to their high sensitivity, immunity to electromagnetic interference, and miniaturization and remote sensing capability. There are several types of OFS sensors, including intensity-based

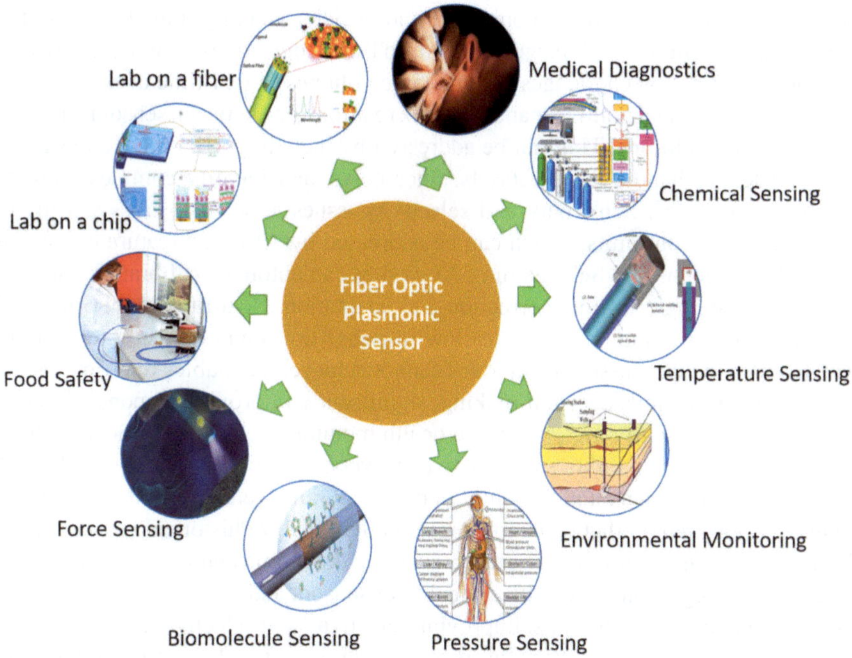

Fig. 2 Applications of fiber optic plasmonic sensor [3]

sensors, wavelength-based sensors, phase-based sensors, and polarization-based sensors. An Optical Fiber Sensing Network (OFSN) is a system of multiple fiber-optic sensors distributed over an area connected to a central processing unit. It enables large-scale monitoring over long distances, real-time data collection and analysis, and distributed sensing for critical infrastructure.

AI is transforming OFS by enhancing data processing, automation, and predictive capabilities. AI solutions include ML, predictive maintenance and fault Detection, smart and adaptive Sensing, reinforcement learning, fuzzy logic and expert Systems, biomedical OFS, and cybersecurity in optical fiber networks. Real-time data processing filters noise and extracts meaningful insights. Predictive maintenance prevents fiber failures before they occur and adaptive sensing tunes sensor parameters for dynamic environments. Biomedical innovation enables early disease detection, while cybersecurity enhances optical fiber networks against cyber threats. AI is transforming OFS and networks by filtering noise, preventing fiber failures, enhancing adaptive sensing, enabling early disease detection, and securing optical fiber networks against cyber threats. This transformation is making OFS and networks smarter, faster, and more resilient.

The integration of AI with OFS and sensing networks addresses several technological gaps that have limited the full potential of these sensors. Key challenges include data processing and interpretation, sensitivity and selectivity issues, calibration and drift compensation, real-time monitoring and decision-making limitations, limited predictive capabilities, and integration with IoT and edge computing. Data processing and interpretation challenges include large volumes of real-time data, noise and signal distortions, and nonlinear and complex relationships between sensor responses and external conditions. This can be addressed by including ML and deep learning (DL) models to handle large datasets, filter noise, and identify complex patterns in sensor responses. Sensitivity and selectivity issues arise from cross-sensitivity and lower detection limits, which can be addressed by AI-based feature extraction and pattern recognition. Self-learning AI models can automate calibration and predict drift, enabling real-time correction and adaptation. Real-time monitoring and decision-making limitations include slow response times and human intervention, which traditional processing techniques cannot achieve. AI enables real-time monitoring and autonomous decision-making, significantly improving response times in applications like smart cities, structural health monitoring, and biomedical sensing. Limited predictive capabilities include reactive versus predictive monitoring, lack of preventive maintenance, and difficulty in detecting rare events in complex environments. AI-powered predictive maintenance models analyze historical sensor data to detect early warning signs of failures, preventing downtime and costly repairs.

Thus the integration of AI with optical fiber sensors addresses key technological gaps by enhancing data analysis, improving sensitivity and selectivity, enabling real-time decision-making, and allowing predictive monitoring. This transformation is driving advancements in healthcare, environmental monitoring, smart infrastructure, and industrial automation.

3 Role of AI in Optical Fiber Sensors

OFS are extensively employed for the monitoring of physical, chemical, and biological characteristics owing to their exceptional sensitivity, resistance to electromagnetic interference, and ability to provide long-distance sensing. AI enhances these advantages by refining data analysis, prediction, and decision-making processes. This leads to enhanced functionality, accuracy, and adaptability across various applications. AI significantly enhances the performance, usability, and versatility of OFS. Utilizing sophisticated algorithms and computing capabilities, AI facilitates efficient data processing, real-time decision-making, and predictive analytics, hence enhancing the performance of these sensors in various applications. It plays a crucial role in analyzing and interpreting data from optical fiber sensors, reducing noise, extracting key patterns, and detecting anomalies. It also provides real-time monitoring and control, enabling applications like health monitoring in wearable devices and industrial processes. AI aids in predictive maintenance, predicting potential failures and extending equipment lifespan. It enhances sensor performance by dynamically adjusting parameters based on environmental changes. AI also combines data from multiple sensors, improving sensitivity and selectivity [4]. AI automates routine tasks, optimizes sensor network configurations, and offers scalability for large sensor networks.

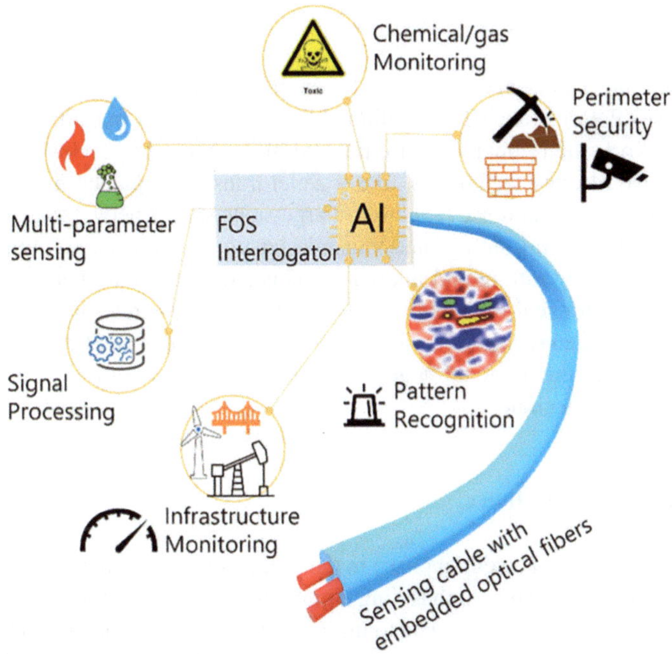

Fig. 3 Schematic representation of diverse uses of AI-enhanced OFS [4]

Lastly, AI can interpret data for personalized healthcare treatments. Figure 3 shows a sensing cable containing embedded optical fibers and linked to an intelligent OFS interrogator utilizing AI technology and can be employed for many applications. The following are the different ways in which AI can improve the system.

3.1 *Data Analysis and Interpretation*

AI markedly enhances data processing and signal refinement in OFS by utilizing sophisticated algorithms to extract pertinent information, reduce noise, and optimize sensor efficacy. It significantly enhances the performance, usability, and versatility of optical fiber sensors. Utilizing sophisticated algorithms and computing capabilities, AI facilitates fast data processing, instantaneous decision-making, and predictive analytics, hence enhancing the performance of these sensors in various applications [5]. The following points discusses some of the important tasks performed by AI that is widely used in various applications.

3.1.1 Noise Reduction

OFS often collect unwanted noise due to environmental factors or system imperfections. AI solutions like ML algorithms like support vector machines (SVM) and Neural Networks help identify and eliminate noise while preserving relevant signals. Adaptive filtering techniques, like wavelet transforms guided by AI, improve signal clarity. It helps in noise reduction. Large-scale sensor networks generate vast amounts of data, making storage and transmission challenging. AI solutions include data compression algorithms and smart data management systems that prioritize critical information for real-time analysis while archiving less relevant data [4, 6, 7]. Optical sensors often require precise spectral analysis, which can be achieved using AI solutions like convolutional neural networks (CNNs) and multi-peak detection and deconvolution techniques.

3.1.2 Feature Extraction

Feature extraction is an essential phase in the processing of data obtained from OFS. It involves identifying and emphasizing significant patterns, attributes, or measurements from unprocessed sensor data to improve analysis and decision-making. In AI systems, feature extraction integrates signal processing techniques with ML and statistical methodologies to identify the most relevant data. ML models identify essential patterns and characteristics within complex datasets, facilitating accurate interpretation of sensor data. Feature extraction for OFS involves several steps, including preprocessing raw sensor data, identifying signal features, and employing advanced AI-driven techniques. Preprocessing involves noise reduction, baseline correction, and normalization to improve data quality. Spectral features, such as peak wavelength shifts and frequency components, are crucial for refractive index sensing or biochemical analysis. Statistical features, such as mean, variance, and standard deviation, assess signal consistency and dispersion. Morphological features, such as peak shapes and slope and curvature, evaluate changes in signal gradient. Advanced AI-driven methods are employed to handle complex and high-dimensional data, such as principal component analysis (PCA), wavelet transform, ML feature engineering, DL feature extraction, and clustering and dimensionality reduction. PCA reduces the dimensionality of the data while retaining the most significant variance, while wavelet transform decomposes signals into different frequency bands for time-frequency analysis. ML feature engineering uses algorithms like decision trees or random forests to rank the importance of features, while DL feature extraction uses CNNs and Recurrent Neural Networks (RNNs) to learn hierarchical and temporal features directly from raw data. Clustering and dimensionality reduction techniques like t-distributed stochastic neighbor embedding (t-SNE) and uniform manifold approximation and projection (UMAP) visualize complex feature spaces and highlight clusters of similar data points. Feature extraction in OFS can be used in various applications, including healthcare, structural health monitoring (SHM), environmental monitoring, industrial automation, and telecommunications. Healthcare uses it to extract biomarkers and analyze respiratory patterns, while SHM detects strain and vibration anomalies. Environmental monitoring identifies pollutants through specific spectral shifts, while industrial automation allows real-time process control based on data.

3.1.3 Anomaly Detection

AI identifies anomalies or deviations in sensor data, which are essential for early warning systems across multiple sectors. AI models can detect subtle deviations in sensor signals, indicating faults or unusual conditions, such as irregular heartbeats or respiratory patterns in health monitoring, providing an advantage in AI technology. AI-based optical fiber systems are revolutionizing anomaly detection by using ML, DL, signal processing, and real-time monitoring. AI helps detect irregularities in optical signals, such as fiber degradation, environmental changes, or abnormal sensor readings, ensuring early fault detection and predictive maintenance. ML models are trained on labeled normal and abnormal data, while DL models use CNNs, RNNs and LSTMs, and autoencoders to detect abnormalities in time-series sensor data. Real-time signal processing includes Fourier transform and wavelet analysis, statistical methods, and edge AI processing to analyze anomalies locally without relying on cloud computing. Examples include AI-powered fiber optic sensors in oil pipelines detecting real-time pressure drops in oil pipelines. AI-powered predictive maintenance and fault detection include predictive analytics, self-healing mechanisms, and AI-optimized maintenance scheduling. For example, AI-enhanced fiber optic sensors detect thermal anomalies in power grids to prevent electrical failures. AI and IoT integration for smart anomaly detection enables real-time anomaly alerts via cloud platforms, AI-driven automation, and big data analytics. For example, AI-driven optical fiber networks detect traffic congestion anomalies in smart cities, optimizing urban mobility. AI revolutionizes optical fiber anomaly detection by enabling real-time, accurate, and automated anomaly detection, reducing false alarms, and enhancing system reliability with self-learning capabilities. This technology is transforming telecommunications, infrastructure monitoring, medical diagnostics, and industrial automation.

AI can help in signal interpretation by detecting patterns or anomalies in raw sensor data and identifying important variables for analysis. Pattern recognition algorithms detect specific patterns or anomalies, while feature extraction techniques like PCA identify the most important variables. AI can distinguish between signals from different analytes in complex mixtures. AI-based super-resolution techniques and predictive models can amplify weak signals produced by OFS, ensuring signal fidelity and enhancing low-intensity signals based on historical data and trends. The issue of environmental changes affecting sensor performance can be addressed by AI solutions that adjust the sensor's sensitivity and wavelength parameters in real-time, ensuring accuracy during temperature fluctuations.

3.2 Real-Time Monitoring and Control

Real-time monitoring and control involve the use of OFS to collect real-time data such as strain, temperature, pressure, or chemical concentrations. This data is used in various applications such as SHM and biomedical measurements. AI algorithms then process the data to extract meaningful features, reduce noise, and interpret results. Techniques used include signal denoising, feature extraction, and pattern recognition. AI systems then analyze the processed data to make informed decisions, such as triggering an alert in environmental monitoring. Based on these decisions, the system takes corrective or adaptive actions, such as adjusting machine operation to maintain optimal performance. DL algorithms process real-time data streams, providing immediate insights and prompt responses. For instance, AI can analyze temperature, pressure, or strain changes in infrastructure monitoring. AI can help the following tasks are easily achieved and can help the system in real time monitoring and control:

1. High Speed Processing: AI systems such as Neural Networks (NN) and CNN can analyze substantial amounts of sensor data within milliseconds. This facilitates prompt identification of changes or anomalies.
2. Adaptive Response: ML models dynamically modify sensor parameters (e.g., sensitivity or wavelength) in accordance with environmental variables. An AI-enhanced optical fiber sensor may adjust its sensitivity according to varying light intensity.
3. Automated Alert Systems: AI identifies critical conditions and activates alarms or alerts instantaneously. In healthcare, an AI-driven biosensor can notify medical personnel of irregular heartbeats or abnormal glucose concentrations.
4. Predictive Capabilities: AI forecasts future trends or failures with current and past data like help anticipating the failure of a bridge component by analyzing real-time strain data.
5. Integration with IoT and Edge Computing: AI enables OFS to interact with other devices inside an Internet of Things (IoT) network. Edge computing processes data locally on the device, hence minimizing latency for real-time applications.

3.3 Predictive Maintenance

Predictive Maintenance (PdM) is a technology that uses artificial intelligence and data analytics to predict failures or performance degradation in optical fiber sensor networks. This system collects real-time data such as signal loss, wavelength shift, and phase and intensity changes from fiber sensors, which are continuously monitored and sent to an AI-based predictive maintenance system. AI-based data processing and anomaly detection are crucial for early detection of fiber degradation before failure occurs. ML and DL models analyze historical sensor data, while anomaly detection algorithms identify sudden changes. Signal processing AI removes noise and false positives for accurate diagnostics. Predictive failure modeling using AI algorithms is essential for predicting exact failure times. Regression-Based ML models, reinforcement learning, and hybrid AI models help predict time-to-failure using past trends. Real-time decision making and automated maintenance are also essential for critical fiber networks. AI-driven edge computing processes sensor data in real time without cloud delays, and AI-controlled alerts and alarms notify technicians before a breakdown occurs. Automated self-healing networks reroute optical signals to bypass faulty fiber sections, resulting in zero downtime and self-healing fiber-optic networks.

AI is playing a significant role in predictive maintenance in OFS by utilizing data analytics, ML, and real-time monitoring techniques. AI algorithms analyze historical and real-time sensor data to detect early signs of degradation or potential failures, using techniques such as supervised learning, unsupervised learning, and DL. AI-driven edge computing processes sensor data in real-time, reducing response time for maintenance alerts and predicting potential failures based on deviations in sensor readings. AI also enables self-healing networks and smart diagnostics, enabling self-correcting algorithms to reroute data in case of failures. AI-driven predictive maintenance models, such as regression models, recurrent neural networks, Long Short-Term Memory (LSTM), random forest, and decision trees, estimate the remaining useful life of optical fibers based on usage patterns. AI has applications in telecommunications, structural health monitoring, healthcare, and biomedicine, ensuring sensor reliability in biosensing applications for real-time patient monitoring. By integrating AI into optical fiber sensor systems, maintenance becomes predictive, reducing costs, improving reliability, and enhancing system longevity. AI algorithms used in PdM include Random Forest, LSTM, CNN, reinforcement learning, and autoencoders. AI-driven Predictive Maintenance ensures smarter, self-repairing, and highly reliable optical fiber sensor networks, reducing costs, preventing unexpected network failures, and ensuring zero downtime.

3.4 Adaptive and Smart Sensing

AI-based OFS utilize adaptive and smart sensing to dynamically adjust to environmental changes, optimize data collection, and enhance decision-making. This is achieved through real-time AI algorithms, ML models, and edge computing. Adaptive sensing involves ML and DL to process historical and real-time data, while neural networks recognize patterns in optical signals and self-adjust detection parameters based on environmental conditions. RL optimizes sensor settings by continuously learning from feedback. For example, in environmental monitoring, AI-powered fiber optic sensors adjust sensitivity based on humidity, temperature, or pressure fluctuations. Real-time data processing and smart decision-making are achieved through edge AI and IoT integration, anomaly detection, self-optimizing sensors, multi-modal and context-aware sensing, predictive analytics, self-healing networks, and self-calibration of optical fiber sensors. AI-enabled fiber networks reroute signals automatically when damage is detected, ensuring long-term stability and reliability. Lastly, AI algorithms continuously calibrate optical fiber sensors without human intervention, reducing the need for frequent manual recalibration. This results in improved accuracy, efficiency, reduced false alarms, enhanced real-time decision-making, and minimal human intervention. AI-powered optical fiber systems are shaping the future of autonomous sensing, predictive diagnostics, and smart infrastructure monitoring.

3.5 Automation and Optimization

With the help of AI industries are getting automated and many of the important decisions are being performed by various AI algorithm. The system is able to take better actions as they are based on the past and present data. Technologies such as ML, DL, real-time data processing, and edge computing enable self-monitoring, self-adjusting, and predictive decision making improved efficiency, accuracy, and performance. AI helps in reducing noise thereby improving the data acquisition and processing. It can also improve the detection of complex signal variations and analyze the incoming optical signals for instant insights. In medical diagnostics the integration of AI in optical biosensors is able to detect the disease biomarkers automatically without the need of manual data interpretation. This avoids not only the time but also the human error that ensures accuracy and fast decision making. AI also makes the system reliable by predicting failure and optimizes the maintenance scheduling. Optimization with AI becomes much more convenient and can be implemented at various points of decision making.

In telecommunication networks, AI identifies optical fiber deterioration and optimizes power levels to avert communication disruptions. AI-driven optical fiber networks adaptively modulate bandwidth to optimize energy consumption in 5G communications [8]. It automates sensor calibration by dynamically altering parameters based on real-time data, so raising accuracy and facilitating continual learning and improvement. In SHM, AI can detect cracks and stress in bridges with manual intervention. With AI optimization of resources can also be achieved reducing the energy consumption, minimizing the false alarm and enhancing the signal processing through AI based Fourier transforms and neural networks [9, 10]. AI-driven control systems modify sensor functions according to predictive models, enhance system efficiency, and interface with IoT devices for complete automation. In oil and gas pipelines AI allows to autonomously identify leaks and initiate preventive measures without human involvement.

4 AI Algorithms

AI improves the capabilities of OFS by employing sophisticated algorithms to interpret data, extract features, identify anomalies, and optimize performance. The choice of the algorithm for an application depends on data being examined and detailed nature of work that requires intricate details of the application. These algorithms significantly improve the performance of the system by adding intelligence into it. It also enables real time data processing, anomaly detection, predictive maintenance, and smart sensing. Machine learning models examine trends in optical sensor data to improve decision-making, precision, and efficiency. Supervised learning methods, such as Decision Trees, Support Vector Machines, and Random Forests, identify normal versus faulty signals in optical fiber networks, whereas unsupervised learning techniques, like K-Means Clustering and Principal Component Analysis, uncover anomalies and hidden patterns without the need of labeled data. Similar to this there is deep learning methods that can extract complex features that can automate decision-making in optical fiber systems. Different methods of DL can help achieve different goals. CNN can extract spatial features whereas Recurrent Neural Networks (RNNs) and Long Short-Term Memory (LSTM) architectures evaluate time-series optical fiber data for trend forecasting, facilitating real-time surveillance of environmental alterations. Reinforcement learning (RL) autonomously optimizes the performance of optical fiber systems, facilitating enhanced tuning of fiber optic sensors for greater efficiency. AI-powered 5G optical networks ensure stable communication. Table 1 provides an overview of the various AI algorithms frequently employed in OFS. The following section discusses the different algorithm used for various applications.

4.1 Machine Learning

A Machine Learning algorithm is a computational technique that allows systems to identify patterns from data, generate predictions, and enhance performance over time without any additional programming [11]. It has boundless applications in fiber optic sensors throughout healthcare, environmental monitoring, industrial automation, and telecommunications, augmenting the intelligence and efficiency of these sensors. ML algorithms can be classified as supervised learning, unsupervised learning, RL, and DL. Each category serves unique purposes, encompassing pattern recognition and autonomous decision-making. Supervised learning is a ML technique that employs labeled data to generate predictions. It encompasses sensor data input, output from established outcomes, training to reduce mistakes, and prediction. This is a very common method to improve sensitivity, precision, and real-time observation. It is employed in optical fiber sensors for predictive modeling, signal processing, anomaly detection, and biosensing applications. These algorithms help find refractive index,

Table 1 Overview of commonly used AI algorithms for various applications

AI algorithm	Use case	References
Linear regression	Predicting RI changes	[11, 12]
Logistic regression	Binary classification of sensor signals	[13]
SVM	Classification	[14, 15]
Random forest	Multi-class classification, optimization	[16, 17]
Decision trees	Fault localization, anomaly detection	[15]
KNN	Real-time monitoring	[17, 18]
ANNs	Disease diagnosis, signal enhancement	[15, 19–21]
CNNs	Defect detection in optical fiber	[22, 23]
RNNs, AEs and LSTM	Predictive maintenance	[24]
Generative adversarial networks (GANs)	Improved AI training	[25]
Deep reinforcement learning (DRL)	Adaptive optical fiber system optimization	[26, 27]
Transformers (BERT, GPT)	Smart traffic bandwidth management	[28]

strain, temperature, and chemical concentrations, mitigate noise, detect faint signals, and identify structural anomalies. It is utilized for applications like chemical detection, biosensing, fault identification, and structural health monitoring.

Supervised learning is a technique that employs labeled data to train models capable of predicting outcomes. Different algorithms are available for distinct purposes in optical fiber sensing and networks. Linear Regression is one of the algorithm that predicts variations in refractive index for chemical sensors, and Logistic Regression is employed for defect detection in optical fiber networks. Support Vector Machines (SVM) are employed for the classification of optical signals in biosensing and fault detection applications and decision trees for structural health monitoring. Table 1 gives the overview of different algorithms and their applications in the discussed domain. Supervised learning enhances automation, predictive analytics, and anomaly detection in optical fiber sensing. Selecting the appropriate method is dependent upon the data type, the complexity of the problem, and the performance criteria. The optimal method selection is dependent upon type of decision or prediction, classification, detection and predictive maintenance. Different aspects are to be considered before finalizing the right method [13]. In AI-Enhanced Optical Fiber Systems, regression algorithms are used for continuous variable prediction, classification models for fault detection and biosensing, and boosting and DL for high-dimensional, real-time processing.

Unsupervised learning is a category of machine learning in which the model is trained on unlabeled data. They are utilized OFS networks detect patterns, clusters, and abnormalities. These techniques facilitate defect identification, signal classification, anomaly detection, and predictive maintenance without requiring pre-labeled datasets. The algorithms employed in optical fiber systems comprise K-Means Clustering, Hierarchical Clustering, DBSCAN (Density-Based Clustering), Gaussian Mixture Model (GMM), Principal Component Analysis (PCA), Independent Component Analysis (ICA), Autoencoders (Neural Networks), and Isolation Forest. The selection of an appropriate unsupervised learning algorithm is dependent upon the data's structure, complexity, and the nature of the problem. The optimal method selection is dependent upon clustering similar sensor signals, recognizing anomalies in fiber networks, detecting patterns in biosensing, managing noisy sensor data, and detecting unidentified problems in networks. Hence Unsupervised learning is used in many optical fiber systems for fault detection, disease detection, SHM and predictive maintenance. In conclusion, unsupervised learning enhances automation, fault detection, and efficiency in optical fiber sensing by using clustering algorithms for signal classification and biosensing, dimensionality reduction for cleaning noisy data, and anomaly detection for fault detection and predictive maintenance.

4.2 Deep Learning

DL is a subset of ML aiming to process complex data patterns. This becomes very useful in analyzing optical fiber sensor data as they are high dimensional. It facilitates automated pattern recognition, achieves high accuracy, enables real-time processing, and supports self-learning and adaptation. The commonly used DL architectures include CNNs, RNNs, Long Short-Term Memory (LSTM), Autoencoders, Generative Adversarial Networks (GANs), Transformers (BERT, GPT), and Deep Reinforcement Learning (DRL). CNNs extract spatial information from fiber optic signal images or spectral data, hence improving optical imaging. LSTM is used to predict network congestion and improve the bandwidth allocation in fiber networks. Patterns in sensors data is identified by autoencoders and with the help of GANs sensor precision is enhanced. Transformers rapidly process sequential data, improving fiber optic signal classification in real-time communication and facilitating automated problem diagnostics in fiber optic networks. DRL enhances optical fiber network performance by automating power regulation in optical fiber connections, improving dynamic routing in fiber-optic communication systems, and autonomously altering sensor sensitivity in intelligent fiber optic biosensors. DRL enhances wavelength allocation in dense wavelength-division multiplexing (DWDM) optical networks. It enhances optical fiber networks by increasing speed, intelligence, and efficiency. It enhances optical fiber networks by increasing speed, intelligence, and efficiency.

These algorithms are employed in AI-Enhanced OFS for defect detection, predictive maintenance, biosensors for disease diagnosis, smart chemical and environmental sensing, SHM, adaptive optical networks, and real-time optimization. With Dl the entire optical fiber sensor system is becoming smart and is utilized in many diverse applications.

4.3 Reinforcement Learning

Reinforcement Learning (RL) is a ML methodology in which an artificial intelligence agent acquires optimal decision-making skills by interaction with an environment, receiving rewards or penalties accordingly. RL is employed to tackle issues including dynamic wavelength allocation in optical fiber communication, adaptive power regulation for optical fiber links, fault detection and predictive maintenance in optical fiber sensors, traffic prediction and load balancing in optical networks, and intelligent optical biosensors for healthcare applications. Dynamic wavelength allocation facilitates faster and interference-free transmission, while adaptive power control enhances the energy efficiency of optical networks. Predictive maintenance guarantees uninterrupted operation in industrial fiber optic networks, whereas traffic balancing optimizes performance in high-speed internet and telecommunications networks. Intelligent biosensors enhance the accuracy of identifying diseases in medical diagnostics. RL is facilitating self-optimizing, adaptable, and intelligent fiber optic systems. RL-powered AI is enhancing optical fiber networks and sensors, rendering them more intelligent, adaptive, and efficient.

4.4 Hybrid AI

Hybrid AI integrates many AI methodologies, including ML, DL, expert systems, and RL, to develop systems that are more precise, adaptable, and efficient.

Hybrid AI is a combination of multiple AI techniques, including ML, DL, Expert Systems, and RL, to create more accurate, adaptive, and efficient systems. It is used in OFS and networks to improve performance in areas such as optical communication networks, fiber-optic sensing and biosensors, predictive maintenance, and fault detection. One of the main challenges in using hybrid AI is signal distortion and latency issues in fiber-optic communication. To address this, ML predicts traffic patterns and adjusts bandwidth in real time, while expert systems use pre-defined network rules to prevent signal degradation. This results in high-speed 5G and 6G fiber-optic networks with real-time traffic optimization. Another challenge is low sensitivity in optical biosensors for disease detection. Hybrid AI solutions include DL to identify disease biomarkers from sensor data and physical-based modeling to enhance signal processing for accurate molecular detection. This leads to highly sensitive biosensors for early disease diagnosis. Predictive maintenance and fault detection are another challenge in optical fiber systems. ML detects early signs of fiber degradation, while RL adjusts network parameters to prevent failures. This results in zero downtime and automated network health monitoring. Adaptive optical fiber sensors for environmental monitoring use IoT and AI algorithms to adjust sensor sensitivity based on environmental conditions. Cybersecurity in optical networks is another challenge, with DL detecting anomalous network behavior and expert systems implementing predefined security protocols for real-time threat mitigation. In conclusion, hybrid AI is revolutionizing optical fiber systems by making them smarter, adaptive, and secure.

The choice of an AI algorithm for AI-enhanced optical fiber sensors depends on several factors, including data characteristics, application requirements, real-time processing needs, accuracy, and computational efficiency. The key considerations for choosing the right AI algorithm include the type of data collected, application-specific requirements, accuracy versus computational complexity, real-time versus batch processing, and adaptability and learning capabilities. Data types include Time-Series Data (RNN, LSTM), Spectral Data (Wavelength shifts, refractive index changes), and Anomaly Detection Data (fiber degradation, signal loss). Application-specific requirements include Signal Optimization in Telecommunications, Structural Health Monitoring, Biosensing and Medical Diagnosis, and Chemical Detection. The best AI algorithm for AI-enhanced optical fiber sensors depends on the type of data, application requirements, accuracy versus complexity trade-off, real-time versus batch processing, and adaptability and learning needs. AI-powered optical fiber sensors are transforming industries with smart automation, predictive maintenance, and real-time monitoring.

5 Applications of AI Enhanced Optical Fiber System

5.1 Healthcare

AI-enhanced optical fiber sensors are transforming healthcare by enabling real-time, highly sensitive, and non-invasive monitoring of various medical parameters. These sensors offer biocompatibility, high precision, and immunity to electromagnetic interference, making them ideal for medical applications. AI-based optical fiber biosensors detect biomarkers for diseases like cancer, diabetes, and infections using ML and DL for early disease detection. They also detect glucose levels non-invasively in sweat, saliva, or interstitial fluid, using AI algorithms for predictive diabetes management and personalized insulin recommendations [12]. Figure 4 shows the comparison of wearable health monitoring to centralized healthcare services as a convenient approach in remote and real-time tailored healthcare systems. It also demonstrates AI instruments and individualized health surveillance for therapeutic interventions and disease identification. Aid-driven fiber-optic ECG patches continuously monitor

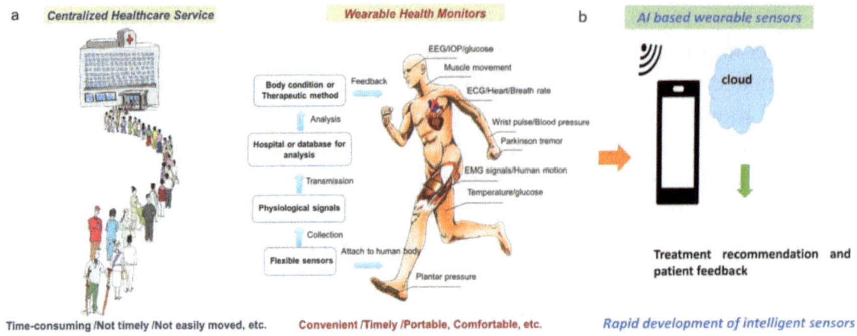

Fig. 4 Demonstrates the different ways the healthcare facilities can be handled from centralized healthcare service to wearable health monitors to AI based wearable sensors [29]

cardiac health with real-time alerts. Neuro-optical fiber sensors track brain activity, neurotransmitters, and temperature changes, using AI deep learning Models for early disease markers and predictive AI models for seizures and cognitive decline. AI-Enhanced optical fiber sensors detect COVID-19, tuberculosis, and influenza at ultra-low concentrations, enabling on-site, real-time diagnostics in rural areas. Edge AI Processing enables on-site, real-time diagnosis without a lab. AI-driven fiber-optic sensors track oxygen levels, pH, and moisture in wounds, predicting infection risk and healing rate based on sensor data. Smart bandages with AI-driven fiber sensors provide real-time wound monitoring.

AI-based optical fiber sensors measure drug concentration in blood and urine, allowing AI-driven predictive models to adjust dosage in real-time for personalized medicine. They track antibiotics, chemotherapy drugs, and opioids for cancer patients. AI-enhanced fiber-optic wearables transmit health data in real-time, analyze sensor data for telemedicine consultations, and ensure low-latency remote patient monitoring through 5G with AI-enabled optical fiber networks. AI has helped revolutionize healthcare with optical fiber sensors, enabling early disease detection, non-invasive monitoring, real-time and predictive analysis, telemedicine and remote care, and personalized medicine. AI-powered OFS are paving the way for the future of smart, predictive, and personalized healthcare.

5.1.1 Biosensing for Disease Diagnosis

The integration of AI with optical fiber biosensing technologies is facilitating more effective and efficient disease diagnostics, enabling early detection and enhanced patient outcomes. It combines the sensitivity of optical fiber technology with the analytical power of artificial intelligence. This integration allows for rapid, accurate, and real-time detection of various biomarkers associated with diseases. Key developments include a fully automated system for multiple cerebrospinal fluid biomarker detection, an AI-enabled device for traumatic brain injury monitoring, and an optical fiber-based biosensor for the label-free detection of salivary interleukin-8 protein. Infectious disease diagnostics have also seen advancements, with optical biosensors used for the detection of various life-threatening viruses and bacteria. The combination of AI and optical fiber biosensing technology has improved the interpretation of complex sensor data, leading to improved diagnostic capabilities [30]. Additionally, the integration of AI in salivary biomarker detection has led to an ultra-low detection limit, promising non-invasive disease diagnostics.

The benefits of AI integration include enhanced sensitivity and specificity, real-time data analysis, and pattern recognition. The fusion of AI with optical fiber biosensing technology is paving the way for more effective and efficient disease diagnostics, offering the potential for early detection and improved patient outcomes. AI-based optical fiber sensors are being developed to detect neurological disorders like Alzheimer's, Parkinson's, and epilepsy. These sensors track brain activity, neurotransmitters, and temperature changes, and use AI deep learning models to analyze EEG data for early disease markers. Predictive AI models can also detect seizures and cognitive decline before symptoms appear. AI-Enhanced Optical Fiber Sensors are being developed for rapid and accurate detection of infectious diseases like COVID-19, tuberculosis, and influenza [31]. These sensors use LSPR-based fibers, AI-driven spectral analysis, and edge AI processing for on-site, real-time diagnostics in rural areas, enabling immediate diagnosis without a lab. There are recent research that introduces the creation of a biofunctionalized fiber-optic ball-resonator biosensor for the real-time identification of the vaccinia virus. The sensor exhibited increased sensitivity and specificity, presenting a viable instrument for fast viral diagnostics [32]. Another work utilized fiber-optic biosensors in cancer diagnosis and treatment. This encompasses the identification of cancer biomarkers, the distinction of malignant cells, the validation of ex vivo tumor models, and the in vivo observation of tumor progression, emphasizing the adaptability of these sensors in oncology [33] (Fig. 5).

5.1.2 Telemedicine and Remote Care

AI is being used in optical fiber-based telemedicine and remote patient monitoring to provide real-time data transmission and analysis. Smart fiber-optic wearables transmit health data in real-time, while AI cloud analytics analyze sensor data for telemedicine consultations. 5G and AI-enabled optical fiber networks ensure low-latency remote patient monitoring. OFS with AI is huge impact on remote healthcare by providing real-time health monitoring, predictive AI algorithms, and remote diagnosis and treatment. These sensors can be used in smart wearables, biosensors, drug monitoring, neurological monitoring, and smart optical fiber networks. AI-powered fiber-optic smart wearables monitor vital signs like ECG, blood pressure, oxygen levels, glucose, and temperature in real-time, detecting irregularities and alerting doctors. They also enable low-latency data transmission to telemedicine platforms. Biosensors detect infectious diseases like COVID-19, TB, and Influenza remotely via saliva/sweat testing and chronic disease biomarkers like diabetes and cancer without blood draws. They use AI cloud processing for instant results via telemedicine

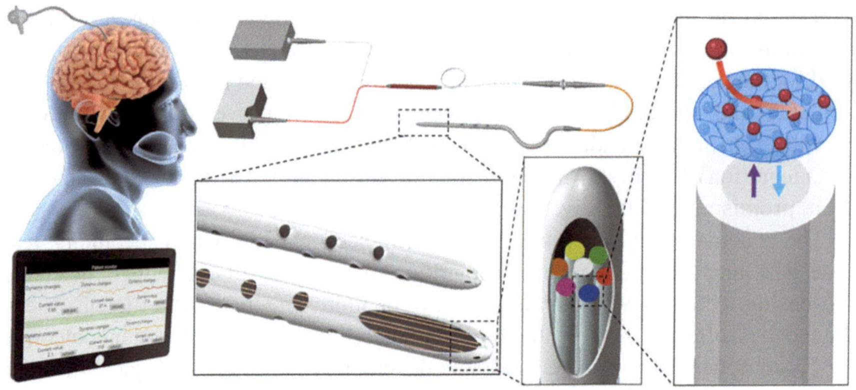

Fig. 5 Multiplexed optical fiber sensing device for the continuous and simultaneous monitoring of six cerebrospinal fluid (CSF) biomarkers utilizing tip-functionalized optical fibers and computational algorithms [34]

apps. Remote drug monitoring and personalized medicine use AI-driven fiber-optic biosensors to track drug concentration levels in real-time, predict personalized dosage adjustments, and optimize treatment. Neurological diseases require continuous monitoring, and AI-powered fiber-optic fiber sensors detect neurotransmitter levels, brain activity, and temperature changes, predicting seizures, Alzheimer's cognitive decline, and Parkinson's tremors [34]. Smart optical fiber networks enhance telemedicine by providing ultra-fast, low-latency, secure medical data transfer, supporting real-time remote surgeries and AI-assisted diagnostics, and integrating AI chatbots and virtual healthcare assistants for patient interaction. AI algorithms used in optical fiber sensors include LSTM, CNN, Random Forest and XGBoost, Reinforcement Learning, and Autoencoders for early disease detection. The future of AI-based optical fiber sensors in remote healthcare includes fully automated, AI-integrated hospitals, real-time disease surveillance, wearable AI healthcare devices, robotic, fiber-optic tele-surgeries, and edge AI processing for healthcare. These sensors enables smarter, faster, and more personalized remote healthcare [35–37] (Fig. 6).

5.2 *Environmental Monitoring*

AI-powered OFS in environmental monitoring is providing real-time, high-precision, and long-distance sensing capabilities. These sensors track air quality, water pollution, climate change, and industrial emissions with AI-driven data analytics, anomaly detection, and predictive modeling. AI-Enhanced OFS offer continuous and instant monitoring of environmental changes, remote and distributed sensing, highly sensitive and accurate detection of chemical and physical changes, and AI-driven pattern recognition for early warnings and disaster prediction. They also work with minimal

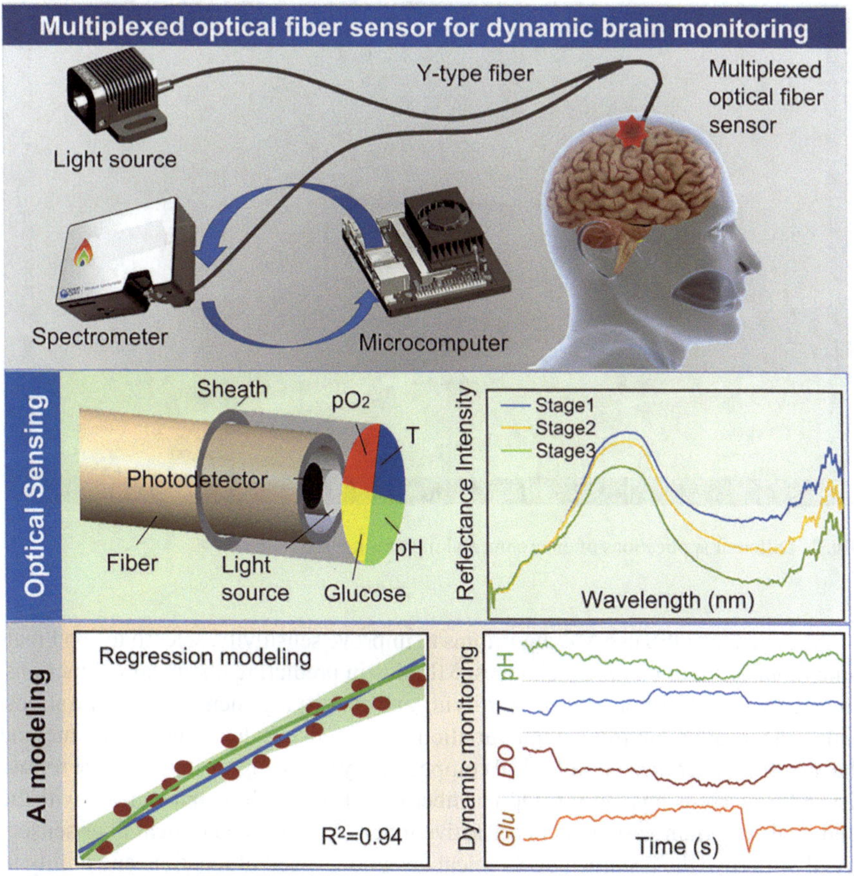

Fig. 6 A multiplexed optical fiber sensor controlled by regression algorithms has been created for the dynamic monitoring of brain pH, temperature, dissolved oxygen, and glucose concentrations [38]

human intervention, working for years with minimal human intervention. AI-based OFS have various applications in environmental monitoring. Figure 7 shows different aspects of environmental monitoring that is widely considered. For example, they can detect CO, CO, SO, NO, CH, O levels in urban and industrial areas, detect heavy metals in drinking water, pathogens and bacteria in rivers and lakes, and chemical pollutants from agriculture and industry. They can also help in wildfire detection and prevention, predicting rising temperatures and dry conditions in forests, and enabling faster firefighting responses. AI-powered optical fiber gas sensors can detect toxic gas leaks in industries and radioactive leaks in nuclear plants for early containment. They can also predict potential failures and alert authorities instantly. AI-powered fiber-optic leak detection systems prevent oil spills in pipelines.

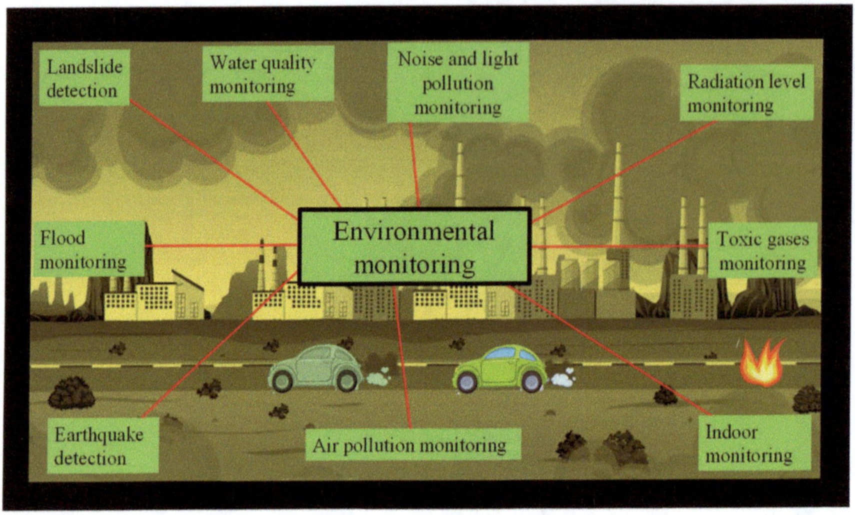

Fig. 7 Different applications of environmental monitoring [42]

AI-enhanced OFS use ML algorithms to improve sensitivity, selectivity, and real-time monitoring of chemical changes. ML aids in predicting chemical interactions, enhancing signal-to-noise ratio, identifying and classifying chemical concentrations, and compensating environmental variations. Refractive Index Sensing is a method that identifies chemical or biological compounds by assessing variations in the refractive index of a sensing medium. Optical fiber sensors utilize light interactions with the analyte to ascertain variations in refractive index. AI enhanced optical fiber sensors greatly improve the accuracy of chemical detection, speed of reaction, and ability to monitor in real time. The integration of ML algorithms is transforming healthcare, environmental safety, and industrial monitoring using these sensors.

There are several chemicals that have been effectively sensed using ML-enhanced optical fiber sensors. ML is being used in various applications, including glucose detection, dissolving oxygen (DO), heavy metal detection, volatile organic compounds (VOCs), liquid identification, and chemical sensing. ML models are used to analyze complex spectral data from fiber-optic sensors, enabling accurate quantification of glucose concentrations [34]. ML algorithms help distinguish overlapping emission spectra in fiber-optic sensors, facilitating precise DO measurement. Fiber-optic nanosensors, combined with ML techniques, analyze spectral data to identify and quantify heavy metal concentrations [39]. Optical fiber sensors detect changes in refractive index caused by VOCs, and ML algorithms classify compounds based on sensor data. Lastly, a simple fiber-optic tip sensor collects data on liquid samples, and ML models classify the liquids by analyzing sensor responses [40]. These applications demonstrate the potential of ML in various applications, such as glucose detection, DO detection, heavy metal detection, VOC detection, and liquid identification. Overall, ML is being used in various applications to improve the efficiency and accuracy of various applications. These examples demonstrate the varied applications of ML in improving the chemical detection capacities of optical fiber sensors. The use of ML facilitates enhanced accuracy, efficiency, and real-time analysis across many domains, such as environmental monitoring, healthcare, and industrial safety. The use of AI in water quality monitoring has been a topic of interest. Linear Regression and Random Forest are used to predict RI shifts for heavy metal detection, such as lead, mercury, and arsenic in water. Biosensing, such as SVM and ANN, classifies RI shifts for vitamin detection, such as ascorbic acid and glucose levels in blood samples. Industrial chemical leak detection uses Gradient Boosting and Neural Networks to predict chemical leaks based on RI changes [41]. This method has been applied to detect toxic gas leaks in factories, improving response time by 30% compared to conventional sensors. Food quality monitoring uses Decision Trees and K-Nearest Neighbors (KNN) to classify contaminants in food, such as pesticides, preservatives, and spoilage in liquids. These AI-based sensors have achieved 95% accuracy in contamination detection.

These AI-driven sensors can track sea level rise, ocean temperature variations, glacier melting and carbon absorption in polar regions, and extreme weather patterns. Examples include fiber-optic networks in Antarctica tracking global warming trends. AI algorithms used in environmental optical fiber sensing include CNN for pollution and climate data detection, LSTM for air quality changes and wildfire risks, reinforcement learning for industrial emissions control, Random Forest and XGBoost for water contaminants and gas leaks, and autoencoders for anomalies in environmental sensor data. The future of AI-based optical fiber sensors includes AI-integrated smart cities, renewable energy monitoring, autonomous environmental drones, global AI climate models, and disaster prevention. These sensors are revolutionizing environmental monitoring, enabling a cleaner, safer, and more sustainable world.

5.3 Smart Infrastructure

AI-powered OFS provides real-time monitoring, predictive maintenance, and automation in critical sectors like transportation, energy, smart cities, and industrial safety. These networks provide high-speed, ultra-sensitive, and distributed sensing to enhance structural safety, energy efficiency, and operational performance. AI-Enhanced OFS are used in smart bridges and SHM, smart transportation and railways, smart cities and infrastructure automation. It also contributes in smart power grids and renewable energy monitoring, industrial Safety and hazardous area monitoring. SHM is achieved through AI-powered fiber-optic strain, vibration, and tilt sensors that detect cracks, stress, and corrosion in bridges, dams, tunnels, and skyscrapers [43]. AI-based damage classification predicts structural failure before it happens, ensuring timely maintenance and reducing repair costs. Real-time monitoring ensures timely maintenance and reduces repair costs. Smart transportation and railways use AI-powered fiber-optic sensors to detect rail track defects, vibrations, and train axle loads, optimize maintenance schedules, and integrate traffic management with AI fiber sensors. Smart power grids and renewable energy monitoring improve power networks by detecting faults, overheating, and load imbalances in power lines, optimizing energy flow in smart grids for reduced wastage, and monitoring solar panels and wind turbines for maximum efficiency. AI-based industrial safety and hazardous area monitoring provide leak detection for oil and gas pipelines, fire, smoke, and toxic gas detection in chemical plants, and predictive safety alerts for workers in hazardous environments. AI-driven fiber sensors in offshore oil rigs detect gas leaks instantly. AI-powered earthquake and disaster early warning systems detect seismic activity for early earthquake warnings, landslide movements, underground water changes, flood levels, and storm surges in coastal cities. AI algorithms used in smart infrastructure include LSTM, CNN, Random Forest and XGBoost, RL, and Autoencoders. The future of AI-based optical fiber sensors in smart infrastructure includes automated, energy-efficient, and climate-responsive skyscrapers, autonomous AI-controlled traffic systems, self-monitoring industrial plants with zero downtime, disaster-resilient cities, and 5G-connected fiber sensor networks for ultra-fast AI data processing for real-time city management. AI-powered optical fiber sensors are building the future of smart infrastructure, ensuring safer, greener, and more efficient cities.

5.4 Telecommunications

AI-powered optical fiber sensors in telecommunications enables high-speed data transmission, real-time network monitoring, fault detection, and adaptive optimization. These smart fiber networks enhance network reliability, reduce latency, and optimize bandwidth, making them essential for 5G, IoT, and future 6G communications. AI-enhanced fiber sensors can detect fiber faults and signal degradations, predict fiber wear, allocate bandwidth based on user demand, enhance low-latency and high-speed data transmission, and protect fiber networks from hacks and failures. This results in faster, smarter, and more reliable global communication networks. AI-driven fiber sensors can optimize network performance by dynamically adjusting bandwidth based on demand, optimizing signal strength for 5G and fiber broadband, and reducing latency for cloud computing and streaming services. They also detect microbends and fiber breaks instantly, predict maintenance needs before failures occur, and reduce downtime and repair costs for telecom providers. AI-based fiber sensors can optimize 5G small cell deployment and signal quality, dynamic frequency allocation for real-time optimization, and intelligent routing for future 6G wireless networks [28]. They can also detect and prevent cyber threats targeting fiber-optic communication systems, such as intrusions, unauthorized access, anomalies in optical signal patterns, and data breaches. AI-powered undersea and space fiber networks can optimize fault detection and repair prediction, satellite communication, space exploration, and ultra-secure fiber data transmission. Examples of AI-enabled networks include Google and Facebook undersea fiber cables. AI algorithms used in optical fiber networks include reinforcement learning (RL), deep neural networks

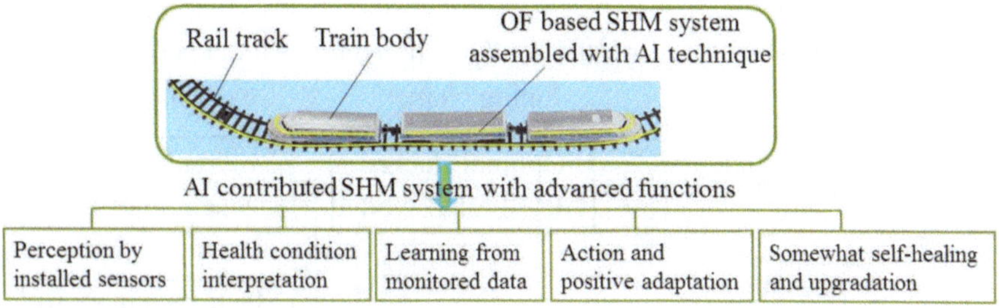

Fig. 8 Utilization of artificial intelligence in the health monitoring of railway structural systems [44]

(DNNs), SVMs, convolutional neural networks CNNs, and LSTM [27]. These algorithms make fiber networks faster, smarter, and more secure. Figure 9 shows a overall system proposed to enhance the accuracy of strain measurement.

The future of AI-based optical fiber networks includes self-healing, ultra-fast, energy-efficient networks, quantum fiber networks for finance and defense, autonomous network operations, AI-integrated smart cities, and space-based optical fiber networks for satellite internet for global access. AI-powered optical fiber networks are shaping the future of global communication and are poised to revolutionize the industry (Fig. 8).

Traditional OFS generate large amounts of raw data, which can be affected by noise and environmental fluctuations. AI techniques such as ML and deep learning are used to extract meaningful features, reduce noise, and improve sensor accuracy. Real-time monitoring and predictive analytics enable applications such as SHM, telemedicine, and disaster prediction. AI-based smart and adaptive sensing helps fiber optic sensors dynamically adapt to changing environmental conditions by self-calibrating in real time. Biomedical sensors assisted by AI adjust sensitivity according to patient conditions for more precise diagnosis of diseases. Detection of anomalies powered by AI in networks uses AI models to identify anomalies, cyber threats, and security breaches. AI-powered optical fiber sensors are increasingly being used in smart cities, industrial automation, and environmental monitoring through IoT connectivity [5] (Fig. 9).

6 Conclusion

AI is revolutionizing optical fiber sensors and sensing networks, making them smarter, more adaptive, and autonomous. Future advancements in ML, IoT, quantum sensing, and energy-efficient AI models will drive innovation in this field. AI-driven self-healing optical networks, quantum sensing and computing, energy-efficient optical fiber systems, next-gen optical biosensors in personalized healthcare, advanced environmental and climate monitoring, and AI-integrated optical fiber sensors in

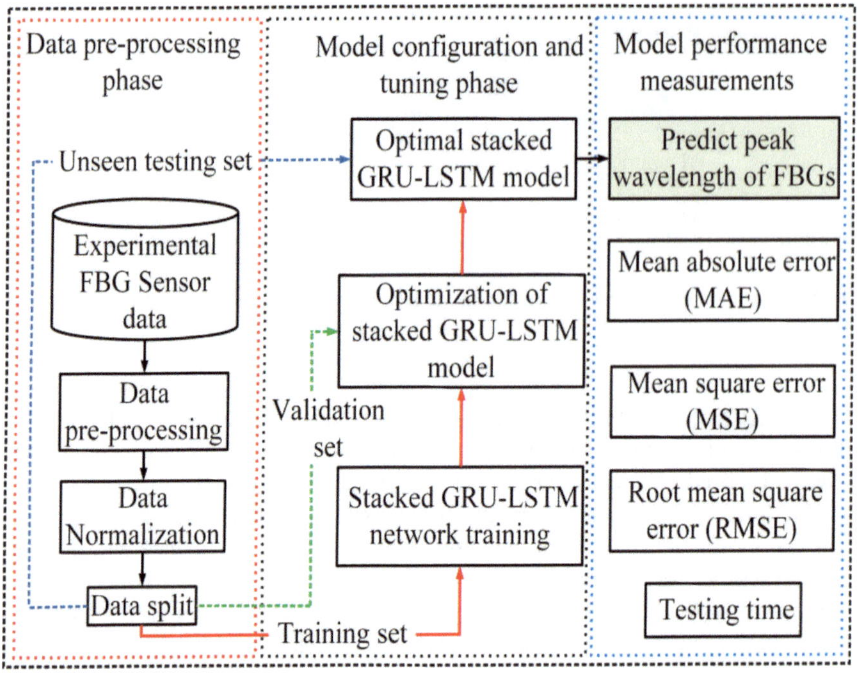

Fig. 9 A proposed system structure with overview of the AI enhanced system telecommunication [45]

space exploration are some of the promising developments. AI algorithms will enable optical fiber networks to self-heal and reconfigure automatically in case of failures, optimizing data routing in 5G and 6G communication networks. Quantum sensing and computing will improve applications in biomedical diagnostics, space exploration, and geophysics. Energy-efficient optical fiber systems will optimize power consumption, making fiber-optic networks more sustainable and cost-effective. Next-generation optical biosensors in personalized healthcare will enable ultra-precise disease diagnosis and real-time patient monitoring, while AI-based sensors will enhance remote sensing for disaster prediction. AI-driven optical fiber sensors will also play a crucial role in autonomous space missions and deep-space communication networks. In conclusion, AI is making far reaching changes in optical fiber sensors and sensing networks, paving the way for next-generation AI-driven technologies in healthcare, environment, smart cities, space exploration, and ultra-fast communication networks.

References

1. A. Bogomolov, A. Evseeva, E. Ignatiev, V. Korneev, New approaches to data processing and analysis in optical sensing. TrAC, Trends Anal. Chem. **160**, 116950 (2023)
2. C. Wang, T. He, H. Zhou, Z. Zhang, C. Lee, Artificial intelligence enhanced sensors-enabling technologies to next-generation healthcare and biomedical platform. Bioelectron. Med. **9**(1), 17 (2023)
3. M.A. Gandhi, S. Chu, K. Senthilnathan, P.R. Babu, K. Nakkeeran, Q. Li, Recent advances in plasmonic sensor-based fiber optic probes for biological applications. Appl. Sci. **9**(5), 949 (2019)
4. A. Venketeswaran, N. Lalam, J. Wuenschell, P.R. Ohodnicki, M. Badar, K.P. Chen, P. Lu, Y. Duan, B. Chorpening, M. Buric, Recent advances in machine learning for fiber optic sensor applications. Adv. Intell. Syst. **4**(1), 2100067 (2022)
5. P. Kamencay, P. Hockicko, R. Hudec, Sensors data processing using machine learning. MDPI (2024)
6. Y. Zhou, Y.-N. Zhang, Q. Yu, L. Ren, Q. Liu, Y. Zhao, Application of machine learning in optical fiber sensors. Measurement 114391 (2024)
7. Z. Sun, M. Fang, D. Huang, H. Yang, C. Zhao, S. Li, F. Liu, An unsupervised learning method to enhance SNR for distributed optical fiber vibration sensor based on phase-sensitive OTDR. Opt. Laser Technol. **180**, 111570 (2025)
8. A.F. Mohammed, J. Lee, S. Park, Dynamic bandwidth slicing in passive optical networks to empower federated learning. Sensors **24**(15), 5000 (2024)
9. U. Jayawickrema, H. Herath, N. Hettiarachchi, H. Sooriyaarachchi, J. Epaarachchi, Fibre-optic sensor and deep learning-based structural health monitoring systems for civil structures: a review. Measurement **199**, 111543 (2022)
10. S. Villalba, J.R. Casas, Application of optical fiber distributed sensing to health monitoring of concrete structures. Mech. Syst. Signal Process. **39**(1–2), 441–451 (2013)
11. Y.S. Dwivedi, R. Singh, A.K. Sharma, A.K. Sharma, Enhancing the performance of photonic sensor using machine-learning approach. IEEE Sens. J. **23**(3), 2320–2327 (2022)
12. M. Shokrekhodaei, D.P. Cistola, R.C. Roberts, S. Quinones, Non-invasive glucose monitoring using optical sensor and machine learning techniques for diabetes applications. IEEE Access **9**, 73029–73045 (2021)
13. D.F. Kandamali, X. Cao, M. Tian, Z. Jin, H. Dong, K. Yu, Machine learning methods for identification and classification of events in ϕ-OTDR systems: a review. Appl. Opt. **61**(11), 2975–2997 (2022)
14. Y. Wang, P. Wang, K. Ding, H. Li, J. Zhang, X. Liu, Q. Bai, D. Wang, B. Jin, Pattern recognition using relevant vector machine in optical fiber vibration sensing system. IEEE Access **7**, 5886–5895 (2019)
15. H. Song, Y. Li, M. Liu, K. Wang, J. Li, M. Zhang, Y. Zhao, J. Zhang, Experimental study of machine-learning-based detection and location of eavesdropping in end-to-end optical fiber communications. Opt. Fiber Technol. **68**, 102669 (2022)
16. Y. Wang, M. Li, X. Li, H. Wang, L. Tian, R. Deng, Identification and classification of water absorption profile of distributed optical fiber vibration signal based on XGBoost algorithm. SN Appl. Sci. **4**(11), 289 (2022)
17. K.K. Soothar, Y. Chen, A.H. Magsi, C. Hu, H. Shah, Optimizing optical fiber faults detection: a comparative analysis of advanced machine learning approaches. Comput. Mater. Contin. **79**(2) (2024)
18. S. Keser, Ş.E. Hayber, Fiber optic tactile sensor for surface roughness recognition by machine learning algorithms. Sens. Actuators, A **332**, 113071 (2021)
19. C. Karapanagiotis, K. Hicke, K. Krebber, A collection of machine learning assisted distributed fiber optic sensors for infrastructure monitoring. tm-Tech. Mess. **90**(3), 177–195 (2023)
20. J. Rana, A.K. Sharma, Y.K. Prajapati, Intervention of machine learning and explainable artificial intelligence in fiber optic sensor device data for systematic and comprehensive performance-optimization. IEEE Sens. Lett. (2024)

21. H.A. Arief, T. Wiktorski, P.J. Thomas, A survey on distributed fibre optic sensor data modelling techniques and machine learning algorithms for multiphase fluid flow estimation. Sensors **21**(8), 2801 (2021)
22. C. Lyu, Z. Huo, X. Cheng, J. Jiang, A. Alimasi, H. Liu, Distributed optical fiber sensing intrusion pattern recognition based on GAF and CNN. J. Lightwave Technol. **38**(15), 4174–4182 (2020)
23. Z. Sun, K. Liu, J. Jiang, T. Xu, S. Wang, H. Guo, Z. Zhou, K. Xue, Y. Huang, T. Liu, Optical fiber distributed vibration sensing using grayscale image and multi-class deep learning framework for multi-event recognition. IEEE Sens. J. **21**(17), 19112–19120 (2021)
24. Q. Zhang, Z. Wang, S. Duan, B. Cao, Y. Wu, J. Chen, H. Zhang, M. Wang, An improved end-to-end autoencoder based on reinforcement learning by using decision tree for optical transceivers. Micromachines **13**(1), 31 (2021)
25. N. Islam, M.M. Shoaib Hasan, I. Hossain Shibly, M.B. Rashid, M.A. Yousuf, F. Haider, R. Ahmmed Aoni, R. Ahmed, Plasmonic sensor using generative adversarial networks integration. Opt. Express **32**(20), 34184–34198 (2024)
26. E. Durech, W. Newberry, J. Franke, M.V. Sarunic, Wavefront sensor-less adaptive optics using deep reinforcement learning. Biomed. Opt. Express **12**(9), 5423–5438 (2021)
27. J. Li, D. Wang, M. Zhang, S. Cui, Digital twin-enabled self-evolved optical transceiver using deep reinforcement learning. Opt. Lett. **45**(16), 4654–4657 (2020)
28. A. Tzanakaki, M. Anastasopoulos, V.-M. Alevizaki, Intent-based control and management framework for optical transport networks supporting B5G services empowered by large language models. J. Opt. Commun. Netw. **17**(1), 112–123 (2024)
29. S. Shajari, K. Kuruvinashetti, A. Komeili, U. Sundararaj, The emergence of AI-based wearable sensors for digital health technology: a review. Sensors **23**(23), 9498 (2023)
30. J. Parmar, S.K. Patel, V. Katkar, A. Natesan, Graphene-based refractive index sensor using machine learning for detection of *Mycobacterium tuberculosis* bacteria. IEEE Trans. Nanobiosci. **22**(1), 92–98 (2022)
31. S.A. Nakhjavani, H. Mirzajani, S. Cararra, M.C. Onbaşlı, Advances in biosensor technologies for infectious diseases detection. TrAC Trends Anal. Chem. 117979 (2024)
32. A. Rakhimbekova, B. Kudaibergenov, K. Seitkamal, A. Bellone, A. Dauletova, M. Sypabekova, M. Olivero, G. Perrone, A. Radaelli, C. Zanotto et al., Rapid detection of vaccinia virus using biofunctionalized fiber-optic ball-tip biosensors. Sci. Rep. **13**(1), 17470 (2023)
33. F. Jin, Z. Xu, D. Cao, Y. Ran, B.-O. Guan, Fiber-optic biosensors for cancer theranostics: from in vitro to in vivo. Photonic Sens. **14**(4), 240415 (2024)
34. Y. Zhang, N. Zhang, Y. Hu, C. Pereira, M. Fertleman, N. Jiang, A.K. Yetisen, Fully automated and AI-assisted optical fiber sensing system for multiplexed and continuous brain monitoring. ACS Sens. **9**(12), 6605–6620 (2024)
35. X. Zhang, C. Wang, T. Zheng, H. Wu, Q. Wu, Y. Wang, Wearable optical fiber sensors in medical monitoring applications: a review. Sensors **23**(15), 6671 (2023)
36. A.G. Leal-Junior, C.A. Diaz, L.M. Avellar, M.J. Pontes, C. Marques, A. Frizera, Polymer optical fiber sensors in healthcare applications: a comprehensive review. Sensors **19**(14), 3156 (2019)
37. X. Li, Y. Li, H. Wei, C. Wang, B. Liu, A review of wearable optical fiber sensors for rehabilitation monitoring. Sensors **24**(11), 3602 (2024)
38. Y. Zhang, Y. Hu, Q. Liu, K. Lou, S. Wang, N. Zhang, N. Jiang, A.K. Yetisen, Multiplexed optical fiber sensors for dynamic brain monitoring. Matter **5**(11), 3947–3976 (2022)
39. V. Matějec, I. Kašík, I. Bartoň, Fiber-optic nanosensors for chemical detection. Chemosensors **11**(10), 521 (2023)
40. W. Naku, C. Zhu, A.K. Nambisan, R.E. Gerald, J. Huang, Machine learning identifies liquids employing a simple fiber-optic tip sensor. Opt. Express **29**(24), 40000–40014 (2021)
41. M. Al Mahfuz, S. Afroj, A. Rahman, M.A. Hossain, M.A. Hossain, M.S. Habib, Ultra-sensitive visible-IR range fiber based plasmonic sensor: a finite-element analysis and deep learning approach for RI prediction. IEEE Access (2024)
42. M.A. Butt, G.S. Voronkov, E.P. Grakhova, R.V. Kutluyarov, N.L. Kazanskiy, S.N. Khonina, Environmental monitoring: a comprehensive review on optical waveguide and fiber-based sensors. Biosensors **12**(11), 1038 (2022)

43. Q. Song, G. Yan, G. Tang, F. Ansari, Robust principal component analysis and support vector machine for detection of microcracks with distributed optical fiber sensors. Mech. Syst. Signal Process. **146**, 107019 (2021)
44. Y.W. Sasy Chan, H.-P. Wang, P. Xiang, Optical fiber sensors for monitoring railway infrastructures: a review towards smart concept. Symmetry **13**(12), 2251 (2021)
45. A.M. Dehnaw, Y.C. Manie, L.-Y. Du, C.-K. Yao, J.-W. Jiang, B.-X. Liu, P.-C. Peng, Integrated sensor-optics communication system using bidirectional fiber and FSO channels and hybrid deep learning techniques. Sensors **23**(20), 8434 (2023)

Generative AI in Spectroscopy

Debarati Nath, Sukanya Ghosh, Sayantani Bhattacharya, Praveenkumar Hiremath, and Debajit Deb

Abstract Generative AI with spectroscopy is an innovative approach that combines artificial intelligence techniques with spectroscopic analysis, enabling researchers to extract valuable insights from complex spectral data more efficiently and accurately. AI can create high-precision synthetic spectra, enhance signal-to-noise ratios, and support spectral reconstruction by employing deep learning techniques like Generative Adversarial Networks (GANs), Graph Neural Networks (GNN), and Variational Autoencoders (VAEs). GenAI models facilitate the simulation of spectra from molecular structures, solving the forward design, whereas reasoning-driven models address the inverse design by predicting molecular structures with greater accuracy. GenAI can be integrated with spectroscopy to achieve various tasks such as improvement of spectral data analysis, rapid advancement in material discovery, automation, closed-loop optimization, etc. These innovations allow faster and more precise material identification, and real-time spectral prediction across diverse fields such as pharmaceuticals, environmental monitoring, and materials science. Despite challenges related to data quality, model interpretability, and computational demands, integrating Generative AI with spectroscopy offers significant potential for driving advancements in scientific research and industrial applications.

Keywords Generative adverrial neworks · Spectroscopy · Molecular structure · Forward design · Inverse design

D. Nath (✉) · S. Ghosh · S. Bhattacharya
Department of ECE, College of Engineering and Technology, Faculty of Engineering and Technology, SRM Institute of Science and Technology, Kattankulathur, Chennai, Tamil Nadu, India
e-mail: debaratn@srmist.edu.in

D. Deb
Department of ECE, Koneru Lakshmaiah Education Foundation, Vijayawada, India

P. Hiremath
Mechanics, Materials and Component Design, Lund University, Lund, Sweden

Y. Vasimalla and S. Kumar (eds.), *Generative AI for Photonic Sensing*, Progress in Optical Science and Photonics 36, https://doi.org/10.1007/978-981-95-1561-5_6

1 Introduction

The latest progress in Artificial Intelligence has largely been driven by the emergence of Generative Artificial Intelligence (known as Generative AI or GenAI). The rapid development of AI and ML is reshaping the field of chemistry, improving the accuracy and efficiency of molecular property prediction [1] and reaction modeling [2]. The use of machine learning in spectroscopy is still a relatively unexplored area. Spectroscopy is a vital analytical tool employed across diverse scientific domains, such as chemistry, pharmaceuticals, materials science, and environmental research. It offers valuable insights into molecular structures, compositions, and chemical interactions. However, obstacles like data limitations, noise disturbances, and the complexity of spectral interpretation can impede accurate analysis. Advancements in artificial intelligence (AI) have led to the adoption of algorithms such as Partial Least Squares Regression (PLS) [3], Random Forest (RF) [4], Support Vector Regression (SVR) [5], Linear Regression, Support Vector Machine (SVM) [6], and deep learning for spectral data analysis. Deep learning methods, in particular, can autonomously extract meaningful features from Raman spectral data, minimizing the need for manual feature selection. Generative AI is revolutionizing spectroscopy by improving data analysis, providing more precise interpretations of spectral data, and aiding in the discovery of novel materials through predictive modelling [7–9]. This technology enhances the identification of chemical compounds while expediting research and development across industries such as pharmaceuticals, materials science, and environmental monitoring. By utilizing advanced algorithms and machine learning techniques, generative AI can uncover patterns in complex datasets that traditional methods might miss, resulting in more streamlined workflows and innovative advancements. The adoption of generative AI in spectroscopy paves the way for real-time monitoring and analysis, allowing researchers to make data-driven decisions instantly rather than relying only on post-analysis assessments. This flexibility not only enhances the reliability of research findings but also speeds up scientific progress, fostering breakthroughs with far-reaching implications across multiple disciplines. Consequently, the synergy between generative AI and spectroscopy is set to revolutionize conventional research approaches, driving remarkable advancements in areas like materials science, pharmaceuticals, and environmental monitoring. The fusion of these technologies is expected to optimize workflows, lower costs, and improve the accuracy of experimental outcomes, ultimately leading to groundbreaking solutions that tackle critical global challenges. The advancement of spectroscopic analysis is accelerating due to progress in machine learning, particularly in generative modeling such as few- and zero-shot learning [10, 11], and large-scale pretraining. Generative AI refers to artificial intelligence systems capable of producing text, images, and other types of media by leveraging generative models. The progress of machine learning in spectroscopy from basic pattern recognition to advanced generative models is shown in Fig. 1.

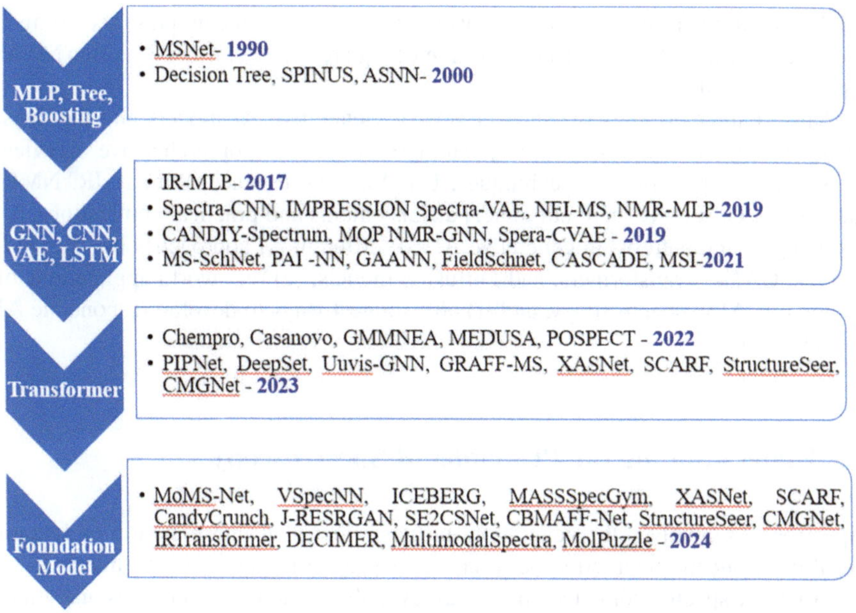

Fig. 1 Progress of machine learning in spectroscopy from basic pattern recognition to advanced generative models

Convolutional neural networks (CNNs) [12] and recurrent neural networks (RNNs) [13] have shown strong performance in tasks such as peak detection, deconvolution, and reaction monitoring. Additionally, transfer learning and pre-trained models [14] help these algorithms generalize across different spectra, reducing the need for extensive retraining. The emergence of foundation models [14] further enhances the Spectroscopy of ML by enabling advanced reasoning and planning for complex tasks like molecular structure elucidation and reaction pathway prediction [15]. Generative Adversarial Networks (GANs) [16] are employed to create realistic synthetic spectral data, enhancing data diversity and supporting more effective model training. Variational Autoencoders (VAEs) [17] play a key role in reducing noise, compressing spectral data, and recognizing patterns. A diffusion model [18] is a generative model that produces data, such as images, audio, or text, by reversing a process that gradually adds noise. It begins with random noise and then refines it to generate highly realistic outputs.

Kwon et al. [19] predict three-dimensional atomic structures from spectroscopic data, especially from (XANES) spectra, using a diffusion model. Tan [20] generates chemical structures from spectroscopic data (IR, UV, and NMR) with high accuracy by utilizing a transformer-based generative AI model (CLAMS). Generative Adversarial Networks (GANs) [21] significantly improve Raman spectroscopy by transforming low-resolution spectral data into high-resolution outputs. Bainbridge and Webb [22] introduce an integrated automated approach for analyzing high-resolution absorption spectra using generative AI techniques, including a genetic algorithm (GVPFIT), non-linear least-squares (VPFIT), and Bayesian Model Averaging (BMA).

This chapter presents (a) an overview of Generative AI models and the latest generative AI techniques, methods, and algorithms, (b) a comprehensive overview of six major spectroscopic techniques, UV–Vis, fluorescence, Raman, IR, NMR, and MS—focusing on methodological advancements and practical applications, (c) challenges, and patterns in generative AI, (d) different AI approaches like Autoencoders, GANs, Transformers, and Diffusion models, (e) real-world applications of generative AI in spectroscopy, and (f) also suggest ways to develop responsible AI in the future.

2 Understanding the Principles of Spectroscopy

Spectroscopy involves the generation, measurement, and interpretation of spectra resulting from the interaction between electromagnetic radiation and matter [23]. A variety of spectroscopic techniques are available to address numerous analytical challenges. These methods vary depending on the type of species being examined (such as atomic or molecular spectroscopy), the nature of the radiation–matter interaction being observed (such as absorption, emission, or diffraction), and the specific region of the electromagnetic spectrum utilized in the analysis. Spectroscopic techniques provide valuable insights and are extensively employed for both qualitative and quantitative analysis. Methods that rely on the absorption or emission of radiation within the ultraviolet (UV), visible (Vis), infrared (IR), and radio frequency (nuclear magnetic resonance, NMR) ranges are among the most commonly used in conventional food analysis laboratories. Each technique is unique and detects different types of molecular or atomic transitions. The different types of spectroscopy techniques and their specific applications in material analysis [23–25] is given below.

2.1 *UV–Vis Spectroscopy*

Electromagnetic radiation in the UV–Vis range has wavelengths between 200 and 700 nm [23–25]. The UV region covers 200–350 nm, while the visible region spans 350–700 nm. The UV range is invisible to the human eye, while the visible range includes colors from violet (short wavelengths) to red (long wavelengths). UV–Vis spectroscopy is generally classified into two types: absorbance spectroscopy and fluorescence spectroscopy, depending on how radiation interacts with matter. The principles of Beer–Lambert Law state that absorbance is directly proportional to the concentration of a substance in solution and the path length. UV–Vis spectroscopy can be used to measure sample concentrations, aiding in analysis. Additionally, UV and visible spectroscopy help identify free electrons and double bonds in molecules while also determining the concentration of a sample.

2.2 *Fluorescence Spectroscopy*

Fluorescence spectroscopy is 1–3 times more sensitive than absorption spectroscopy [23–25]. It works by measuring the light emitted by a substance as it returns from an excited state to its normal state. The substance first absorbs UV or visible light, gaining energy. During the process, excitation and emission happen at the same time. Each substance has a specific wavelength for absorbing light and a longer wavelength for emitting fluorescence, which depends on its chemical properties.

2.3 *Raman Spectroscopy*

It examines vibrational energy levels by measuring inelastic light scattering, making it valuable for material science and chemical identification. This technique studies molecular vibrations and rotations through chemical bonds, helping to understand a material's structure, composition, and interactions. It is widely used because it does not damage the sample and provides detailed molecular information. A key benefit is that no sample preparation is needed, making it useful for in vivo studies. The Raman spectrum measures vibrational modes, which help identify materials. By analysing the Raman fingerprint, doctors can distinguish between benign and cancerous tissues [23–25].

2.4 *Infrared (IR) Spectroscopy*

Infrared (IR) radiation falls between visible light and microwaves in the electromagnetic spectrum. IR spectroscopy utilizes wavelengths from 0.8 to 100 μm, which are classified into three ranges: near-IR (0.8–2.5 μm), mid-IR (2.5–15 μm), and far-IR (15–100 μm). IR spectroscopy operates on the principle that molecules absorb infrared light, causing their bonds to stretch and bend as they vibrate. In this technique, IR light passes through a sample, and when molecules absorb energy, their dipole moment changes due to this motion. This absorption can be detected only if the IR frequency matches the vibration frequency of the molecule's chemical bonds, generating a unique spectrum [23–25].

2.5 Nuclear Magnetic Resonance (NMR) Spectroscopy

Nuclear Magnetic Resonance (NMR) spectroscopy [23–25] is an advanced analytical method used to examine molecular structure, composition, and interactions. It works by exposing atomic nuclei to an external magnetic field and radiofrequency (RF) radiation. By analyzing how nuclei respond to these conditions, NMR provides precise details about molecular architecture, functional groups, and chemical surroundings. The technique is based on the fact that certain atomic nuclei, when subjected to a strong magnetic field, absorb and emit RF energy due to their intrinsic property called spin, which generates a tiny magnetic field around them.

2.6 Mass Spectrometry (MS)

Mass spectrometry (MS) is a highly effective analytical method for identifying the molecular weight, composition, structure, and chemical characteristics of substances [23–25]. It operates by ionizing molecules, sorting them according to their mass-to-charge ratio (m/z), and measuring their concentration. MS is extensively utilized in fields such as chemistry, biochemistry, pharmaceuticals, environmental studies, and forensic investigations. The fundamental concept of MS is that ions, or charged particles, respond consistently when exposed to an electric or magnetic field. The technique follows three key steps: ionization, where molecules are converted into ions; mass analysis, which separates ions based on their mass-to-charge ratio; and detection, which measures their abundance.

Therefore, spectroscopy is an important technique used in science to study how light interacts with matter. It helps scientists understand the structure, composition, and properties of different substances. Over time, advancements in spectroscopy have made it more accurate and useful in fields like research, industry, and medicine. As technology continues to improve, spectroscopy will keep playing a key role in scientific discoveries and innovations.

3 Transformers Based Models (TRMs)

TRMs are one type of deep neural network architecture implemented in 2017. These use self-attention mechanisms to manipulate and produce serial data proficiently [26]. The transformers also can develop input data in parallel, leading to a significant reduction in processing and training time. The TRMs networks consist of an encoder and decoder for transforming input into a continuous representation and generating output based on the representation, respectively. This revolutionary architecture paved the way for numerous tasks, such as machine translation and language generation, and continues to shape the evolution of neural network designs [26]. These models are widely used in various fields, especially in natural language processing (NLP), computer vision, and time series analysis. Generative Pre-trained Transformers (GPTs) models and Bidirectional Encoder Representations from Transformers (BERT) are widely used in NLP tasks such as language translation, contextually relevant text generation, and question answering [26]. Generative Pre-trained Transformers (GPTs) employ autoregressive methods to generate text sequences using advanced language models [27]. BERT analyzes input sequences from both directions, enabling a deeper understanding of contextual relationships. The Vision Transformers (ViT) approach is used for image classification under computer vision by processing images as sequences of patches. ViTs are a class of deep learning models that apply the transformer architecture, originally designed for natural language processing, to computer vision tasks [28]. Multimodal AI transformers combine various data types, such as text and images, enabling applications like visual question answering and text-based image generation. Time-series and speech processing transformers are used for anomaly detection in time-series data, and also for speech synthesis and speech recognition.

3.1 Variational Autoencoders (VAEs)

VAEs are a generative model used to generate new data samples that resemble the input data, enabling tasks like data generation, denoising, and feature learning. VAE has become an essential technique for both generative modeling and representation learning. VAEs consist of an encoder and a decoder as shown in Fig. 2. The encoder's goal is to transform the input into a lower-dimensional representation known as the latent space [26]. It extracts the mean and variance of the latent variables, which define its properties, from the data using a neural network [29]. The decoder adds Gaussian noise to decode the latent output of the encoder into its original input data or newly generated data. Throughout this process, variation is incorporated into the latent space using a standard Gaussian distribution. The primary objective is to obtain an output with a mean and variance similar to the input after incorporating the variance. In order to optimize the model parameters, VAEs use a reconstruction loss function and the Kullback–Leibler (KL) divergence loss. The reconstruction loss function deals with how well the decoder reconstructs the original input. KL divergence loss promotes the latent space to conform to a normal distribution. The VAEs offer a structured approach to learning meaningful data representations and generating new samples from the same distribution [26].

3.2 Diffusion Models (DMs)

The DMs model is implemented to enhance the performance of the simple generative adversarial network [26]. The diffusion model learns the distribution of key

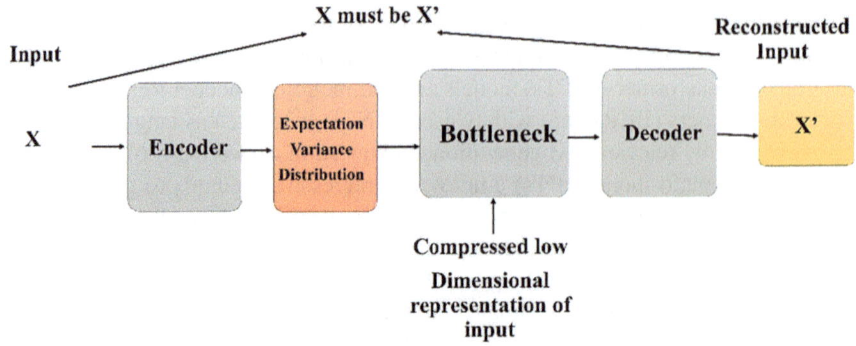

Fig. 2 The working principle of VAEs architecture

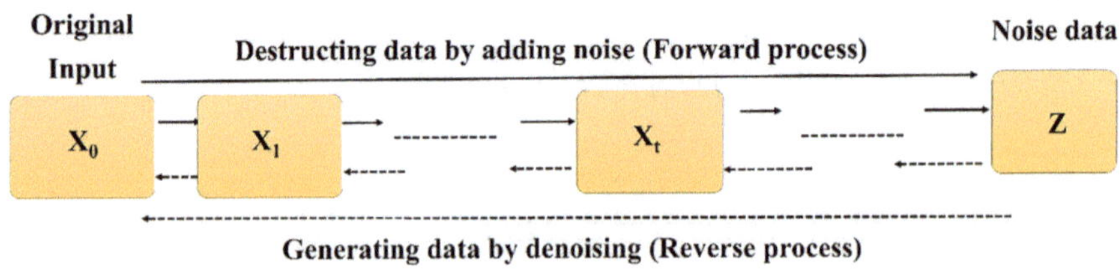

Fig. 3 The working principle of DMs architecture

parameters using a neural network. It is used to generate data similar to the data it is trained on. It is especially known for generating high-quality images and is widely used in fields such as image synthesis, text-to-image conversion, and other applications. During the training phase, the model involves a forward pass process in which Gaussian noise is gradually added to the original input data as shown in Fig. 3. At each stage, the parameters undergo iterations through a Markov process until the data is converted into a state that closely resembles pure Gaussian noise. It also involves a reverse process which undergoes a steady removal of noise added in the forward process, leading to the generation of new data. The model's parameters are continuously refined to enhance the similarity between the generated and original data, resulting in a steady improvement in output quality.

3.3 Advancements in GenAI

Advancements in generative AI have led to a wide range of applications, including image translation, video synthesis and generation, knowledge graph generation, text generation, artistic creativity, and scientific simulations [30, 31]. These innovations are transforming industries by enabling realistic content creation, enhancing automation, and accelerating research in fields such as healthcare, material science, and design.

4 Intersection of Generative AI and Spectroscopy

GenAI is revolutionizing spectroscopy by enhancing material design, refining spectral analysis, and improving sensor performance. Spectroscopy is a fundamental technique to analyze the interaction between matter and electromagnetic radiation, producing a huge amount of complex data. The data reproduce audio signals in the peaks, shifts, and patterns format that encode molecular information [28]. GenAI, capable of recognizing patterns and producing realistic synthetic data, significantly advances spectroscopy applications. GenAI models facilitate the simulation of spectra from molecular structures, solving the forward design, whereas reasoning-driven models address the inverse design by predicting molecular structures with greater accuracy.

4.1 Forward Design

It is used to predict the spectra from the molecular structure and composition-related data, as shown in Fig. 4. Although spectroscopy devices can produce spectra from molecular samples, utilizing AI-driven forward design is highly beneficial and provides several key advantages. This technology uses a computational tool such as neural networks, transformer models. Generally, Density Functional Theory (DFT) and Finite-Difference Time-Domain (FDTD), were used to incorporate forward design in spectroscopic analysis, however, these methods can be computationally expensive. AI-driven forward design can synthesize realistic spectra, reducing the cost and time-consuming experimental work. Moreover, it helps to identify the relationship between molecular structure and spectrum behavior. Due to the lower concentration and poor synthesis process, some molecules are hard to analyze, however, with the help of the GenAI concept, the spectra behavior of corresponding molecules can be predicted. Such GenAI-driven forward design is widely used for detecting biomarkers in medical diagnostics [32].

The inputs of forward design are in the form of molecular graphs, vector-based molecular parameters, etc. ML models used for forward design are Feed Forward Networks (FFN), Graph Neural Networks (GNN), and Foundation Models [28, 30, 33–36] with the concept of encoding prediction framework, as shown in Fig. 5. The models are usually combined with regression or classification models to estimate continuous properties, such as ^{1}H and ^{13}C chemical shifts or spectral features [30, 33, 35, 36]. The output can consist of complete spectra across various modalities, including MS, NMR, IR, Raman, and UV–Vis [28].

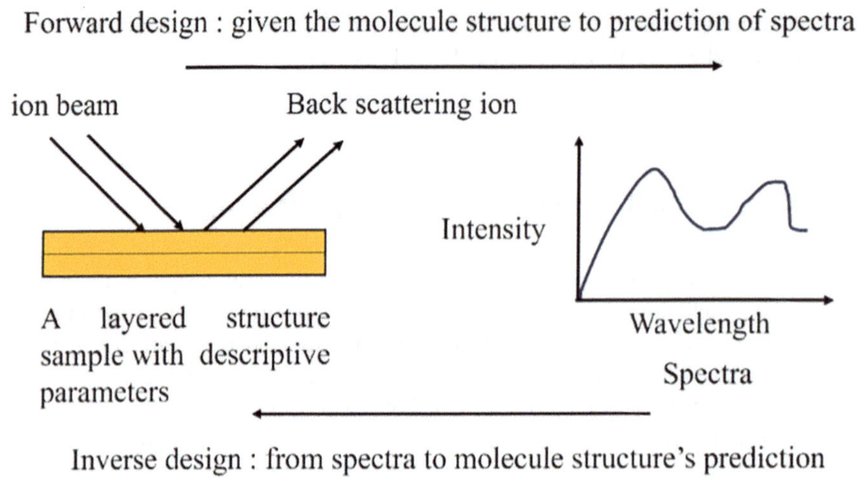

Fig. 4 Overview of forward and inverse design

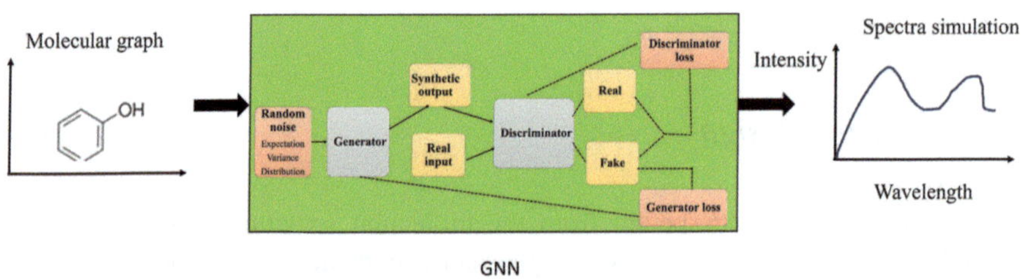

Fig. 5 Overview of spectrum generation using forward design

4.2 Inverse Design

Inverse design from GenAI is considered as a transformative role in integrating GenAI with spectroscopy by allowing the formation of tailored materials and devices with the help of AI-driven optimization of spectral properties [28]. The Inverse design in spectroscopy involves reconstruction of the molecular design or composition from a given experimentally obtained spectral response, as shown in Fig. 5. Such analysis helps researchers to determine unknown material with proper composition ratio, eventually evolving material discovery and applications. Identifying molecular structure from spectra using manual interpretation is time-consuming and most challenging. Such a traditional method is also susceptible to misinterpretation due to signal overlap and sample impurities. The GenAI method minimizes human error and accelerates high-throughput analysis in spectroscopy applications. The method also speeds up advancements in spectroscopic applications, ranging from sensor design to molecular analysis. The GenAI-driven spectroscopy analysis can be incorporated into biomedical diagnostics, environmental monitoring, etc.

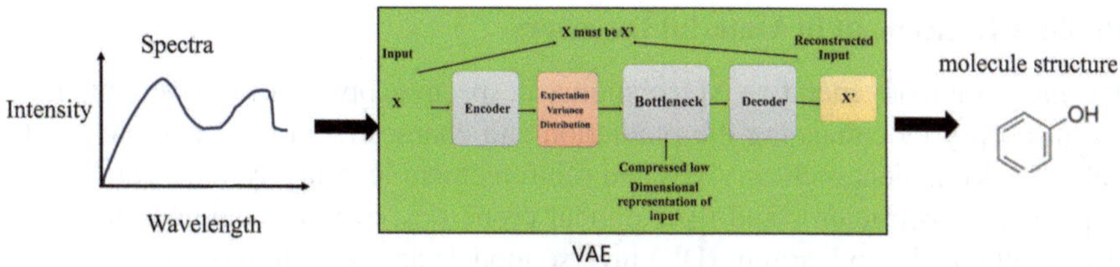

Fig. 6 Spectrum image-based molecule structure group identification using inverse design

In the inverse design, inputs are in the form of sequential data, one-dimensional (1D) NMR signals, high-dimensional spectral vectors, etc. [28]. GenAI modes are generally utilized in either an encode-decode scheme or an encode-predict framework. Random forest, transformer-based model, and VAEs model are used to encode 1D NMR signals into full molecular structures and substructure arrays, enabling the reconstruction of data [37, 38], as depicted in Fig. 6. The output consists of the predicted molecular structure, typically represented as either a sequence of tokens for SMILES strings or a molecular graph [39–42].

GenAI can be integrated with spectroscopy to achieve the following tasks:

Improvement of Analysis of Spectral Data

Generative AI models, such as diffusion models and generative adversarial networks (GANs), are being employed to improve the explanation of spectral data. These models can reduce noise, and reconstruct, and refine spectral signals, allowing for more precise identification of chemical compositions and molecular structures. GANs and VAEs are being explored for generating novel molecules from spectral data, potentially eliminating the need for traditional trial-and-error material development [26]. Large language models (LLMs) reveal unexpected issues like column degradation or solvent interference from raw spectral images [28]. New platforms are implemented to make AI-driven spectroscopy analyses, allowing users to obtain probable molecular structures through structure-spectrum predictions from user input spectra. A new data augmentation technique was implemented utilizing Generative Adversarial Networks (GANs) to tackle imbalanced spectral data in spectroscopy, improving classification accuracy by generating realistic samples that enhance material differentiation and ensure balanced training data [43]. The integration of a GANs with low-resolution Raman spectroscopy has been reported, enhancing spectral analysis by producing high-resolution data, minimizing noise, and refining feature extraction for more precise molecular classification and identification [21]. The spec2struct framework was integrated, leveraging multimodal embeddings and evolutionary algorithms to enhance spectral data analysis. This approach effectively replicates expert chemists' methodologies, offers reliable confidence estimates, and improves the structure elucidation process in spectroscopy [44].

Rapid Advancement in Material Discovery

The integration of generative AI techniques in spectroscopy is revolutionizing material discovery by enhancing the prediction and characterization of complex materials. GenAI models such as VAEs and reinforcement learning systems can be used to predict nanostructures and their optical properties, avoiding outdated trial-and-error methods. Deep learning (DL) inverse models are used in plasmonic switches to achieve design parameters with the highest prediction accuracy with losses [28]. A framework is introduced, combining Diffusion Models from GenAI with XANES spectroscopy to predict 3D atomic structures of disordered materials, demonstrating rapid advancements in material discovery by tailoring atomic arrangements and bridging materials characterization with atomic structure determination [45]. GenAI is used for discovering porous oxide materials in energy storage, specifically using the Crystal Diffusion Variational Autoencoder and machine learning techniques [46]. An unsupervised deep learning framework is included for classifying relaxation dynamics in X-ray photon correlation spectroscopy, facilitating rapid exploration of large datasets and correlating microscopic dynamics with macroscopic properties, thus advancing material discovery without prior physical knowledge [47].

Mapping Spectra to Molecular Structures

DL models use simulation data from the spectra analyzer, such as finite-difference time-domain results for training purposes. Then, the model tries to establish a relationship between the transmission spectra and the device geometry. This facilitates the rapid development of devices like nonlinear plasmonic ring resonators for the application of ultra-fast optical switching [48]. A GAN is integrated with low-resolution Raman spectroscopy to enhance spectral resolution, enabling precise molecular feature identification and mapping spectra to molecular structures, focusing on organic and pharmaceutical drug classification [21]. A machine learning model is developed to predict infrared spectra directly from 3D molecular structures, improving accuracy and efficiency over traditional approaches while effectively mapping spectra to molecular structures and accounting for an harmonic effects [49].

Closed-Loop Optimization

AI-powered workflows combine forward prediction, which determines spectral response from geometry, and inverse design, which derives geometry from target spectra. GANs produce synthetic spectral datasets to train robust inverse models, overcoming data insufficiency in the design of novel materials. AI-driven automated closed-loop approach is developed for scientific discovery, emphasizing closed-loop approaches that could enhance data generation and hypothesis exploration in fields like spectroscopy [50].

Automation and Optimization

GenAI enables the automatic extraction of spectral features and pattern recognition, thereby minimizing manual effort while enhancing efficiency. It can also minimize experimental conditions by predicting the ideal parameters for spectroscopy, reducing time and resource consumption.

5 Utilization of Spectroscopy in the Field of Chemistry

Spectroscopy is the exploration of how substance interacts with electromagnetic radiation, producing data similar to audio signals through shifts, peaks, and patterns that represent molecular data [51]. In contrast, spectrometry focuses on evaluating chemical interactions to understand molecular configurations and characteristics. Frequently used spectroscopic methods include:

- **Raman and Infrared (IR) Spectroscopy**: Identify functional groups present in a compound.
- **Mass Spectrometry (MS)**: Determines the molecular weight and composition of a compound, along with its structural characteristics, by analyzing pieces generated when the molecule disintegrates.
- **Ultraviolet/Visible (UV–Vis) Spectroscopy**: Delivers insights into substances featuring interconnected double bonds.
- **Nuclear Magnetic Resonance (NMR) Spectroscopy**: Offers details regarding atomic cores, particularly the carbon-hydrogen structure. Cutting-edge NMR methods like 2D and 3D-NMR allow for the characterization of intricate compounds, including proteins, nucleic acids, and natural products.

These spectral informations serve as a "molecular microscope," enabling researchers to investigate the hidden aspects of chemistry and biology. The results are typically depicted as charts or diagrams that illustrate the connection between intensity and variables such as wavenumber, wavelength, or mass-to-charge ratio (m/z). Spectroscopic data analysis generally falls into two main categories: the forward problem and the inverse problem, as we discussed in previous sections.

6 SpectraML Development Trajectory

ML has transformed the analysis of spectroscopic data, providing innovative avenues to derive deeper insights, streamline workflows, and reveal patterns beyond human capabilities [51]. Traditionally, computational methods in spectroscopy were confined to basic tasks like pattern recognition and property prediction. However, the introduction of deep learning and advanced ML frameworks has revolutionized the entire spectrum analysis process. For example, Convolutional Neural Networks

(CNNs) thrive in activities such as deconvolution [52] and peak detection [12], similar to feature identification in images. Recurrent Neural Networks (RNNs) and transformers are adept at processing sequential spectral data [13], akin to analyzing audio signals, making them ideal for monitoring reactions and conducting active research. One notable advancement is CASCADE, which accelerates NMR chemical shift predictions by 6000 times compared to the fastest Density Functional Theory (DFT) methods, allowing instantaneous forecasts derived from basic molecular representations [33]. As spectroscopic datasets have expanded in dimensions and intricacy, ML has shown remarkable scalability and flexibility. The evolution from initial forecasting models to contemporary generative and analytical structures—like attention-driven transformers and foundational models—has broadened the range of spectral examination. Generative models can simulate spectral patterns derived from chemical compositions, tackling the direct challenge, while logic-guided frameworks enhance accuracy in predicting molecular structures for the inverse problem. These advancements have introduced unmatched accuracy and swiftness across various applications, through molecular analysis to predicting response pathways. For instance, IMPRESSION forecasts NMR parameters with accuracy approaching that of quantum chemical methods while reducing processing duration from days to mere seconds.

6.1 Synopsis of Spectra ML Methodologies

In this section, an in-depth exploration of machine learning techniques has been discussed that tackle two main challenges: the forward problem (transforming molecules into spectra) and the inverse problem (deriving molecules from spectra). Present discussion is structured around four key elements.

6.2 Data Formats and Preparation Techniques

The effectiveness of spectral analysis hinges on the representation and preprocessing of molecular and spectral data. SpectraML uses vectors, sequences, or images for spectral data and vector-oriented representations, 2D graphs, SMILES strings, or 3D coordinates for molecular structures. This variety accommodates the complexities of molecular interactions. However, heterogeneity, high dimensionality, baseline drift, noise, and instrument variability necessitate strong pre-processing pipelines because standard normalization and alignment fall short in preserving chemical information. Integrating domain knowledge like physics-guided standardization and customized feature extraction within pre-processing [53, 54]. Recent work focuses on constructing large spectral datasets and employing transformer models to map spectral data's latent representations, enabling robust and generalizable spectral analysis for both forward and inverse tasks [55, 56]. These improvements ensure resilience to experimental artifacts and maintain chemical relevance, which establishes a solid base for spectral analysis.

6.3 Molecule-to-Spectrum

The forward problem in SpectraML focuses on forecasting spectral data straight from recognized molecular configurations, offering a more effective substitute to costly quantum-chemical simulations and tedious empirical assessments. This approach is already discussed in previous section. Vector-based inputs often use MLPs, while molecular graphs employ GNNs to capture structural information [30, 33–35, 57]. Coordinate-based features, used in physics-informed neural networks, help predict oscillatory characteristics. The output prediction is primarily regression-based, mapping molecular features to spectral properties like chemical shifts, fragment m/z values, or vibrational properties. Some methods predict discrete spectral features or sequences of spectral tokens using generative models [58]. Beyond spectral prediction, these methods extend to property-focused tasks like reaction behavior analysis and physicochemical parameter prediction, supporting applications in drug discovery and materials design [59].

6.4 Spectrum-to-Molecule

The inverse problem in spectral analysis focuses on deducing molecular structures directly from their observed spectra, offering a supplementary method to conventional structure determination techniques, already explained in previous section. In these models, encoding techniques include transformers, convolutional neural networks [37], Monte Carlo Tree Search, and graph-based models to capture critical spectral patterns. Decoding methods involve generative approaches like SMILES string prediction or iterative molecular graph construction using frameworks such as transformers and Markov decision processes [35]. Additional constraints enhance accuracy through candidate retrieval systems or combined analyses of spectral and structural data. These innovations enable precise elucidation of molecular structures from spectral inputs across diverse applications [42, 60, 61]. The encoding-prediction framework provides an option to generative methods by mapping spectral data directly to discrete structural elements like functional groups and substructures, facilitating applications from metabolite identification to forensic analysis. Techniques like MEDUSA [62], CANOPUS [63], Candy-Crunch [64], CANDIY-spectrum [65], and transformer-based networks analyze MS, IR, and NMR data to predict these elements efficiently [52]. Extending inverse SpectraML beyond full structure elucidation enhances rapid compound identification through sequence-to-sequence models and hybrid systems, reducing reliance on manual verification across various scientific domains.

6.5 Integrated Frameworks and Multi-modal Fusion

Recognizing the shared foundational chemical principles underlying forward and inverse tasks, recent advancements have focused on creating integrated frameworks to address both simultaneously. Foundation models, pre-trained on extensive and diverse spectral datasets, lead this effort by utilizing cross-domain learning to identify common features across modalities like NMR, MS, IR, and Raman. This enables capabilities such as zero-shot and few-shot learning [14]. Additionally, physics-guided generative models, including GAN-based super-resolution techniques and diffusion models, have been developed to generate high-quality spectra while adhering to established chemical constraints [58]. Hybrid architectures that merge the relational modeling strengths of graph neural networks (GNNs) with the sequential modeling abilities of transformers show significant promise for unified spectrum analysis. Furthermore, foundation models are advancing reasoning-driven spectrum analysis [66], addressing complex tasks like peak assignment [67], spectral deconvolution [15, 68], and consistency validation [56]. These models leverage previous chemical understanding to deduce likely molecular configurations, clarify overlapping spectral characteristics, and forecast absent spectral areas. By integrating forward and inverse tasks into a single framework, these methods mitigate data scarcity through synthetic data generation while improving model robustness and interpretability [69]. This unified perspective is driving innovation in fields such as materials science, drug discovery, and environmental monitoring.

6.6 Obstacles and Prospects

Data Integrity, Limitations, and Sophistication

SpectraML encounters several interconnected obstacles stemming from the intrinsic characteristics of experimental spectral information and the constraints of current machine learning (ML) methodologies. A primary challenge is the inconsistency in data integrity, as experimental spectra are often affected by noise, baseline fluctuations, and variations between instruments. These factors lead to discrepancies in intensity and spectral resolution, complicating model training and diminishing predictive accuracy, especially when preprocessing methods lack robustness. Additionally, the limited availability and uneven distribution of high-quality annotated datasets, especially for uncommon or intricate compounds, intensify these issues. Limited data availability restricts model generalization across diverse chemical domains and heightens the likelihood of overfitting, requiring solutions like data enhancement and knowledge transfer. The complexity of spectral data further intensifies these challenges. Spectral measurements often display high dimensionality and overlapping peaks, which complicates the process of feature extraction. Present ML models frequently find it challenging to grasp the intricate, complex patterns embedded in such data, resulting in suboptimal performance for activities like peak identification and feature discrimination. Additionally, numerous architectures fail to fully utilize domain-specific information contained within spectroscopic data, restricting their capacity to harness fundamental chemical and physical concepts. Another significant obstacle is integrating data from various spectroscopic methods (e.g., MS, IR, NMR), each defined by unique scales, formats, and noise characteristics. Developing efficient integration approaches to unify these modalities is complex. Most ML models are fine-tuned for inputs of a single modality and struggle to seize inter-domain connections essential for precise spectral analysis. Finally, attaining model comprehensibility—providing meaningful insights into underlying chemical phenomena—requires balancing model complexity with transparency.

7 Prospects and Evolving Frameworks

7.1 Artificial Data Creation and Physics-Guided Approaches

AI-driven generative models [70], such as LLMs and diffusion models, offer a powerful solution to the challenges of limited and variable spectral data by generating synthetic spectra that mimic real data while capturing complex nonlinear relationships [15, 68, 70–74]. Incorporating physics-based constraints into these models ensures chemical validity and enhances interpretability. This hybrid approach addresses data scarcity, reduces reliance on costly experimental methods, and outperforms traditional simulation techniques like DFT. These advancements promise to accelerate breakthroughs in fields like drug development and materials science.

7.2 Foundation Models: A Transformative Approach for SpectraML

Foundation models [14, 75] revolutionize SpectraML by leveraging cross-domain learning from diverse datasets across IR, MS, NMR, and Raman modalities. They excel at tasks like spectral generation and structure elucidation through few-shot learning while embedding domain-specific constraints for chemical validity. These models enhance spectral analysis with advanced reasoning capabilities, multi-modal integration, error detection, and iterative learning. Despite their transformative potential, addressing challenges like hallucination [75] and robustness [76] is crucial for reliable real-world applications.

SpectraML introduces a vital link between ML and spectroscopy. This section offers a detailed and unified roadmap of Approaches utilizing various spectroscopic methods and highlighting progress in forward and inverse challenges. By addressing trends like generative modeling and foundation models, alongside releasing an open-source repository, present section aims to support researchers in AI and chemistry, promoting collaboration and creativity in spectral analysis.

8 Conclusion

Generative AI can revolutionize spectroscopy by enhancing data analysis, improving spectral resolution, and predictive modeling, and enabling more precise interpretations of complex spectra and complex chemical compositions. Utilizing machine learning models like deep neural networks and transformer-based architectures, AI can aid in reducing spectral noise, forecasting molecular structures, and accurately simulating spectral data. This advancement can lead to significant breakthroughs in various fields, including pharmaceuticals, environmental monitoring, materials science, and biomedical diagnostics by providing faster and more reliable spectral insights and material discovery. The integration of generative AI with spectroscopy introduces innovative techniques capable of revealing hidden patterns, thereby revolutionizing our understanding of materials at a molecular level. However, challenges such as data quality, model interpretability, and the need for extensive training datasets must be tackled for broad adoption. As AI and spectroscopy continue to evolve, the integration of generative AI presents significant opportunities to expand the frontiers of analytical science.

References

1. Z. Guo, C. Zhang, W. Yu, J. Herr, O. Wiest, M. Jiang, N.V. Chawla, Few-shot graph learning for molecular property prediction, in *Proceedings of the Web Conference* (2021), pp. 2559–2567
2. C.W. Coley, W. Jin, L. Rogers, T.F. Jamison, T.S. Jaakkola, W.H. Green, R. Barzilay, K.F. Jensen, A graph-convolutional neural network model for the prediction of chemical reactivity. Chem. Sci. **10**(2), 370–377 (2019)
3. A. Höskuldsson, PLS regression methods. J. Chemom.Chemom. (1988). https://doi.org/10.1002/cem.1180020306
4. L. Breiman, Random forests. Mach. Learn. **45**, 5–32 (2001). https://doi.org/10.1023/A:1010933404324
5. H. Drucker, C.J. Burges, L. Kaufman, A. Smola, V. Vapnik, Support vector regression machines. Adv. Neural. Inf. Process. Syst. **9**, 155–161 (1996)
6. B.E. Boser, I.M. Guyon, V.N. Vapnik, A training algorithm for optimal margin classifiers, in *COLT '92: Proceedings of the Fifth Annual Workshop on Computational Learning Theory* (1992), pp. 144–152. https://doi.org/10.1145/130385.130401
7. L.B. Sanchez, G.A. Aspuru, Inverse molecular design using machine learning: generative models for matter engineering. Science **361**(6400), 360–365 (2018)
8. L. Shitong, S. Chence, X. Minkai, T. Jian, Predicting molecular conformation via dynamic graph score matching, in *35th Conference on Neural Information Processing Systems (NeurIPS 2021)* (2021), pp. 1–12
9. C. Litao, Z. Wentao, N. Zhiwei, L. Shunning, P. Feng, Generative models for inverse design of inorganic solid materials. J. Mater. Inform. **1**(4), 1–15 (2021)

10. W. Yaqing, Y. Quanming, J.T. Kwok, Lionel, M. Ni, Generalizing from a few examples: a survey on few-shot learning. ACM Comput. Surv. (CSUR) **53**(3), 1–34 (2020)
11. Y. Xian, C.H. Lampert, B. Schiele, Z. Akata, Zero-shot learning—a comprehensive evaluation of the good, the bad and the ugly. IEEE Trans. Pattern Anal. Mach. Intell. **41**(9), 2251–2265 (2019)
12. K. O'Shea, R. Nash, An introduction to convolutional neural networks. arXiv preprint arXiv: 1511.08458v2 (2015)
13. M. Schuster, K.K. Paliwal, Bidirectional recurrent neural networks. IEEE Trans. Signal Process. **45**(11), 2673–2681 (1997)
14. R. Bommasani, D.A. Hudson, E. Adeli, R. Altman, S. Arora, S. von Arx, M.S. Bernstein, J. Bohg, A. Bosselut, E. Brunskill et al., On the opportunities and risks of foundation models. arXiv preprint arXiv:2108.07258 (2021)
15. K. Guo, B. Nan, Y. Zhou, T. Guo, Z. Guo, M. Surve, Z. Liang, N.V. Chawla, O. Wiest, X. Zhang, Can LLMs solve molecule puzzles? A multimodal benchmark for molecular structure elucidation, in *38th Conference on Neural Information Processing Systems 37 Datasets and Benchmarks Track (NeurIPS 2024)* (2024), pp. 134721–134746
16. I. Goodfellow, J. Pouget-Abadie, M. Mirza, B. Xu, F.D. Warde, S. Ozair, A. Courville, Y. Bengio, Generative adversarial networks. Commun. ACM **63**(11), 139–144 (2020)
17. D.P. Kingma, M. Welling, Auto-encoding variational Bayes. arXiv:1312.6114 (2014)
18. J. Ho, A.N. Jain, P. Abbeel, Denoising diffusion probabilistic models. Adv. Neural Inf. Process. Syst. **33**, 6840–6851 (2020)
19. H. Kwon, T. Hsu, W. Sun, W. Jeong, F. Aydin, J. Chapman, X. Chen, V. Lordi, M.R. Carbone, D. Lu, F. Zhou, T.A. Pham, Spectroscopy-guided discovery of three-dimensional structures of disordered materials with diffusion models. Mach. Learn. Sci. Technol. **5**(4), 045037 (2024)
20. X. Tan, A transformer based generative chemical language AI model for structural elucidation of organic compounds. https://doi.org/10.48550/arXiv.2410.14719 (2024)
21. V. Yadav, A. Tiwari, S. Siddhanta, Machine learning driven high-resolution Raman spectral generation for accurate molecular feature recognition. https://doi.org/10.48550/arXiv.2407.01597 (2024)
22. M.B. Bainbridge, J.K. Webb, Artificial intelligence applied to the automatic analysis of absorption spectra. Objective measurement of the fine structure constant. Mon. Not. R. Astron. Soc. **468**(2), 1639–1670 (2017). https://doi.org/10.1093/mnras/stx179
23. M.H. Penner, Basic principles of spectroscopy, in *Food Analysis*, ed. by S. Suzanne Nielsen (Springer Science + Business Media, LLC, 2010), pp. 375–385
24. I.H. Ali, H.O. Saheb, L.S. Alhiti, A.A. Al-Fahham, Spectroscopy: types, principles and clinical uses. Int. J. Health Med. Res. **3**(7), 469–472 (2024)
25. S. Sinha, C. Jeyaseelan, G. Singh, T. Munjal, D. Paul, Spectroscopy—principle, types, and applications, in *Basic Biotechniques for Bioprocess and Bioentrepreneurship*, ed. by A.K. Bhatt, R.K. Bhatia, T.C. Bhalla (Elsevier Inc., 2023), pp. 145–164. https://doi.org/10.1016/C2017-0-01594-X
26. S.S. Sengar, A.B. Hasan, S. Kumar, F. Carroll, Generative artificial intelligence: a systematic review and applications. Multimed. Tools Appl. (2024). https://doi.org/10.1007/s11042-024-20016-1
27. A. Radford, K. Narasimhan, T. Salimans, I. Sutskever et al., Improving language understanding by generative pre-training (2018)
28. K. Guo, Y. Shen, G. Montiel, Y. Huang, Y. Zhou, M. Surve, Z. Guo, P. Das, N. Chawla, O. Wiest, X. Zhang, Artificial intelligence in spectroscopy: advancing chemistry from prediction to generation and beyond. https://doi.org/10.48550/arXiv.2502.09897 (2025)
29. Y. Liu, Z. Yang, Z. Yu, Z. Liu, D. Liu, H. Lin, M. Li, S. Ma, M. Avdeev, S. Shi, Generative artificial intelligence and its applications in materials science: current situation and future perspectives. J. Mater. **9**(4), 798–816 (2023)
30. Y. Kwon, D. Lee, Y.S. Choi, M. Kang, S. Kang, Neural message passing for NMR chemical shift prediction. J. Chem. Inf. Model. **60**(4), 2024–2030 (2020)

31. J. Fine, J.K.-Y. Liu, A. Beck, K.Z. Alzarieni, X. Ma, V.M. Boulos, H.I. Kenttämaa, G. Chopra, Graph based machine learning interprets and predicts diagnostic isomer-selective ion–molecule reactions in tandem mass spectrometry. Chem. Sci. **11**(43), 11849–11858 (2020)
32. P. Mandelare, D. Adpressa, E. Kaweesa, L. Zakharov, S. Loesgen, Coculture of two developmental stages of a marine-derived *Aspergillus alliaceus* results in the production of the cytotoxic bianthrone allianthrone A. J. Nat. Prod. **81**(4), 1014–1022 (2018)
33. Y. Guan, S.S. Sowndarya, L.C. Gallegos, P.C.S. John, R.S. Paton, Real-time prediction of ^{1}H and ^{13}C chemical shifts with DFT accuracy using a 3D graph neural network. Chem. Sci. **12**(36), 12012–12026 (2021)
34. Z. Guo, K. Guo, B. Nan, Y. Tian, R.G. Iyer, Y. Ma et al., Graph-based molecular representation learning. arXiv preprint arXiv:2207.04869. https://doi.org/10.48550/arXiv.2207.04869 (2022)
35. E. Jonas, S. Kuhn, Rapid prediction of NMR spectral properties with quantified uncertainty. J. Cheminform. **11**, 1–7 (2019)
36. S. Kang, Y. Kwon, D. Lee, Y.-S. Choi, Predictive modeling of NMR chemical shifts without using atomic level annotations. J. Chem. Inf. Model. **60**(8), 3765–3769 (2020)
37. Z. Huang, M.S. Chen, C.P. Woroch, T.E. Markland, M.W. Kanan, A framework for automated structure elucidation from routine NMR spectra. Chem. Sci. **12**(46), 15329–15338 (2021)
38. L. Yao, M. Yang, J. Song, Z. Yang, H. Sun, H. Shi, X. Liu, X. Ji, Y. Deng, X. Wang, Conditional molecular generation net enables automated structure, elucidation based on ^{1}C NMR spectra and prior knowledge. Anal. Chem. **95**(12), 5393–5401 (2023)
39. E. Jonas, Deep imitation learning for molecular inverse problems, in *Proceedings of the 33rd International Conference on Neural Information Processing Systems*, Article 449 (Curran Associates Inc., Red Hook, NY, USA, 2019), pp. 4990–5000
40. H. Sun, X. Xue, X. Liu, H.-Y. Hu, Y. Deng, X. Wang, Cross-modal retrieval between ^{1}C NMR spectra and structures based on focused libraries. Anal. Chem. **96**(15), 5763–5770 (2024)
41. Z. Tian, Y. Dai, F. Hu, Z. Shen, H. Xu, H. Zhang, J. Xu, Y. Hu, Y. Diao, H. Li, Enhancing chemical reaction monitoring with a deep learning model for NMR spectra image matching to target compounds. J. Chem. Inf. Model. **64**(14), 5624–5633 (2024)
42. X. Zheng, B. Tang, P. Xu, Y. Wang, B. Di, Z. Hua, M. Su, J. Liao, CBMAFF-Net: an intelligent NMR-based nontargeted screening method for new psychoactive substances. Anal. Chem. **96**(47), 18672–18680 (2024)
43. J. Chung, J. Zhang, A.I. Saimon, Y. Liu, B.N. Johnson, Z. Kong, Imbalanced spectral data analysis using data augmentation based on the generative adversarial network (2024). https://doi.org/10.21203/rs.3.rs-3876820/v1
44. A.H. Mirza, K.M. Jablonka, Elucidating structures from spectra using multimodal embeddings and discrete optimization (2024). https://doi.org/10.26434/chemrxiv-2024-f3b18
45. H. Kwon, T. Hsu, W. Sun, W. Jeong, F. Aydin, J. Chapman, X. Chen, V. Lordi, M.R. Carbone, D. Lu, F. Zhou, T.A. Pham, Spectroscopy-guided discovery of three-dimensional structures of disordered materials with diffusion models. Mach. Learn. Sci. Technol. (2024). https://doi.org/10.1088/2632-2153/ad8c10
46. J. Datta, D. Datta, A. Nadimpally, N. Koratkar, Generative AI for discovering porous oxide materials for next-generation energy storage. https://doi.org/10.48550/arxiv.2410.06433 (2024)
47. J.P. Horwath, X. Lin, H. He, Q. Zhang, E.M. Đufresne, M. Chu, S.K.R.S. Sankaranarayanan, W. Chen, S. Narayanan, M.J. Cherukara, AI-NERD: elucidation of relaxation dynamics beyond equilibrium through AI-informed X-ray photon correlation spectroscopy. Nat. Commun. **15** (2024). https://doi.org/10.1038/s41467-024-49381-z
48. Z. Pan, X. Pan, Deep learning and adjoint method accelerated inverse design in photonics: a review. Photonics **10**(7), 852 (2023)
49. S.A. Al, A. Allouche, Neural network approach for predicting infrared spectra from 3D molecular structure. https://doi.org/10.48550/arxiv.2405.05737 (2024)
50. H. Zenil, J. Tegnér, F.S. Abrahão, A. Lavin, V. Kumar, J.G. Frey, A. Weller, L. Soldatova, A. Bundy, N.R. Jennings, K. Takahashi, L. Hunter, S. Džeroski, A.M. Briggs, F.D. Gregory, C.P. Gomes, C.K.I. Williams, J. Rowe, J.A. Evans, H. Kitano, J.B. Tenenbaum, R.D. King, The

future of fundamental science led by generative closed-loop artificial intelligence. https://doi.org/10.48550/arxiv.2307.07522 (2023)
51. J.E. Elias, F.D. Gibbons, O.D. King, F.P. Roth, S.P. Gygi, Intensity-based protein identification by machine learning from a library of tandem mass spectra. Nat. Biotechnol. **22**(2), 214–219 (2004)
52. F. Hu, M.S. Chen, G.M. Rotskoff, M.W. Kanan, T.E. Markland, Accurate and efficient structure elucidation from routine one-dimensional NMR spectra using multitask machine learning. ACS Cent. Sci. **10**(11), 2162–2170 (2024)
53. M. Gastegger, J. Behler, P. Marquetand, Machine learning molecular dynamics for the simulation of infrared spectra. Chem. Sci. **8**(10), 6924–6935 (2017)
54. W. Gerrard, L.A. Bratholm, M.J. Packer, A.J. Mulholland, D.R. Glowacki, C.P. Butts, IMPRESSION—prediction of NMR parameters for 3-dimensional chemical structures using machine learning with near quantum chemical accuracy. Chem. Sci. **11**(2), 508–515 (2020)
55. M. Alberts, T. Laino, A.C. Vaucher, Leveraging infrared spectroscopy for automated structure elucidation. Commun. Chem. **7**(1), 268 (2024)
56. M. Alberts, O. Schilter, F. Zipoli, N. Hartrampf, T. Laino, Unraveling molecular structure: a multimodal spectroscopic dataset for chemistry. Adv. Neural. Inf. Process. Syst. **37**, 125780–125808 (2024)
57. S. Kang, Y. Kwon, D. Lee, Y.S. Choi, Predictive modeling of NMR chemical shifts without using atomic-level annotations. J. Chem. Inf. Model. **60**(8), 3765–3769 (2020)
58. M. Cordova, P. Moutzouri, B. Simões de Almeida, D. Torodii, L. Emsley, Pure isotropic proton NMR spectra in solids using deep learning. Angew. Chem. Int. Ed. **62**(8), e202216607 (2023)
59. A. Leniak, W. Pietrus, R. Kurczab, From NMR to AI: designing a novel chemical representation to enhance machine learning predictions of physicochemical properties. J. Chem. Inf. Model. **64**(8), 3302–3321 (2024)
60. H. Sun, X. Xue, X. Liu, H.Y. Hu, Y. Deng, X. Wang, Cross-modal retrieval between ^{13}C NMR spectra and structures based on focused libraries. Anal. Chem. **96**(15), 5763–5770 (2024)
61. Z. Tian, Y. Dai, F. Hu, Z. Shen, H. Xu, H. Zhang et al., Enhancing chemical reaction monitoring with a deep learning model for NMR spectra image matching to target compounds. J. Chem. Inf. Model. **64**(14), 5624–5633 (2024)
62. D.A. Boiko, K.S. Kozlov, J.V. Burykina, V.V. Ilyushenkova, V.P. Ananikov, Fully automated unconstrained analysis of high-resolution mass spectrometry data with machine learning. J. Am. Chem. Soc. **144**(32), 14590–14606 (2022)
63. K. Dührkop, L.F. Nothias, M. Fleischauer, R. Reher, M. Ludwig, M.A. Hoffmann et al., Systematic classification of unknown metabolites using high-resolution fragmentation mass spectra. Nat. Biotechnol. **39**(4), 462–471 (2021)
64. J. Urban, C. Jin, K.A. Thomsson, N.G. Karlsson, C.M. Ives, E. Fadda, D. Bojar, Predicting glycan structure from tandem mass spectrometry via deep learning. Nat. Methods **21**(7), 1206–1215 (2024)
65. J.A. Fine, A.A. Rajasekar, K.P. Jethava, G. Chopra, Spectral deep learning for prediction and prospective validation of functional groups. Chem. Sci. **11**(18), 4618–4630 (2020)
66. A. Young, B. Wang, H. Röst, MassFormer: tandem mass spectrum prediction for small molecules using graph transformers. arXiv preprint arXiv:2111.04824 (2021)
67. Y. Su, X. Wang, Y. Ye, Y. Xie, Y. Xu, Y. Jiang, C. Wang, Automation and machine learning augmented by large language models in a catalysis study. Chem. Sci. **15**(31), 12200–12233 (2024)
68. N.M. Ralbovsky, I.K. Lednev, Towards development of a novel universal medical diagnostic method: Raman spectroscopy and machine learning. Chem. Soc. Rev. **49**(20), 7428–7453 (2020)
69. Z. Tan, D. Li, S. Wang, A. Beigi, B. Jiang, A. Bhattacharjee et al., Large language models for data annotation and synthesis: a survey. arXiv preprint arXiv:2402.13446 (2024)
70. S. Wu, Y. Huang, C. Gao, D. Chen, Q. Zhang, Y. Wan et al., UniGen: a unified framework for textual dataset generation using large language models. arXiv preprint arXiv:2406.18966 (2024)

71. X. Chen, T. Wang, T. Guo, K. Guo, J. Zhou, H. Li et al., Unveiling the power of language models in chemical research question answering. Commun. Chem. **8**(1), 4 (2025)
72. J. Chen, K. Guo, Z. Liu, O. Isayev, X. Zhang, Uncertainty-aware yield prediction with multimodal molecular features. Proc. AAAI Conf. Artif. Intell. **38**(8), 8274–8282 (2024)
73. D. Chen, Y. Huang, S. Wu, J. Tang, L. Chen, Y. Bai et al., GUI-world: a dataset for GUI-oriented multimodal LLM-based agents. arXiv preprint arXiv:2406.10819 (2024)
74. Y. Huang, J. Shi, Y. Li, C. Fan, S. Wu, Q. Zhang et al., MetaTool benchmark for large language models: deciding whether to use tools and which to use. arXiv preprint arXiv:2310.03128 (2023)
75. Z. Ji, N. Lee, R. Frieske, T. Yu, D. Su, Y. Xu et al., Survey of hallucination in natural language generation. ACM Comput. Surv. **55**(12), 1–38 (2023)
76. Y. Huang, Y. Wang, Z. Xu, C. Gao, S. Wu, J. Ye et al., Breaking focus: contextual distraction curse in large language models. arXiv preprint arXiv:2502.01609 (2025)

Generative AI for Environmental and Atmospheric Sensing

Sudeep Rathore, Ajay Yadav, Rahul Mukherjee, Manoj Sharma, and Robin Kalyan

Abstract The application of generative artificial intelligence (AI) has witnessed significant growth in addressing environmental concerns, encompassing climate modeling and conservation initiatives. This groundbreaking technology possesses the transformative potential to revolutionize the approach to environmental challenges. Through the utilization of sophisticated algorithms and machine learning techniques, generative AI can effectively analyze extensive datasets and generate predictive models to facilitate climate forecasting and the development of effective mitigation strategies. This research explores the transformative potential of generative AI, particularly techniques like GANs, VAEs, and diffusion models, in revolutionizing environmental and atmospheric sensing. By addressing critical challenges posed by climate change and environmental degradation, generative AI enhances the understanding of environmental phenomena. Key applications include data augmentation for rare event modeling, super-resolution of satellite imagery, synthetic climate scenario generation for improved forecasting, pollution mapping, and extreme weather event simulation for better preparedness. Furthermore, it enables applications in biodiversity modeling, anomaly detection, and atmospheric sensing calibration. However challenges such as data quality, ethical considerations, and computational demands exist, generative AI offers innovative solutions for climate modeling, prediction, and

S. Rathore (✉) · A. Yadav · R. Mukherjee · M. Sharma
ECE Department, School of Engineering and Applied Science, Bennett University, Greater Noida, Uttar Pradesh, India
e-mail: E22SOEP0005@bennett.edu.in

A. Yadav
e-mail: ajayyadavbsz@gmail.com

R. Mukherjee
e-mail: rahul110.iitkgp@gmail.com

M. Sharma
e-mail: manoj.sharma1@bennett.edu.in

R. Kalyan
Department of High Performance IC Design, Institute of Microelectronics, Agency for Science, Technology and Research, Singapore, Singapore
e-mail: robin_kalyan@ime.a-star.edu.sg

Y. Vasimalla and S. Kumar (eds.), *Generative AI for Photonic Sensing*, Progress in Optical Science and Photonics 36, https://doi.org/10.1007/978-981-95-1561-5_7

environmental management. These advancements have the potential to significantly improve climate change mitigation strategies, enhance weather forecasting accuracy, and ultimately enable more effective environmental protection efforts.

Keywords Generative AI · GANs · VAEs · Environmental sensing · Atmospheric sensing

1 Introduction

Environmental and atmospheric monitoring is useful for understanding and responding to climate change, air quality issues, and biological changes. It has been partially effective in discerning these processes due to traditional environmental monitoring methods, such as satellite imaging, terrestrial sensors, and numerical simulations. Yet these traditional methods often provide insufficient insights owing to practical constraints of availability of data, resolution, and computational intensity [1].

Conventional sensing techniques, for instance, terrestrial stations, satellites, and aircraft, face the disadvantages of: • Data Sparsity: Limited spatial and temporal coverage owing to very high overcome or tend to create and maintain the sensing networks. • Heterogeneity: Inconsistent data quality and formats among various players present challenges to data integration and analyses. • Bias: Possible biases dawning from unevenness at sampling, sensor deficiencies, and human factors. • Computational Cost: Processing and analyzing great amounts is relatively computationally and costly time consuming.

Environmental and atmospheric sensing are vital for understanding and addressing climate change, air quality crises, and biological transformations. Traditional environmental monitoring methods, including satellite imaging, land-based sensors, and numerical models, have provided useful insights into these processes. Nevertheless, such traditional approaches often run into limitations regarding data access, resolution, and computational cost.

The rise of artificial intelligence (AI), particularly generative AI, has resulted in revolutionary approaches for environmental data collecting, analysis, and predictive modelling. Generative AI refers to a type of machine learning model that creates new data instances that match real-world observations. Models such as Generative Adversarial Networks (GANs) [1], Variational Autoencoders (VAEs) [2], and transformer-based architectures [3] have distinct advantages for handling complicated datasets. In environmental and atmospheric sciences, AI-driven methodologies allow researchers to synthesise high-quality data, improve predictive modelling, and replicate real-world environmental conditions that would otherwise be difficult to record [4].

As an illustration, generative AI could help substitute data in network sensors for synthetic data in areas with sparse monitoring infrastructures. This becomes pivotal in regions where the logistical or financial deployment of a physical sensor is impractical. Also, they can be used in climate modeling by producing high-resolution simulations while obtaining low-resolution inputs, reducing costs of computation while increasing the simulation accuracy, besides helping in disaster management through simulations of extreme weather events like hurricanes and floods while aiding in preparation for responses [5, 6].

Although generative AI holds great promise in environmental and atmospheric sensing, its application is not free of problems. Among them are data quality issues, ethical quandaries, and computation costs, which must be prudently handled for proper and effective application of these technologies. That is to say, concerns emerge in regards to the reliability of synthetic data and its potential for nefarious uses-the worst-case scenario being its influence on climate change messaging and policy-making.

The systematic observation together with measurement and evaluation of natural environmental components and their facets defines environmental monitoring as an essential process. The main purpose of environmental monitoring consists of two parts: to evaluate environment conditions and to detect potentially dangerous alterations affecting ecosystems or public health [7]. Traditional environmental monitoring relies mainly on manual sampling along with laboratory analysis of samples using statistical modeling techniques which have been standard procedures for multiple decades. These assessment methods face limitations because they come with both elevated costs and labor-intensive protocols as well as insufficient precision levels. Manual sampling combined with laboratory analysis creates prolonged delays in decision-making periods, emergency responses both during natural disasters and pollution emergencies [8–10].

Artificial Intelligence has become an innovative technology which improves environmental monitoring by providing new solutions beyond traditional methods during recent years. The environmental monitoring systems gain accuracy together with efficiency and extended scale capacity through the application of Generative AI and especially AI technology. Generative AI represents a group of AI models which create fresh information from fundamental data points. GANs together with VAEs represent AI models which generate synthetic data for training other AI systems to simulate environmental conditions and estimate future environmental trends [1].

The combination of AI systems including Generative AI enables environmental monitoring to process large datasets that originate from a combination of satellite imagery and IoT sensors and historical records. The system allows real-time environmental changes tracking while making predictions about environmental conditions for better environmental management practices. Through the use of AI models scientists can produce more precise and fast solutions regarding natural disaster monitoring and water/air quality assessment together with pollution detection beyond what human experts could achieve [11]. The synthetic datasets produced by generative AI represent environmental conditions accurately which leads to valuable training models for situations that lack accessible and obtainable real data according to [12].

The breakthroughs of AI do not stop the environmental monitoring sector from facing obstacles during its practical implementation. Environmental organizations encounter difficulties due to a shortage of experts trained in AI for environmental work which affects areas that struggle to build their technological fundamentals [13]. Management difficulties in environmental data protection arise from problems regarding access to data and privacy regulations along with control procedures. AI technology ethical protection together with sensitive information protection require proactive data governance establishment from governments and organizations according to [14].

Future progress in environmental monitoring AI appears promising because continued research on data collection techniques combined with improved computing capabilities and AI algorithm development will strengthen the accuracy and operational efficiency of pollution detection systems. Through generative AI technology environmental monitoring can undergo transformation because it enables researchers to create authentic simulated systems along with predictive analytics which support officials and scientists render decisions based on facts [15]. AI technologies will continue their development towards greater involvement in environmental challenges alongside sustainable development activities [16].

This chapter explores the role of generative AI in environmental and atmospheric sensing, highlighting its applications, challenges, and future directions. We begin with an overview of key generative AI techniques, followed by a discussion of their use in air quality prediction, climate modeling, and disaster management. By leveraging the power of generative AI, researchers and policymakers can gain new insights into environmental systems, improve the accuracy of predictive models, and develop more effective strategies for mitigating the impacts of climate change and pollution. However, realizing this potential will require interdisciplinary collaboration, robust ethical frameworks, and continued advancements in AI research.

2 Generative AI: An Overview

Generative Artificial Intelligence (AI) is a class of machine learning algorithms whose (unspoken) goal is to generate new data instances that resemble a given dataset. Discriminative models, like those used in most current AI applications, focus on classifying or predicting outcomes based on the input data they are given. But generative models learn the underlying distribution of the data and can create new samples from this distribution. This ability gives the generative AI a huge advantage when data is scarce, hard to collect, or expensive.

The area of research known as generative AI has advanced quickly in recent years. This has come about through the application of deep learning, greater computing power, and the availability of enormous quantities of data. The main ways of accomplishing generative AI include Generative Adversarial Networks (GANs), Variational Autoencoders (VAEs), and Diffusion Models. Each of these techniques is good at some things and not at others, with the result that they are each fit for different types of tasks associated with environmental and atmospheric sensing.

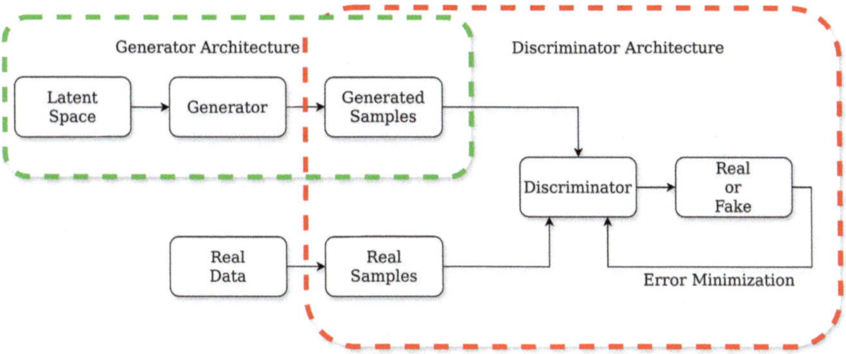

Fig. 1 Generative adversarial networks (GAN) architecture

2.1 Generative Adversarial Networks (GANs)

The widely employed generative AI method named Generative Adversarial Networks (GANs) was invented by Goodfellow et al. [1] in 2014. Two neural networks combine as components of a GAN system: the generator alongside the discriminator which learn through simultaneous adversarial training, as shown in Fig. 1. The generator produces artificial data examples that the discriminator analyzes against genuine dataset content. Real data generation capabilities of the generator enhance progressively alongside the discriminator's capacity to differentiate authentic samples from synthetic ones. This paper summarizes the adversarial training approach through the following steps:

- The generator accepts random noise to generate synthetic samples from it.
- The discriminator works with both synthetic and real data samples before attempting to assign them the labels real or fake.
- The generator receives an update to maximize the discriminator's incorrect real/fake classification while the discriminator seeks to become more accurate at its real/fake evaluation.

GANs operate successfully within different application areas like image synthesis together with video generation and natural language processing tasks. GANs produce synthetic data to augment sensor network data gaps while also simulating rare weather conditions when used in environmental and atmospheric sensing. The application of GANs produces high-resolution satellite imagery from low-resolution inputs which results in improved environmental change monitoring capability.

The success of GANs encounters three main obstacles comprising training instability alongside mode collapse problems and ethical concerns linked to the improper use of synthetic data. The resolution of such problems demands strategic planning regarding network design structures along with optimized training processes and optimization techniques.

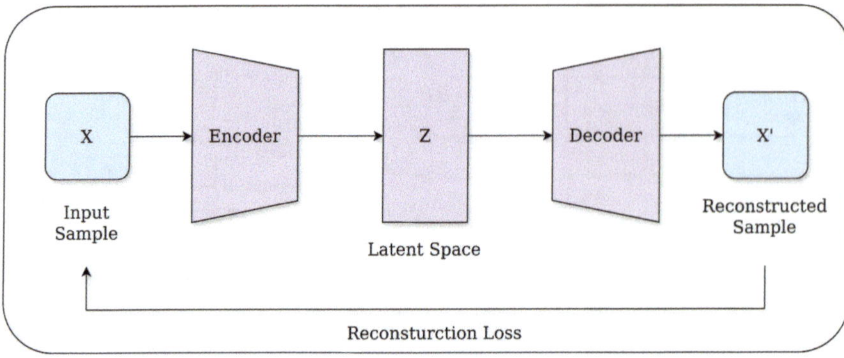

Fig. 2 Variational autoencoders (VAEs) architecture

2.2 *Variational Autoencoders (VAEs)*

VAEs [2] represent a second well-known generative artificial intelligence approach. VAEs, as shown in Fig. 2, serve as probabilistic models that discover hidden input data representations from which new samples may be created. The main objective of VAEs differs from GANs because they seek to model the hidden probability distribution instead of generating realistic data.

A VAE consists of two main components:

The encoder, which maps the input data to a latent space representation.

The decoder, which reconstructs the input data from the latent representation.

VAEs achieve their main advancement through their probabilistic technique for learning latent variable spaces. The encoder defines distributions rather than static point assignments to place input data within the latent space and usually generates Gaussian distributions. The VAE can produce different samples through latent space sampling because of this probabilistic approach definition.

The training process for VAEs led to maximizing evidence lower bound (ELBO) as it maintains a balance between decoder reconstruction quality and latent space uniformity. The well-organized nature of the latent space becomes possible due to this training process.

VAEs operate in different fields including image generation along with anomaly detection and data compression applications. VAEs demonstrate exceptional suitability in environmental and atmospheric sensory applications to solve data imputation tasks along with both anomaly detection and dimensionality reduction problems. Researchers have employed VAEs to complete missing air quality data measurements from underpopulated sensor systems which leads to better pollution level forecasting.

The main drawback of VAEs involves less sharp and less realistic sample generation than GANs. The primary objective of VAEs lies in developing effective models for data distribution no matter how their sample generation capabilities perform.

Nowadays VQ-VAEs (Vector Quantized VAEs) together with Hierarchical VAEs succeed in enhancing sample quality through probabilistic modeling techniques.

2.3 Diffusion Models

Ho et al. [17] developed diffusion models which represent a modern generative AI technology that produces top-quality samples. The process of diffusion models begins with input data which gets progressively noised before training models to reverse the process for sample generation. Non-equilibrium thermodynamics serves as inspiration through this method despite its evolution starting as orderly before moving into disordered phases and returning to order.

The training steps for diffusion models consist of two essential procedures.

- The process of forward diffusion consists of corrupting the input data through progressively increasing Gaussian noise coverage across time.
- The trained neural network learns to undo noise addition steps through the reverse diffusion process which allows it to create new data samples.

The advantages of diffusion models surpasses both GANs and VAEs because they present stable training capabilities and high scalability and outstanding performance in sample generation. Diffusion models show the best fit for environmental and atmospheric data generation applications because they deliver accurate results. The application of diffusion models achieved realistic climate simulation tasks and better resolution of meteorological data [5].

The computational complexity of diffusion models becomes challenging due to their need for precise parameter adjustments through noise schedule selection and number of steps control. Scientists are dedicated to developing diffusion models that become more efficient and scalable for extensive environmental applications.

2.4 Relevance to Environmental and Atmospheric Sensing

Generative AI systems show great promise to solve important problems which occur in environmental monitoring and atmospheric observation. For example: Generating synthetic data through artificial intelligence models enhances limited datasets so machine learning models function better during air quality forecast and climate simulation.

The capability of generative models includes filling gaps in sensor networks by creating plausible data values to achieve better environmental condition monitoring.

Simulation by generative models allows the creation of complex environmental systems simulations to help forecast environmental changes along with decision-making support.

Through generative modeling techniques users achieve higher resolution in satellite imagery and climate data which delivers improved environmental observation details.

Research and practice applications using the advantages of GANs VAEs and diffusion models allow scientists to create progressive solutions for environmental concerns. The successful application of generative AI for this field needs the solution of technical obstacles and ethical problems regarding synthetic data usage and training stability and computational performance.

3 Applications of Generative AI in Environmental and Atmospheric Sensing

Generative AI functions as a powerful instrument in environmental sensing which provides advanced solutions to three key issues including sparse data sets and complicated calculations and requirements for detailed simulations. The transformation of natural world understanding through environmental monitoring involves generative AI operations that produce synthetic data and enhance predictive models and simulate complex environmental systems. The section analyzes three essential uses of generative AI environmental sensing such as forecasting air quality and conducting climate modeling and managing disasters.

3.1 Air Quality Prediction

Public health with ecosystems and climate change stands as essential factors related to environmental sensing because of the critical need to predict air quality conditions. Earth-based air quality sensors together with satellite systems currently monitor air quality yet leave gaps in data collection mainly in unfrequented and less developed territories. Generative AIs use GANs and VAEs to tackle data sparsity problems through synthetic air quality data production and predictive model accuracy improvement [18].

Data Augmentation and Imputation

The main use of generative AI technology in air quality prediction involves generating synthetic data to expand restricted or deficient datasets. The application of GANs produces realistic air pollution maps through gap completion in sensor networks. The synthetic datasets provide training materials for machine learning models which enhances their predictive capabilities when dealing with areas with minimal monitoring systems [19].

The technique of generative models operates through VAEs to carry out data imputation tasks specifically for air quality datasets that require missing value estimation. VAEs learn data distributions to generate realistic air quality data values for missing measurements which enhances both accuracy and fullness of air quality evaluation.

High-Resolution Air Quality Mapping

The usage of generative AI systems helps enhance air quality data resolution which leads to deeper insights about pollution distribution patterns. GANs produce high-resolution air pollution maps through the processing of low-resolution satellite imagery. High-resolution maps generated through this technology assist policymakers alongside researchers by pinpointing pollution hot spots which enables them to create specific interventions for air quality enhancement.

3.2 *Climate Modeling*

Generative AI achieves critical advancements through its work in climate modeling. The predictive understanding of Earth's climate system depends on climate models because they enable scientists to predict future climate conditions as well as support policy implementations. Correcting this situation is that traditional climate models require huge computational power but fail to simulate small-scale climate systems accurately. Diffusion models together with GANs aid climate research by producing detailed climate simulations while making climate simulations more efficient.

High-Resolution Climate Simulations

The resolution of climate simulations can receive an improvement through generative AI systems by transforming coarse-resolution inputs into detailed outcomes. The application of diffusion models generates high-resolution climate data starting from low-resolution global climate models. Nonetheless the high-resolution simulations deliver improved local climate process representation for better predictions about extreme weather events together with their regional climate effects.

Downscaling Climate Data

Climate data at coarse resolutions gets translated through a mechanism which produces finer spatial and temporal scales through downscaling. Generative AI techniques specifically GANs provide a solution for downscaling tasks to generate high-resolution climate data which includes realistic weather patterns. The downsampled climate information serves various purposes in both regional climate assessments and water resource management and agricultural planning systems.

Uncertainty Quantification

Generative AI models enable scientists to determine the degree of uncertainty inside climate prediction models thus establishing their accuracy reliability. VAEs serve as tools to produce multiple climate simulation sets which present distinct future climate condition scenarios. Climate change ensembles supplied by these systems give policymakers accessible information that enables them to make decisions even with unknown risks.

3.3 Disaster Management

Generative AI stands out as a crucial technology for disaster management since it helps forecast and handle natural disasters along with their response plans. Generative AI techniques serve disaster preparedness and response by creating extreme weather simulations and synthetic disaster models and high-resolution satellite images which results in better preparedness measures.

Synthetic Disaster Scenarios
Generative AI technology develops synthetic disaster simulations which model the consequences of extreme weather disasters particularly hurricanes and floods. Through GANs researchers have generated realistic satellite images that show the aftermath of disasters which cover flood zones as well as storm surge incidents and wildfire scattering. Synthetic disaster scenarios developed by these artificial intelligence systems serve both teams preparing for emergencies and the testing of contingency plans and future disaster impact evaluation.

Damage Assessment
Generative AI systems perform assessments of damage that occurs from natural disasters. Through their usage researchers have developed high-resolution before-and-after picture models to assists disaster assessment and recovery operations. The high-resolution images enable emergency responders to better understand damage extent, thus they can make more effective resource distribution decisions.

Early Warning Systems
Generative AI can enhance early warning systems by improving the accuracy of disaster predictions. For example, diffusion models have been used to generate high-resolution weather forecasts, enabling more accurate predictions of extreme weather events such as hurricanes and tornadoes. These improved forecasts provide critical lead time for evacuations and other preparedness measures, reducing the impacts of disasters on communities.

3.4 Biodiversity and Ecosystem Monitoring

Generative AI helps biodiversity and ecosystem surveillance through the creation of artificial data as well as better species distribution modeling and ecologic operation emulation. GANs produce synthetic satellite imagery which enhances deforestation and habitat loss and species population monitoring capabilities. These synthetic datasets serve important functions for biodiversity protection programs as well as biodiversity evaluation activities.

3.5 Water Resource Management

The water resource management industry receives help from generative AI technology. Hydrological process simulation along with water resources prediction and allocation optimization utilize generative models for their solution. VAEs function as an essential component for simulated streamflow data creation that leads to better river basin water availability prediction results. Climate change alongside growing demand requires accurate water resource predictions that represent the key to proper management.

4 Generative AI Use Cases in Environmental and Atmospheric Sensing

Environmental Signal Reconstruction Learning with Generative Adversarial Network: [20] ESR-GAN represents a new framework described in the paper which employs Generative Adversarial Networks (GANs) to generate detailed environmental signals through sparse monitoring website data [20]. ESR-GAN combines three-layer CNN for its generative model process and a discriminative model to measure reconstruction accuracy while establishing end-to-end signal mapping between low- and high-quality signals, as shown in Fig. 3. The authors created an impact map to handle the uneven monitoring site distribution which enables adversarial training to enhance reconstruction quality within remote areas. ESR-GAN delivers superior performance than benchmark techniques Natural Neighbor Interpolation, Kriging, and SRGAN when used for air quality monitoring of PM2.5 and PM10 and NO_2 in Beijing thus proving its capability for sparse data processing and detailed environmental surveillance. The framework demonstrates successful generalizations through various regions and sampling conditions which leads to future research efforts focused on integrating temporal data to boost performance in urban environmental signal reconstruction.

The authors processed data through spatial interpolation for model training within the ESR-GAN network. Surface interpolation through MATLAB scattered Interpolant produced high-quality Z_h sensing signals from existing measurements throughout a Beijing site with fourteen monitoring points until they could act as environmental signal ground truth. Their simulation of sparse monitoring used random measurement removal on this area and employed identical interpolation for

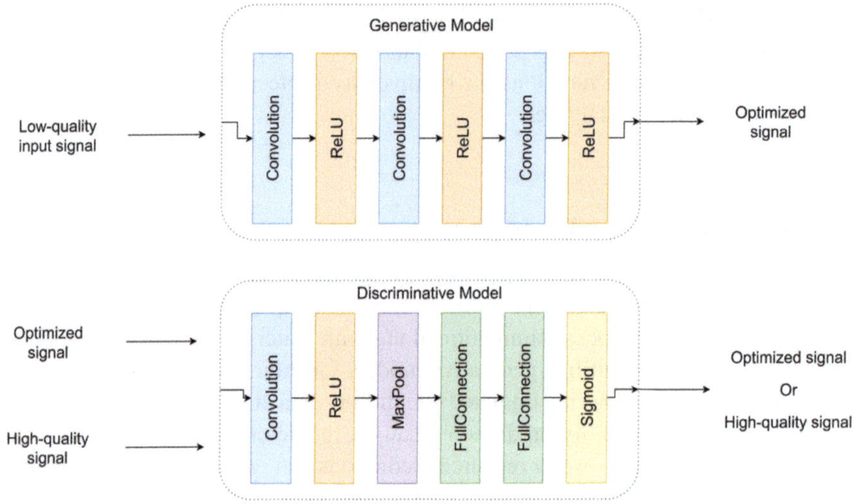

Fig. 3 Architecture of generative model and discriminative model used in environmental signal reconstruction

creating low-quality signals (Z_l). The training data consisted of low-quality and high-quality input-output signal pairs that enabled the generative model to discover relationships between sparse data and refined data and simultaneously facilitated discriminative model reconstruction accuracy enhancement. The method allowed the ESR-GAN to effectively manage the issues arising from unbalanced monitoring site locations.

Air Pollution Prediction with Multi-modal Data and Deep Neural Networks [21]: A new method of air pollution forecasting is introduced through the combination of deep neural networks with multi-modal data consisting of camera images along with meteorological data [21]. The authors implemented four network architectures by combining CNNs and GANs to resolve class imbalance issues present in their dataset, the process flow of the suggested approach is shown in Fig. 4. The research adopts Generative AI technology specifically through CGANs to produce synthetic images for improving the representation of minority class images (polluted images) in the dataset. Transmission through the CGAN method allows datasets to become balanced while non-polluted images remain dominant due to the seasonal distribution of air pollution throughout Skopje North Macedonia. The trained CGAN uses minority class images to produce new images that build a more uniform training set. The model attains better results from this augmentation method in conjunction with traditional image rotation and flipping strategies. The Inception model employing weather data proved most efficient for air pollution prediction as it provided the best performance from the enhancement of pre-trained model usage with weather information. The research demonstrates how wide availability of camera images can substitute sparse air pollution sensors to monitor air quality in real-time.

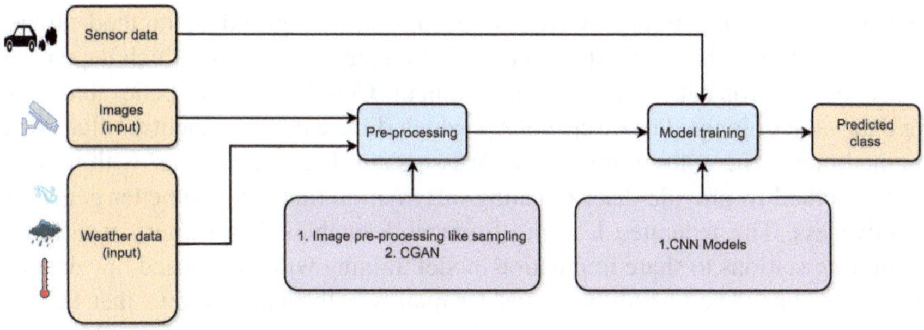

Fig. 4 Process flow of the suggested approach for predicting air pollution

The author develops a new air pollution forecasting system that combines camera images alongside weather data which uses four deep learning models for evaluation purposes. A static Vodno mountain camera near Skopje North Macedonia captures images together with meteorological data from a local weather station that serves as the input data. The system generates predicted Air Quality Index (AQI) data which then gets evaluated through PM2.5 sensor measurements. The research presents four different model architectures which include the basic Convolutional Neural Network (CNN) and Residual Network (ResNet) and Inception model and a custom pre-trained Inception model. Merging convolutional blocks with pooling layers makes up the basic CNN structure but ResNet brings residual blocks as a solution to tackle deep network gradient vanishing issues. The Inception blocks of Inception models combine different filter sizes but the modified Inception network adds weather data processing through the model's additional sub-path. The authors apply data augmentation strategies to handle the extreme class disproportion in the dataset which comprise random oversampling and undersampling together with standard image transformations (flipping and rotation) and Conditional Generative Adversarial Networks (CGANs). The CGAN receives images from the minority class during training to create synthesized samples that improve the dataset balance. Part of the model runs training operations through TensorFlow and Keras using findings from grid search optimization. The custom pre-trained Inception model working with data augmentation from CGAN achieves the best prediction results because it proves the strength of combining generative AI techniques with multi-modal environmental data.

Federated Conditional Generative Adversarial Nets Imputation Method for Air Quality Missing Data [22]: A new method for air quality missing data imputation uses Federated Conditional Generative Adversarial Networks (FCGAI) that integrates Generative Adversarial Networks (GANs) with federated learning to solve the issues of data privacy and non-independent identically distributed (non-IID) data [22]. The authors concentrate their analysis on developing a method to impute air quality data from monitoring stations whereby confidentiality and security along with industrial competition thwart data exchange. The proposed method utilizes Conditional GANs (CGANs) to create realistic data imputations which depend on present observation data. The generator within the CGAN generates believable missing data values alongside a discriminator which differentiates authentic values with imputed ones. The authors implement Wasserstein distance together with a "Hint mask" method to provide discriminating information that leads to better generator effectiveness. The federated learning framework enables data owners at different monitoring stations to share imputation model training with unexposed raw data for privacy reasons. The testing on actual air quality information shows that FCGAI delivers superior performance compared to conventional imputation methods and local GAN models through more accurate outcomes and training process stability. The application of generative AI along with federated learning capabilities enables the development of authentic synthetic data completion for missing fields and secured training across various information sources respectively.

The author presents FCGAI as a framework that implements federated learning with CGAN-based imputation systems to solve air quality monitoring network data imputation issues. The approach deals with situations that require privacy protection by allowing monitoring stations to operate independently. The approach employs a conditional GAN-based imputation model that trains the generator to develop realistic data imputations using observed information to which the discriminator assigns integrity classifications. The authors enhance GAN stability and performance by implementing Wasserstein distance together with their "Hint mask" implementation. Through the hint mask mechanism the discriminator receives specific information to improve generator performance while stopping mode collapse from occurring. Under the federated learning structure several data owners who maintain separate monitoring stations work together on model training by sharing model parameters instead of their actual data. Each participant in this system trains their specific model from individual data before delivering only the model parameters (such as weights) to a central server where the aggregated results are sent back to all participants. Repetition runs until the model reaches convergence. Federated learning secures both model intelligence from combined participant data without revealing someone's confidential information. The paper outlines CGAN structure through an explanation of its generator and discriminator units and presents optimization methods that employ gradient penalties and Wasserstein distance minimization for stable training procedures. The proposed method achieves successful data imputation using actual air quality records while maintaining full confidentiality of the underlying information.

Using Artificial Intelligence to Visualize the Impacts of Climate Change [23]: The authors suggest adopting Generative Adversarial Networks (GANs) within generative AI systems to develop customized visualizations about climate change effects for raising public interest combined with action promotion. The tool developed by authors enables users to submit addresses for artificial intelligence to generate photo-realistic visualizations showing their future locations in 2050 under climate change conditions including rising waters and storm damage and wildfire destruction. The tool employs conditional GANs to develop these images and aims to transform vague climate change end results into concrete pictures that matter to each viewer. The two components of the GAN model form a Masker for selecting flood zones and a Painter for producing authentic water effect simulations in those areas. The authors create training data by utilizing a simulated environment developed from Unity 3D game engine since actual world data is missing. The target application will establish an interactive website which enables viewers to observe potential climate change effects on their locations through visual representations supported by scientific data and methods for preventing those impacts. This research project employs generative AI while seeking to unite scientific information with public understanding thus helping people bypass misconceptions to take real steps in fighting global warming.

The author establishes how the AI Climate Impact Visualizer develops through Generative Adversarial Networks (GANs) to produce authentic visualizations of climate change effects by showing flood areas most specifically. The authors built their GAN system with two fundamental parts known as Masker and Painter. The Masker component identifies flooding zones in images and the Painter produces authentic water effects with reflections along with textures. The training of the GAN became difficult because actual real-world data was insufficient so the authors built their simulation environment in Unity 3D. Through their simulated world they yielded diverse training data sets consisting of before-and-after flood scenarios that included urban and suburban places for the GAN model development. The simulated data contained three-dimensional spatial information and depth capabilities designed to help the model process flooding in its proper environment. The GAN received training from a mix of real images and simulated data to generate outcomes that remained authentic yet maintained correct geographical settings. GAN outputs personalized realistic images showing flooded locations that users can specify according to their specifications. The research team aims to embed their GAN model within an interactive website system which allows users to provide an address so they can receive visualized predictions of 2050 climate change effects combined with scientific knowledge and actionable prevention steps. Using generative AI brings abstract climate change consequences into tangible forms that enhance personal connection to enable public involvement and awareness.

A Multi-modal Deep-Learning Air Quality Prediction Method Based on Multi-station Time-Series Data and Remote-Sensing Images: Case Study of Beijing and Tianjin [24]: The author develops a new multi-modal deep-learning model called Res-GCN that optimizes air quality predictions by connecting remote-sensing images with multi-station time-series data. The model connects a visual feature extraction machine of Residual Network (ResNet) with a dynamic spatio-temporal graph convolution network (DSTGCN) for discovering spatio-temporal patterns from time-series data. Dynamical time warping (DTW) distance serves in the DSTGCN model for building variable air quality data graphs that help the model handle shifting time patterns. A parameter matrix merges both modality features which leads to predictions from fully connected and convolutional layers. The researchers used Beijing and Tianjin real-world datasets as they evaluated the model which showed better prediction accuracy for multi-step air quality forecasting as opposed to conventional baseline solutions. The research paper incorporates deep-learning techniques ResNet and DSTGCN but it does not explicitly use generative AI for prediction generation. Instead the methods use advanced neural network structures to extract complex multi-modal data patterns which increase prediction accuracy. The study demonstrates how joining multiple datasets with modern deep-learning algorithms can optimize predictive capabilities in environmental surveillance operations.

The author presents Res-GCN as a deep-learning multi-modal framework which combines time-series data from multiple stations with remote-sensing images for air quality prediction. The model includes three essential units: (1) ResNet performs spatial feature extraction from remote-sensing images, (2) DSTGCN analyses spatio-temporal relationships in time series data and (3) a feature integration procedure which unites outputs from both input modalities. The ResNet module extracts remote-sensing image details with the help of its residual block structure for identifying subtle pollutant spread patterns. The DSTGCN builds dynamic graphs through dynamic time warping (DTW) distance adjustment for air quality data temporal fluctuations thus allowing the model to identify changing spatial station dependencies. The DSTGCN utilizes GCNs to extract spatial features together with TCNs to understand temporal dependencies while simultaneously using temporal attention for precision time-step weighting. The predictive capabilities of both networks benefit from a parameter matrix fusion technique that joins their output results. Future air quality predictions are obtained by passing the fused features through fully connected layers together with convolutional layers. Air quality forecasting benefits from both multi-modal data fusion and a system for dynamic graph development according to this approach.

DisasterGAN: Generative Adversarial Networks for Remote Sensing Disaster Image Generation [25]: The authors present DisasterGAN which utilizes GAN-based technology to synthesize images from satellite observation platforms for developing better assessments regarding disaster detection and building damage evaluation. Both models called Disaster Translation GAN and Damaged Building Generation GAN were created by the authors to tackle the training data limitations of existing datasets like xBD. The Disaster Translation GAN produces disaster images containing several disaster types from one model and the Damaged Building Generation GAN edits building damage zones specifically while preserving the rest of the image. Through their GAN implementation the two models generate realistic disaster imagery for data augmentation purposes that improves building damage assessment models. The research evaluates these models by conducting qualitative and quantitative evaluations which prove that the synthesized images improve damage classification precision especially for challenging categories. GANs operate under Generative AI to produce virtual disaster images that duplicate real-world conditions in order to solve the insufficient data and uneven labeled data problems faced by remote sensing programs.

The authors presents DisasterGAN which contains two models called Disaster Translation GAN and Damaged Building Generation GAN for generating realistic remote sensing disaster images during data augmentation. A Disaster Translation GAN built after StarGAN can transform pre-disaster images into post-disaster images with different disaster types through a unified model architecture. The system utilizes a generator and discriminator framework which enables the generator to transform pre-disaster pictures together with target disaster information into post-disaster pictures after receiving input from the discriminator for distinguishing real from generated images and identifying disaster attributes. The training process utilizes adversarial loss together with attribute classification loss and reconstruction loss to produce realistic images with maintained original image structure. The Damaged Building Generation GAN draws inspiration from SaGAN to modify particular damaged architectural areas in images without affecting their surrounding sections. The code incorporates mask-guided components that enable precise manipulation of affected building areas without changing the non-damaged sections of the background. The model receives adversarial loss training and attribute classification loss training together with reconstruction loss training. High-quality synthetic disaster images produced by the models utilize GANs to expand training data for building damage assessment models while managing scarce data and imbalanced classes in remote sensing datasets.

Artificial Intelligence in Groundwater Management: Innovations, Challenges, and Future Prospects [26]: The author examines the adoption of Artificial Intelligence (AI) with generative AI capabilities in groundwater management since it demonstrates its revolutionary nature and also discusses existing obstacles and expected upcoming developments. Groundwater management receives a revolution from AI-driven innovations including predictive modeling and real-time monitoring and data integration and decision support systems and optimization algorithms which create precise forecasting and efficient resource allocation and proactive contamination detection capabilities. Through generative AI systems researchers create artificial data which helps predictive analytics succeed better while running simulations when real data possesses limitations or small availability. Better predictions regarding groundwater behavior under different environmental situations including climate change become feasible through this capability. Data quality issues and low data quantity alongside model interpretability difficulties and specialized expert requirement are among the obstacles identified in the paper. Open-source development will produce improved AI models for hydrological forecasting as well as risk evaluation capabilities and communication systems aimed at scientist and community interaction.

The novel usage of AI in groundwater management is detailed, with emphasis on five major aspects: predictive modeling, real-time monitoring, data integration, decision support systems, and optimization. In predictive modeling, AI helps to develop a history of groundwater levels and statistical models to predict future levels, which in turn helps in effective planning and management of the scarce resource. Real time monitoring uses IoT sensors and AI algorithms to periodically examine the groundwater quality and detect any contamination early enough to take action quickly. In data integration, AI is used to reconcile different data sources including geological surveys, satellite imagery and climate data to generate accurate simulation of groundwater flow models. Decision support systems enabled by AI provide valuable recommendations in the form of suggestions and what if analyses to policy makers and water managers for appropriate groundwater management. Last, optimization algorithms supported by AI determine the best pumping regimes to minimize energy use and environmental effects in the extraction of groundwater. However, the paper also notes some limitations of these methods, including the demand for accurate data, the ability to interpret AI models, and technical expertise. However, the integration of AI, including generative AI, has the potential to greatly enhance the management of groundwater by enhancing the prediction, utilization, and interaction of available resources.

5 Conclusion

The emergence of generative AI represents a groundbreaking tool for environmental and atmospheric sensing because it provides new solutions to solve key problems that involve sparse data and high computational demands as well as requirements for detailed simulations. The combination of GANs, VAEs and diffusion models shows substantial ability to enhance the prediction of air quality as well as climate modeling and disaster management functions while helping monitor biodiversity and manage water resources. These technologies enable the creation of synthetic data and better predictive accuracy and complex environmental system simulation which results in superior decisions for environmental management. Generative AI implementation faces different challenges due to quality issues in data input and concerns about ethics as well as significant requirements for computing power. The necessary solutions depend on combined efforts between different academic fields together with solid moral principles and better artificial intelligence exploration programs. Generative AI development shows promise for environmental science to achieve better climate change strategies along with improved weather forecasting accuracy while protecting ecosystems to create sustainable and resilient outcomes.

References

1. I. Goodfellow, J. Pouget-Abadie, M. Mirza, B. Xu, D. Warde-Farley, S. Ozair, A. Courville, Y. Bengio, Generative adversarial nets. Adv. Neural Inf. Process. Syst. **27** (2014)
2. D.P. Kingma, M. Welling, Auto-encoding variational Bayes [online] (2013)
3. A. Vaswani, N. Shazeer, N. Parmar, J. Uszkoreit, L. Jones, A.N. Gomez, Ł. Kaiser, I. Polosukhin, Attention is all you need. Adv. Neural Inf. Process. Syst. **30**
4. M. Reichstein, G. Camps-Valls, B. Stevens, M. Jung, J. Denzler, N. Carvalhais, F. Prabhat, Deep learning and process understanding for data-driven earth system science. Nature **566**(7743), 195–204 (2019)

5. S. Rasp, P.D. Dueben, S. Scher, J.A. Weyn, S. Mouatadid, N. Thuerey, WeatherBench: a benchmark data set for data-driven weather forecasting. J. Adv. Model. Earth Syst. **12**(11), e2020MS002203 (2020)
6. L. Huang, F. Antolini, A. Mostafavi, R. Blessing, M. Garcia, S.D. Brody, High-resolution flood probability mapping using generative machine learning with large-scale synthetic precipitation and inundation data. arXiv preprint arXiv:2409.13936 (2024)
7. J.F. Artiola, I.L. Pepper, M.L. Brusseau (eds.), *Environmental Monitoring and Characterization* (Academic Press, 2004)
8. M.C. Thomson, S.J. Connor, S.E. Zebiak, M. Jancloes, A. Mihretie, Africa needs climate data to fight disease. Nature **471**(7339), 440–442 (2011)
9. P. Ceccato, K. Fernandes, D. Ruiz, E. Allis, Climate and environmental monitoring for decision making. Earth Perspect. **1**, 1–22 (2014)
10. S.A. Dressing, D.W. Meals, J.B. Harcum, J. Spooner, J.B. Stribling, R.P. Richards, C.J. Millard, S.A. Lanberg, J.G. O'Donnell, *Monitoring and Evaluating Nonpoint Source Watershed Projects*, vol. 40 (United States Environmental Protection Agency, Washington, DC, 2016)
11. S. Subramaniam, N. Raju, A. Ganesan, N. Rajavel, M. Chenniappan, C. Prakash, A. Pramanik, A.K. Basak, S. Dixit, Artificial intelligence technologies for forecasting air pollution and human health: a narrative review. Sustainability **14**(16), 9951 (2022)
12. L. Zhao, Z. Li, L. Qu, A novel machine learning-based artificial intelligence method for predicting the air pollution index PM2.5. J. Clean. Prod. **468**, 143042 (2024)
13. D.B. Olawade, O. Fapohunda, O.Z. Wada, S.O. Usman, A.O. Ige, O. Ajisafe, B.I. Oladapo, Smart waste management: a paradigm shift enabled by artificial intelligence. Waste Manag. Bull. (2024)
14. M. Coeckelbergh, AI for climate: freedom, justice, and other ethical and political challenges. AI Ethics **1**(1), 67–72 (2021)
15. M. Zaresefat, R. Derakhshani, Revolutionizing groundwater management with hybrid AI models: a practical review. Water **15**(9), 1750 (2023)
16. R. Nishant, M. Kennedy, J. Corbett, Artificial intelligence for sustainability: challenges, opportunities, and a research agenda. Int. J. Inf. Manage. **53**, 102104 (2020)
17. J. Ho, A. Jain, P. Abbeel, Denoising diffusion probabilistic models. Adv. Neural Inf. Process. Syst. **33**, 6840–6851 (2020)
18. K. Wang, T. Yang, S. Kong, M. Li, A generative AI-based deep learning model for air quality index prediction
19. A. Shammasi, Enhancing data imputation with generative AI: transforming time series through image processing. Doctoral dissertation, Macquarie University, 2024
20. X. Kang, L. Liu, H. Ma, ESR-GAN: environmental signal reconstruction learning with generative adversarial network. IEEE Internet Things J. **8**(1), 636–646 (2020)
21. J. Kalajdjieski, E. Zdravevski, R. Corizzo, P. Lameski, S. Kalajdziski, I.M. Pires, N.M. Garcia, V. Trajkovik, Air pollution prediction with multi-modal data and deep neural networks. Remote Sens. **12**(24), 4142 (2020)
22. X. Zhou, X. Liu, G. Lan, J. Wu, Federated conditional generative adversarial nets imputation method for air quality missing data. Knowl.-Based Syst. **228**, 107261 (2021)
23. A. Luccioni, V. Schmidt, V. Vardanyan, Y. Bengio, Using artificial intelligence to visualize the impacts of climate change. IEEE Comput. Graph. Appl. **41**(1), 8–14 (2021)
24. H. Xia, X. Chen, Z. Wang, X. Chen, F. Dong, A multi-modal deep-learning air quality prediction method based on multi-station time-series data and remote-sensing images: case study of Beijing and Tianjin. Entropy **26**(1) (2024)
25. X. Rui, Y. Cao, X. Yuan, Y. Kang, W. Song, DisasterGAN: generative adversarial networks for remote sensing disaster image generation. Remote Sens. **13**(21), 4284 (2021)
26. M. Shaikh, F. Birajdar, Artificial intelligence in groundwater management: innovations, challenges, and future prospects. Int. J. Sci. Res. Arch. **11**(01), 502–512 (2024)

Medical and Biological Sensing with AI Driven Photonic Sensors

Sneha Sharma and Dharmendra Kumar

Abstract Photonic sensing devices are gaining significant importance across diverse fields, including medicine, healthcare, early disease diagnosis, agriculture, environmental monitoring and automotive industrial applications. Using light, it detects changes in the optical properties of a material influenced by the specific physical, chemical or biological parameter being measured. Photonic based sensor offers high sensitivity and specificity, fast response time, able to detect minor changes with high precision, non-invasive, non-destructive and provides real time monitoring and remote sensing capability features. Recent years there is a significant demand of rapid and accurate detection of biomarkers for disease diagnosis using miniaturized portable and wearable devices based on photonics sensors. This chapter explores various medical and biological sensing methods including surface plasmon resonance, optical fiber, photonic crystal, optical waveguide, integrated photonic and microwave photonic based sensors. It also highlights the challenges and limitations associated with photonic sensors, providing a comprehensive overview of their applications and the obstacles that need to be addressed for future advancements. Integrating artificial intelligence (AI) in photonic sensors is revolutionizing by enabling highly accurate, fast and real-time data monitoring. The data from the sensors can further process and interpret using AI algorithms such as machine learning and deep learning models. By classifying and analysing periodic patterns of data based on previous events predictive model can be developed which correlates with particular health symptoms and provide early signs of diseases. The primary challenge in AI driven photonic sensors are data security and privacy. Despite this limitation, intelligent photonic sensors using AI offers high potential in developing advance medical biosensing devices for precise detection of critical diseases. Photonic crystal-based sensors with advanced modern technology is increasing rapidly by providing continuous monitoring of health, offering early disease diagnosis to provide good health for

S. Sharma (✉)
Department of Electronics and Communication Engineering, Dayananda Sagar University, Bangalore, India
e-mail: sneha.ece@gmail.com

D. Kumar
Department of Electronics and Communication Engineering, Guru Ghasidas Vishwavidyalaya, Bilaspur, Chhattisgarh, India

Y. Vasimalla and S. Kumar (eds.), *Generative AI for Photonic Sensing*, Progress in Optical Science and Photonics 36, https://doi.org/10.1007/978-981-95-1561-5_8

the growing population. Also, photonic based devices are able to monitor environment pollution by monitoring greenhouse gases, remote sensing to observe climate change and thus reducing carbon footprint which emits due to industrial gas emissions. Advanced 2D material-based sensors are often used to provide clean and green energy which are useful for energy harvesting and photonic based solar panels to increase the efficiency. By integrating AI with the photonic sensors, it is revolutionizing the current infrastructure by offering smart way of health monitoring the structures for industrial and medical applications.

Keywords Photonic sensors · Biosensors · Disease diagnosis · Artificial intelligence · Photonic crystal · Optical sensors

1 Introduction

Conventional method of disease diagnosis is quite expensive and time consuming, so in resource limited areas more affordable, fast, highly sensitive and specific early disease detection devices are required. The advantage of using portable lab-on-a-chip biomedical sensing device provides detection and prevention of critical illness for patients deprived of medical care facilities and reducing frequent hospital visits [1]. Biomedical sensors are devices that detect various chemical or biomolecule reactions and changes its output signal in accordance with the quantity of the analyte concentration in the reaction and is widely used for diagnosis of various diseases in healthcare and medical analysis [2] as shown in Fig. 1. Accurate and non-invasive method of determining diseases and measuring various metabolites for clinical analysis using body fluid other than blood has shown significant growth [3, 4]. These types of methods are majorly required for patients who needs to monitor health parameters like glucose, urea, glycemia, etc. frequently. There is an increasing demand of biosensors for the detection and monitoring of particular biomedical analytes by less invasive method which can reduce the pain of patients and give fast, reliable and accurate results [5].

Any biosensor mainly contains two major parts that is receptor layer and transducer shown in Fig. 2. Biorecognition layer is present in the receptor part which detects the presence of biological analytes. Any interaction between the receptor and the analyte produces the biochemical reaction or changes the certain output characteristics of the device which is converted into readable electrical signals by transducer [6, 7]. For diagnosis of diseases classification of various biological cells and its fundamental function is necessary. Biomedical parameter detection using sensors is changing the health care industry. Further integrating this device with artificial intelligence and making future smart biosensors is *revolutionizing* the trend of health care sector [8–10]. With wide range of availability of data, the machine learning algorithms can be applied for classification of cells or analysing the data and therefore advancing the detection methodology automatically without the physical need of expert intervention [11].

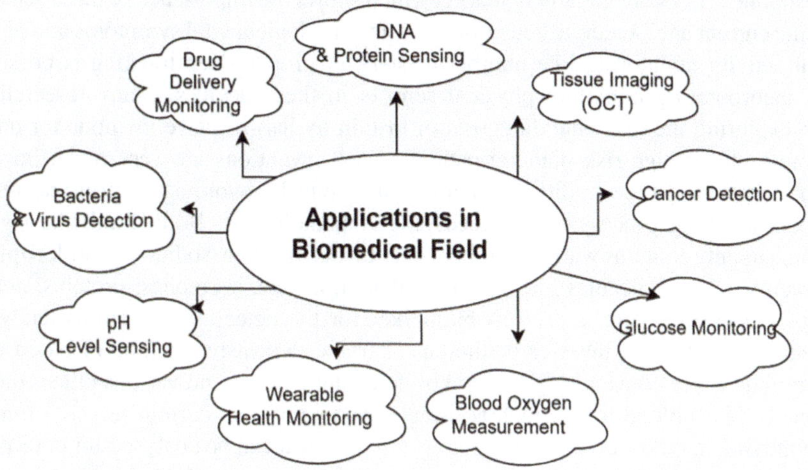

Fig. 1 Applications of bio-sensors in medical domain

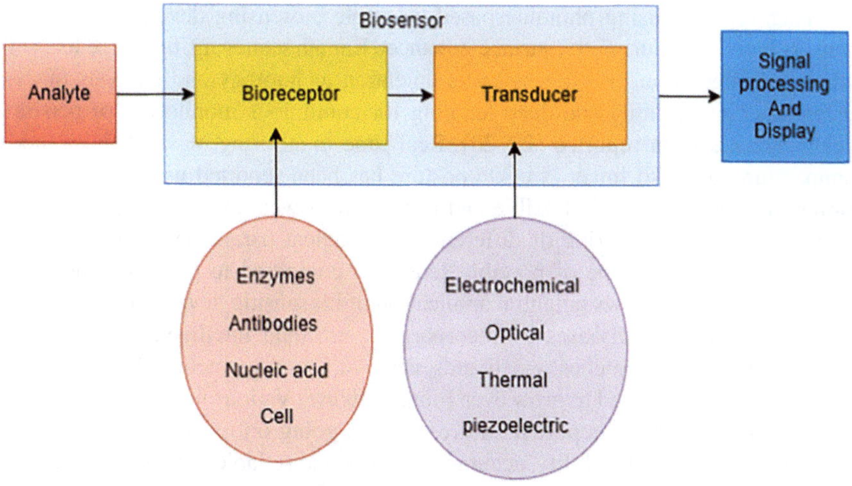

Fig. 2 Biosensor and its major components

Recent years real time and continuous monitoring of various biological and physical activity tracking by non-invasive and minimally invasive techniques using bodily fluids in wearable and flexible biosensors have shown great potential [12, 13]. Development and commercialization of these sensors are also possible with the advancement of nano-fabrication technology [14, 15] which shows great impact in the society by providing early diagnosis of various medical and biological problems and improving the living conditions. Non-invasive detection of physical activity using wearable devices are already in market which shows the importance of these sensors in the current age. Accurate detection of various biological vital symptoms and physical activity monitoring like number of steps or proper sleep tracking is possible by incorporating miniature physical sensors in these devices. Many researchers are exploring the real time diagnosis of insulin by less-invasive methods for daily monitoring of high-risk diabetic patients. Such inventions are very much important compared to the traditional method of testing by avoiding painful and risky blood sampling process which needs expert pathologists. Body fluid like sweat which mainly contains water and other constituents like urea, sodium chloride, lipids, proteins, small molecules like glucose, ethanol, lactate, hormones, proteins, nutrients, electrolytes can be used as biomarker for the detection of various analytes continuously by non-invasive method as alternate to invasive method of blood test sampling. Since conventional method of detecting glucose and various other conditions involves blood tests which is invasive and painful so current research trends are moving into non-invasive technology where sweat can be analysed for detection of several metabolites. Authors have reported glucose level, pH, lactate and several other biological conditions using sweat based biomedical sensors. Recently many researches have reported non-invasive sweat sensor for detection of glucose, ethanol, pH, calcium, lactate, cystic fibrosis [16–18]. Monitoring of cortisol in human sweat has been reported using photonic based wearable biosensing device [19]. Optical sensors based on colorimetry, surface-enhanced Raman scattering, fluorescence, with advanced integration of wireless communication technology and analysis of data using machine learning and deep learning for continuous monitoring of personal healthcare has been reported [20–23]. Real time monitoring of wrist pulse, body temperature and also finger knuckle posture has been reported using optical fiber Bragg grating as sensor in flexible and personalized wearable device [24]. Therefore, continuous monitoring of different physiological parameter is possible with the invention of this type of portable sensor as compared to previous traditional methodology. This shows potential application of biosensing devices and automatic diagnosis of various diseases by incorporating artificial intelligence in wearable gadgets like in smartwatches, wristbands, earwear, smart garments, shoes and even in eye glasses is expected to grow even more in future. Various non-invasive biological sensing devices are reported and research is going on to improve the various parameters in terms of stability, accuracy, validity and reliable data with low power consumption and ability to integrate advanced wireless communication technology with secure personal data.

There is a tremendous growth in the use of photonic based sensors in various domains like monitoring the quality of food, detection of toxic gases for industrial and environmental applications, biomedical sensors for diagnosis of diseases, monitoring of various physical and chemical parameters for space and defence applications. To measure or sense any physical, chemical or biological analytes, it utilizes photons or light particles and observes the changes in optical behaviour such as amplitude, frequency, phase shift or change in polarization mode. Photonics sensors has the ability to revolutionize medical and various industrial field due to their efficient, quick, precise and highly sensitive characteristics. Photonic based waveguides have transformed medical and healthcare industry where it can quickly detect early symptoms of various deadly diseases by less non-invasive method and photonic based waveguides for supercontinuum generation for optical coherence tomography application which improves resolution and sensitivity of cross-sectional medical images of biological tissues. For high power industrial applications photonics sensors are more preferable compare to electronic sensors since it is not prone to electromagnetic interference as light energy is used instead of electrical energy. Intensity of light or wavelength changes as refractive index of the photonic sensors changes when it comes in contact with any analyte thus providing precise and accurate results. Sensors based on optical fiber have the advantage of using it for longer distance and distributed sensing applications as there is less attenuation of light. Photonic sensors are extremely sensitive and resistant to various physical parameters. With the current fabrication technology most of the photonic sensors can be integrated into chips and thereby can be used as highly sensitive, compact and portable sensors. Researchers have shown great interest in photonic sensors for determining the various vital symptoms of diseases in health care devices. Several photonic sensors have been reported like subwavelength grating ridge waveguide, single slot and double slot waveguide microring resonator with a very high sensitivity of 840 nm/RIU [25]. The most common type of photonic sensors in biomedical industry are Mach–Zehnder interferometer (MZI), micro ring resonator structures and surface plasmon resonance techniques. Biomolecular interactions between antibody and antigen plays an important role for identification of various health related symptoms. Accurate detection of biomolecules like protein is one of the important parameters for medical applications. Plasmonic waveguide in integrated MZI has been used for the investigation of protein conformational change in biomolecular interaction and as shown great potential in biochemical sensing device [26]. A very highly sensitive of 21,500 nm/RIU silicon photonic based biosensor has been reported using MZI and ring resonator for monitoring of immunoglobulin G pairs [27]. Non-invasive real time monitoring and data processing for detection of tuberculosis by testing human urine is developed using photonic sensor chip integrated in MZI with a detection limit of 475 pg/ml [28]. Such point-of-care biomedical devices offers accurate, fast, reliable, high sensitivity and specificity for regular monitoring of various signs of early deadly diseases.

The sensors face various challenges like noise and drift that cause variations in the outcomes, unable to track useful information from large amount of data generated from sensor, lack of feedback and control signals and mainly open loop system. Intelligent biosensing devices are possible in the era of industry 4.0 by integrating advanced artificial intelligence (AI) algorithms and IoT in the healthcare devices to form a complete closed loop system for the continuous real time health monitoring of patients by collecting and automatically analysing data and offers more convenient way of detecting diseases with secured personal data. Machine learning (ML) which is the subset of AI can solve many complicated problems without being explicitly programmed and learn from previous experiences and patterns. Recent advances in ML algorithms like deep learning and convolution neural network has the ability to automatically find the abnormalities from the big amount of data generated by the sensors and thus provides real time monitoring of various vital signs like cardiac disease, heart beat rate, blood pressure, breathing rate, sleeping disorder and various early symptoms of diseases with comfort and easily accessible to the people. The fluid characterization in photonic sensors has been performed by using machine learning techniques reducing the decision making and energy consumption with same accuracy [29]. Continuous wearable cardiovascular monitoring system has been developed using ML algorithms with high SNR, response time 40 ms and sensitivity of 0.21 $\mu A\ kPa^{-1}$ [30]. The accuracy of photonic crystal biosensor has been analysed for the early detection and investigation of tumor in human body using various ML algorithms like support vector machine, logistic regression, k-nearest neighbour and artificial neural network [31]. Smart garment has been developed using polymer based optical fiber and merging AI algorithm such as k-nearest neighbours classifier to enhance the decision-making process and provides an accuracy of 98.14% for remote monitoring of health [32].

This review highlights the potential of AI/ML algorithm techniques in photonic sensors for enhancing its functionality by optimizing sensor design with advanced material selection, efficient data collection and signal processing with improved accuracy, sensitivity and adaptability for wide range of medical applications. AI with photonic sensors into the hardware level for edge computing poses challenges like energy consumption, interpreting data and privacy, stability of sensor, algorithmic optimization and scalability for commercial application. Despite this challenges AI has the ability to revolutionize the medical and health care sector providing patient medical data collection and real time disease diagnosis for early treatment. The basic concepts of photonic sensors for various medical and biological applications will be discussed with its advantages and limitations. For enhancing the sensors performance various machine learning and AI techniques for data processing and pattern recognition will be outlined. The challenges, opportunities and future research direction of AI enhanced smart photonic sensors for biological and medical applications will be highlighted with various case studies for disease detection.

2 Principles of Photonic Sensors for Biomedical Applications

Photonic based devices which uses light in the form of photons are used for many different industrial and healthcare applications. For medical and biological sensing, the biomolecules interact with light energy through various methods like absorption, scattering and fluorescence to detect the presence of defects in the molecules. The main principle of sensing using this method is that as the biomolecules interacts with light the particular characteristics of the device or property of the light changes in terms of wavelength shift, polarization, phase change or intensity which is analysed to find the abnormality [33, 34]. Classification of various photonics sensors shown in Fig. 3 based on operating principle methodology, structural geometry and based on various application. Recent years integrated photonic sensors has gain specific interest in the various domains of medicine, biology, automotive and industrial fields a several optical components or circuits can be easily integrated on single chip using existing CMOS technology. The performance of this type of sensors can be greatly enhanced in terms of its sensitivity and selectivity, ability of its compactness with various electronic devices. In the field of photonic sensing technologies, surface plasmons resonance and ring resonators are the integrate features which is gaining popularity in waveguide based integrated photonic sensor devices. Photonic sensors such as optical waveguide, surface plasmon resonance-based prism biosensor, photonic crystal-based waveguide and optical fiber based waveguide with its architecture is shown in Fig. 4, which are available for biological or medical sensing applications. Optical waveguide can be used for biosensing applications where the biomolecules interacts with the light wave evanescent from the waveguide and the resonance peak red shifts [35]. This device can be further modified with polymer based optical biosensors which offers easier fabrication methods with lower cost but may suffer from poor sensitivity than compare to silicon-based waveguides. Ring resonator-based waveguide can be selected for improving the performance of the device as it provides high quality factor and good confinement ability.

Photonic sensing based on surface plasmon resonance has attracted the recent sensing technology as it offers real time and label free detection method. These sensors are widely used for DNA hybridization, protein analysis or early disease detection. When the density of electron wave propagates at the interface between metal and dielectric surface plasmons occurs. Surface plasmon wave excites when the incident photon energy oscillates with the free electrons near the metal area and form the resonance. When any biomolecular target is present near the sensor surface the refractive index changes due to the absorption of target molecules on the surface and thus shifts in resonance peak occurs. This sensor uses angular interrogation system where the significant angle shift of SPR wave is analysed. The sensitivity and performance can be enhanced further by optimizing the device and using 2D material and silicon nanosheet authors reported the highest sensitivity [36].

Photonic crystal-based waveguide sensors attracted the sensing technology due to its ability to control light in the wavelength scale. For various applications like biomolecular, bacterial or refractive index changes it can be used. Photonic crystal-based sensor is used to identify the biological molecules in the refractive index range from 1.33 to 1.40. The device is further optimized and analysed the sensitivity and FOM with values of 185.117 nm/RIU and 9.717 RIU^{-1}, respectively [37]. A photonic crystal-based ring resonator has been investigated for detection of various blood components, concentration of glucose in urine, cancer biomarker detection and change in refractive index and Q factor is identified [38–41]. A highly sensitive two-dimensional photonic crystal based microcavity resonator is proposed and fabricated on silicon on insulator chip. The device is further coated internally with various proteins like glutaraldehyde and bovine serum albumin and observed variation in resonance redshift with different sizes [42]. This type of device has also been used for classification of various brain tissues based on change in refractive index during robotic surgery with a very high sensitivity, quality factor and limit of detection [43].

Fig. 3 Classification of photonic sensors

Sensors based on optical fiber uses principles of absorption, surface plasmon resonance, refractive index method. This sensing method has attracted great interest in chemical and biomolecular detection applications as it has several advantages such as immunity to electromagnetic interference, low limit of detection capability and distributed and multiplexed abilities. The surface of the fiber is generally coated with various sensitive materials which interacts with the target samples and thus changes the phase, polarization, intensity and refractive index properties of optical

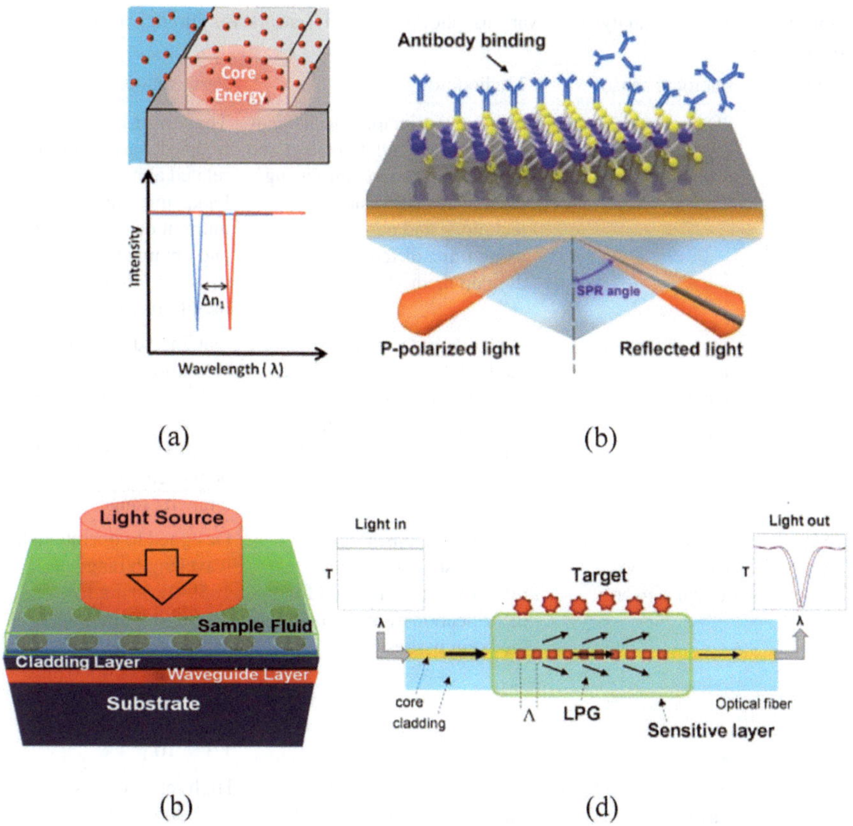

Fig. 4 **a** Optical waveguide. Reprinted/adapted with permission from [35] © Optical Society of America. **b** Surface plasmon resonance-based biosensor [36]. **c** Photonic crystal based sensor [37]. **d** Optical fiber based sensor [38]

fiber. Various types of optical fiber are explored like fiber Bragg grating sensors, single mode fiber, multi core fiber, microstructured and photonic crystal fibers for healthcare applications. Hollow core Optical fiber biosensor for DNA detection has been proposed by observing the redshift in the fiber transmission spectrum [44]. Optical fiber based on fluorescence has been developed for detection of *Salmonella typhi* in food samples [45]. Tapered optical fiber based on frequency domain reflectometer principle has been proposed for the detection of biochemical substances which is significant to cancer or cell abnormality with a very high sensing resolution of 680 μm [46]. Table 1 shows the comparative analysis of several photonic based sensors for bio medical applications.

Photonics sensors offers various advantages to be used in wide area of industrial and medical applications. However, there are certain challenges involved in utilizing photonic based sensors that need to be addressed for real time monitoring applications such as sensitivity, selectivity, specificity, scalability, precision, calibration,

Table 1 Comparative analysis of various photonic based sensors for biological and medical applications

References	Photonic sensor	Methodology	Performance
[47]	2D photonic crystal-based ring resonator	Using the principle mechanism of photonic band gap effect and simulating the structure using plane wave expansion technique and performance using finite difference time domain method	The sensor has been used to detect various blood related abnormality like leukaemia, thalassemia, nutrient deficiency, etc. and obtained sensitivity of 166 nm/RIU with quality factor of 3702
[48]	Photonic crystal resonator based optical biosensor	Variation in the refractive index method. Investigation of the performance using finite difference time domain method for identifying cancer cells in the blood	Obtained detection resolution and limit of 10^{-4}, quality factor of 190 with high sensitivity of upto 900. Also offers better sensitivity in the terahertz range
[43]	Two-dimensional photonic crystal optical sensor	The optical sensor detecting mechanism is based on refractive index method for identification of abnormality in the brain tissues	Designed to operate in the wavelength range of 1650–1800 nm for classifying tumor and cancer affected cells with sensitivity of 4600 nm/RIU, quality factor 570 and detection limit of 1.3×10^{-3}
[49]	Photonic crystal fiber sensor	Refractive index based for early detection of brain cancer	High relative sensitivity 98%, low loss confinement and material loss in the terahertz region of 1–3 THz
[50]	Silicon on insulator based photonic wire waveguide	Evanescent field-based sensor using Mach–Zehnder interferometer-based device	High sensitivity, high variation in effective refractive index 0.31 RIU^{-1} with the change in bimolecular solution
[51]	One dimensional photonic crystal	A defect layer created for sensing oxygen concentration in the haemoglobin sample. Transfer matrix method is used to analyse the transmission spectra in case of variations	Obtained a sensitivity of 1500 nm/RIU
[52]	Magneto plasmonic sensor based on one dimensional photonic crystal and ferromagnetic layer	Variation in the refractive index due to change in magneto optical Kerr effect in spectrum	Sharper magnet based plasmonic resonance with angular width of 0.02° and limit of detection of the order of 10^{-6}

(continued)

Table 1 (continued)

References	Photonic sensor	Methodology	Performance
[53]	Eight symbolic shaped resonator coupled with plasmonic waveguide	Effect on the transmission spectrum is analysed using finite difference time domain method for detection of basal cancer cells	Sensitivity of 1200 nm/RIU is obtained with Q factor of 270 at wavelength of 1180 nm. The structure is suitable for biosensing and filtering operation
[54]	Tapered optical fiber based sensor	Based on evanescent waves and change in refractive index of the surrounding medium the device is used for chemical and biological sensing applications. The structure is modified with ZnO based nanoparticles for enhanced performance	Detection of ethanol solution in the concentration range of 0–100% with sensitivity of 4 a.u/%. Identification of creatinine solution with sensitivity of 0.11 a.u/μM
[55]	Nanostructure photonic crystal-based sensor	Identification of biomarkers or virus from biological data by observing the shift in the resonant peak wavelength	Obtained less than 10 pm resolution in capturing the shift in wavelength. Detected HIV-1 virus spiked with concentration in the range of 10^4–10^8 samples/ml
[56]	Photonic crystal fiber sensor	Surface plasmon resonance methodology-based device for detection of bacteria presence in the blood. Finite element method has been used to investigate the performance of the sensor	Received wavelength sensitivity 20,000 nm/RIU and intensity sensitivity 1400/RIU with spectral resolution of 10^{-6} for the variation in refractive index of the analyte
[57]	Photonic crystal fiber based optical sensor	Surface plasmon resonance-based sensor with Au, $BaTiO_3$ and graphene coating. To investigate the performance finite element method is used	Detection of COVID-19 virus with amplitude sensitivity of 15/RIU, 2300 nm/RIU wavelength sensitivity and resolution of 4.2×10^{-4}

portability and signal processing. Sensors must be able to detect low concentrations of analytes and ability to distinguish between target samples and any interference present in the device to provide stable and reproducible results. Progress in fabrication technology process enabled portable sensors by integrating lab on chip device and as wearable and implantable device for easy diagnosis. Due to its multiplexing nature these devices are expected to grow in remote and resource deprived areas for health monitoring applications. The advancement in the technology of IoT and AI further incorporated in these devices for better signal processing and networking in large scale. Also, photonic sensors are further researched in terms of sustainability and overcome the challenges involved in mass manufacturing process due to at present high cost development process. Further introducing high speed wireless connectivity and IoT the security of the patient data is a challenge which needs to addressed. Also, integration of AI for data processing needs more memory and power consumption also need to be explored. Self-generating power for wearable sensors based on piezoelectric or triboelectric effect are reported. Also, the sensor interfacing and data can be noisy so need to address the robust techniques to extract the required useful information from the output data [58].

3 AI and ML Concepts for Medical Applications

Recent emerging technologies like AI and ML has become a powerful tool in analysing the problems in advance. These technologies are widely used in various applications like natural language processing, computer vision, enhancing the image and video processing, in industrial and medical applications for advance diagnosis and automation in the treatment, etc. Using various of AI/ML algorithms machine are able to learn from the available data and also giving smart outcomes. AI enabled devices are capable of learning and behaving similar like human intelligence through rigorous training the model with large number of required datasets using various algorithmic techniques. ML and deep learning is a part of AI which is shown in Fig. 5 with its various classifications for applications in medical optics [59–61]. Subset of AI is ML which have many techniques and gives the machine to learn itself without any set of programming rules. ML provides various data driven methods and techniques for AI and gives the machine to learn itself without any set of programming rules.

For biomedicine applications of disease detection and diagnosis ML uses the algorithms which learns from the dataset generated by the large number of healthcare devices and make prediction or classify the problems. Various AI and ML algorithms and its concepts and applications are summarized in Table 2. DL is again the advanced techniques that uses artificial neural network techniques which is inspired by human brain neurons and recognizes patterns. Performance of DL is better than ML but requires large amount of data during training process and is expensive and more computational time required. Whereas ML is useful when smaller dataset is available and comparatively faster in processing. The various classification of ML which can used for such applications are supervised, unsupervised and reinforcement learning. Supervised learning uses labelled data during training to solve problems. In this approach model learns to recognize the patterns and predicts the result. In this technique the algorithms find the correlation between input pattern and the target outcome. For health care applications supervised learning is used to train the machine and identify the pattern and relation between the health parameters of patient and medical abnormality. A properly trained model diagnoses the presence of diseases early. This is further classified into classification and regression methods. Using classification various problem can be solved by using logistic regression that gives

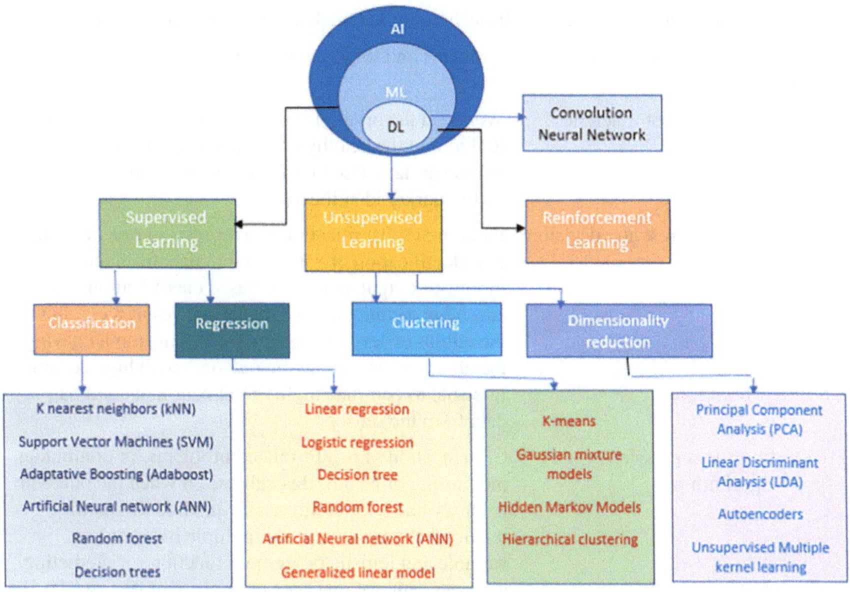

Fig. 5 The hierarchical relation between artificial intelligence (AI), machine learning (ML) and deep learning (DL) and various categories of ML and DL algorithm

the normal or abnormal conditions. Naïve Bayes approach uses probabilistic determination method and generally works on large set of data. Support vector machine (SVM) is very popular method when input features are highly classified and also able to handle non-linear relationship between the data. Forest and decision trees are ensemble process that is useful to improve the accuracy of the predicted results and also to solve the problem of overfitting. Identification of various health related risks can be pre-identified using statistical based regression method that finds the relation between several health parameters. When the nature of health parameter is continuous in characteristics then linear regression is applied. For identifying presence or absence of early disease logistic regression is applicable that gives binary outcome. The k-nearest neighbour algorithm is a classification technique that categories the data by allocating it to the nearest sample point. Regression algorithm is used to predict the outcome by finding the correlation between dependent and independent parameters. Linear regression is the commonly used method to analyse the disease using best fit linear straight line to find the relation between data obtained from sensor and the analyte quantity being used.

Unsupervised technique uses unlabelled data for analysis of the problem and determines the patterns and correlation between the available medical data without having any previous understanding of the disease being determined. Two categories under supervised techniques are clustering and dimensionality reduction methods are useful for biomedical applications. Clustering algorithm groups similar medical data available and identify the nature of data to find the abnormality without using any

Table 2 Summary of several AI and ML techniques for medical diagnosis applications

S. No.	Various AI/ML algorithms	Concepts and its applications in medical
1	Support vector machine	Works on the principle of binary classification a powerful tool to find the healthy or diseased data from the biosensor data. Used for cancer genomic detection based on biomarker identification
2	Random forest and decision trees	It combines classification and regression tree methods. For identification of various anomalies from the biosensor output using rule-based classification and ensemble techniques. For diagnostic result it combines the output of several decision trees. This model having the ability of predicting the outcome with high accuracy. It is able to correlate highly nonlinear input and output data from the sensors
3	Logistic regression algorithm	Classification of health-related problems for continuous monitoring of health. Basically, used when the outcome is binary and uses multivariable method to find the relation between single and multiple independent variable and logistic or sigmoid function for predicting the probability of presence or absence of diseases for the data obtained from medical history
4	K-nearest neighbours	Useful for classification and processing of biomedical signals. Medical diagnosis for prediction of heart, lung diseases or prediction of tumors based on the medical data collected from patients. It calculates the distance between the query point and all other point in the available data set and identifies the closest data points in the features and predicts or classifies based on the distance of the data. By comparing the new data with the available existing medical data of the patient it has the potential to diagnose the disease
5	Unsupervised algorithms like K-means clustering and principal component analysis	Effective in minimizing noise from the medical sensor data based on the concept of dimensionality reduction and feature extracting methods. For grouping similar data sets into clusters K-means algorithm is utilized. By combining both methods, it identifies patterns in medical data for precise diagnosis of the health problems
6	Deep learning-based CNN	Extraction of features and analysing the biomedical images from the sensors. For automatic diagnosis this method provides high accuracy in the medical decision-making process. It works on the principle of feature extraction, pattern recognition, classification of diseases by analysing medical images and useful for personalized medicine

labelled data. K-means clustering is very useful in finding hidden complexities and is widely used technique. Identification of important features of the patient data can be simplified using the dimension reduction algorithm. Using this method large and complex data set can be simplified for pre-processing. Principal component analysis is a type of dimension reduction method being used for converting the sample into reduced form to hold the fluctuations among the available sample data. DL can be applied to get clearer refined outcome at any level of model training. The challenges faced during training is overfitting and underfitting that need to be addressed during the process.

4 AI for Sensor Performance Enhancement and Disease Prediction

AI and ML has the ability to enhance the sensor outcome by enabling real time monitoring and early detection of diseases and decision making on the sensors data for various healthcare and industry applications. Figure 6a shows the process to enhance the sensor outcome by enabling AI and ML methodologies. The output signal generated by sensor is processed and digitalized using AI is illustrated in Fig. 6b. Using various different ML algorithms diseases are identified by training the system and has been reported. Infectious diseases detection has been diagnosed using the available routine test of blood components of the patients using supervised machine learning (SVM) algorithm [62]. Classification of infectious and non-infectious disease prediction is developed based on the availability of the data such as C-reactive protein, white cell count, bilirubin, creatinine, ALT and alkaline phosphatase from various individual patients. A supervised machine learning algorithm embedded within photonic sensors for automatic diagnosis support system and disease detection can be incorporated. Various bacterial or viral infections are identified from the spectrum of white blood samples using SVM and dimensional reduction technique like linear discriminant analysis (LDA) and principal component analysis (PCA) method with a very high sensitivity of 93% sensitivity and specificity of 85% [63]. Deep learning method like convolution neural network (CNN) is used to identify rapid label-free bacterial identification and data set has been taken from Raman optical spectroscopy and achieved a very high accuracy of 99.7% including noisy spectra [64]. Deep learning technique has been used to predict the bacterial contaminant like *Escherichia coli* in water from the output generated by AI-based lab on chip device using photonic crystal based optical biosensor with high accuracy of 95% [65]. Neural network based automatic antimicrobial resistance detection has been reported [66] where the output has been collected from the array of optical fiber. A highly sensitive silicon dioxide coated silver nanorod based surface enhanced Raman scattering device embedded ML algorithms like SVM, kNN and random forest has been used for pathogen detection with a very high accuracy of

99%. Such classification and quantification technique from saliva enhances importance of non-invasive diagnosis in medical field. Neural network and PCA techniques are used for developing highly sensitive and intelligent MEMS sensing device for its application in industry and as medical equipment [67]. A real diagnosis system has been reported utilizing AI and deep neural network for advanced heal AI based deep neural network for real time advanced diagnosis system healthcare system [68]. Also, the process of enhancing signal processing by integrating various AI and ML techniques for the data generated by optical transducers has been reported [69] by applying feature selection and extraction method and using various machine learning algorithms like classification and regression methods and predicting the outcome in digital form by smartphone or digital camera. AI empowered optical biosensor for antibacterial resistance diagnosis [70] and lung cancer prediction from the output of surface enhanced Raman scattering (SERS) based biosensing device [71] has been reported shown in Fig. 7.

Regression analysis has been applied to determine the figure of merit which evaluates sensors performance. Various ML models such as linear regression, decision tree, random forest, Gaussian process regressor and XGBoost techniques are investigated on the data set. The performance of the optical biosensor depends on the sensitivity and accuracy which predicts its figure of merit. The sensors ability to differentiate the parameter from the diverse range of parameters is called sensitivity and how precisely the sensor is able to measure it is called accuracy. To enhance the performance parameter of fiber optic surface plasmon resonance (SPR) based photonic sensor by using various machine learning algorithms is reported in [72]. MoS_2 monolayer is used as metal layer in the SPR sensor. Several numerical simulations need to perform to get the optimum sensor design by varying the parameters or proper material selection. Considering the precise step size during simulation requires very long time and also consume more memory and hence proper optimization is difficult during simulation process. Therefore accuracy, sensitivity and optimized parameters during fabrication

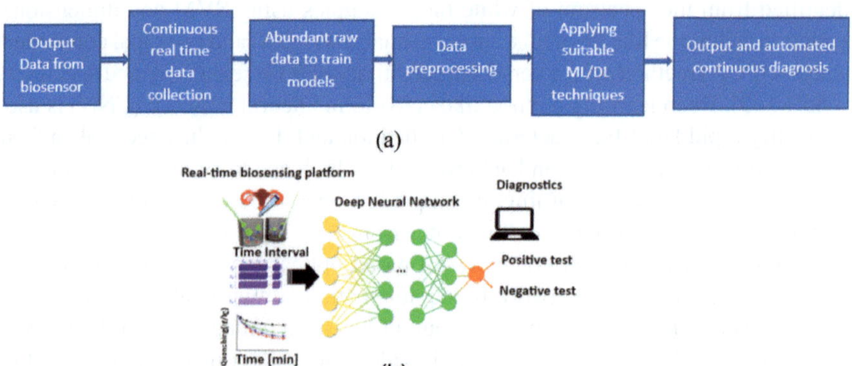

Fig. 6 **a** Flowchart to enhance sensor performance using AI/ML based techniques. **b** AI based deep neural network for real time advanced diagnosis system [68]

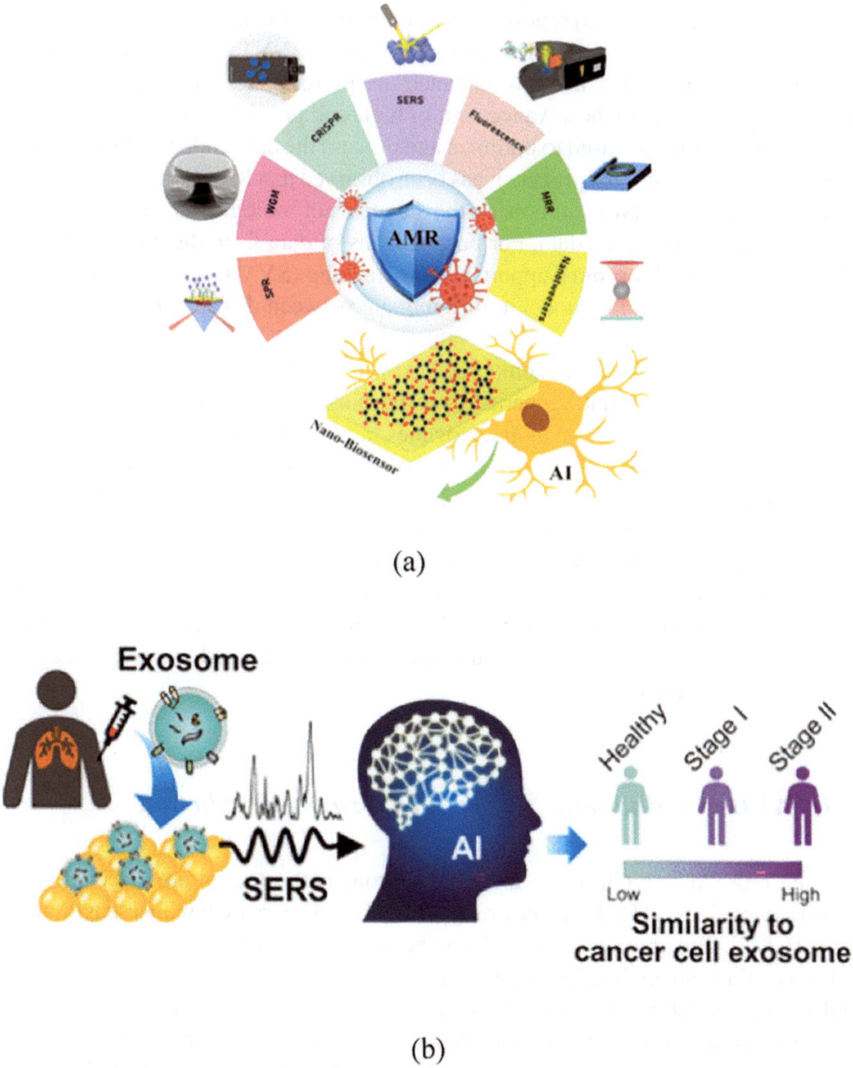

Fig. 7 **a** AI enabled photonic biosensor [70]. **b** Deep learning-based lung cancer diagnosis using SERS based sensor [71]

process is difficult to achieve. To overcome these issues several ML and AI techniques are being explored by researchers. ML algorithms has significantly contributed in handling data processing, enhancing signal to noise ratio, validation and testing of large data set generation in optical devices. Various methods like principal component analysis and linear discriminant techniques have been used for feature selection of the raw data generated by the various sensors.

In this work authors [73] reported the development of nano-enabled impedimetric electrochemical biosensor for detection of dengue DENV1 and DENV2 serotypes. The device is enabled with AI to enhance the performance of biosensor as post-processing tool. Due to the advancement of high-end processors nowadays machine learning algorithms are used to optimize and process the data collected by the device. Compared to conventional process it has been observed that ML has the ability to offer more accurate and robust solutions in the diagnosis of biological data and capability of reducing the erroneous data. For post analysis of data from the biosensor one of the most popular algorithm Support Vector Machine (SVM) has been used here to develop the classification model to differentiate between different DENV infection with accuracy of 94%.

Several ML algorithms are compared [74] which are useful to analyse and extract information from raw data generated by the biosensor to detect the presence of foodborne illness like listeria monocytogenes in milk and other dairy products. The basic steps include the collection of data, processing data and building the model with various ML classifiers and then evaluation of the model. After labelling the data set as 1 or 0 respectively in case in the presence or absence of infection, different ML classifiers such as SVM, KNN, logistic regression, gradient boost, AdaBoost, random forest, decision tree, Naïve Bayes and SGD classifier are applied to the dataset for training of the model. Five-fold cross validation method is used to check the validity of the model. Performance of various classifiers are evaluated using accuracy, F1 score, precision and recall.

4.1 AI Based Photonic Sensor for Hormone Prediction

Women reproductive hormones like progesterone and estradiol are detected in the blood sample in the THz frequency range using hollow core PCF with a sensitivity of 99.88% [75]. The predicting the performance of the sensor ML algorithms are used. ML algorithms are embedded in photonic sensor devices for predicting, forecasting and recommendations and even to save the computation time. Linear regression algorithm is used for predicting the relative sensitivity of the biosensor. To get the optimal structure and save the simulation time ML algorithms have been implemented to predict the results in less time. Locally weighted Linear regression method has been used to determine the weighting and prediction function. The loss function and R^2 score has been estimated to predict the accuracy of the model. The actual and predicted relative sensitivity value of the device has been compared with the weight hyperparameter tuning of 0.02. It has been observed here that with the ML model the predicted data fits the actual data in the graph suitably. For optimization of algorithm the weighting hyperparameter value is tuned and varied from 20 to 0.02. It has been observed that increasing weighting parameter decreases the prediction accuracy as per actual value. Also, the observed results comparatively same with the R^2 score for the reproductive hormone of various concentration level analysed. The highest R^2 value of 1 is achieved for all level of concentration with hyper tuning parameter of 0.02. Authors suggested this type of hollow core PCF [75] design as shown in Fig. 8 can be fabricated using the existing methods like stack and draw, sol–gel and 3D printing techniques.

One dimensional photonic crystal waveguide has been proposed for determining the concentration level hormone responsible for women reproductive system using transfer matrix method [76].

A refractive index variation technique has been proposed based on gold and graphene metasurfaces sensor to identify the reproductive hormones with a sensitivity of 375 GHz/RIU and good detection limit and figure of merit performance in the terahertz range [77]. ML algorithms like KNN regression model has been used to optimize the systems performance further. ML enabled plasmonic waveguide has been explored for prediction of estrogen hormone. SVM algorithm has been compared with all ML model and offered a very high sensitivity and accuracy of more than 85% with F1 score of 0.874 [78].

4.2 AI and Photonic Based Analyte Detection

Sensing of biological or chemical mixture in water is investigated in the mid-infrared region based on nanophotonic waveguide [79]. In this study spectrum collected from the device has been further classified and analysed using ML algorithms with higher accuracy. Also, convolution regression method has been used to detect the concentration level of each analytes. Three layer of convolution neural network model has been used with two fully connected layer for classification of the mixture spectrum and obtained accuracy of more than 96% and able to differentiate 27 types of chemical from the spectrum. This AI integrated smart sensing device is compact, portable and consumes less power and offers multi analyte detection process. AI based CNN algorithm enabled localized surface plasmon resonance-based gold nanoparticles has been studied for the monitoring and predicting the concentration of chemical and biological substances [80] as shown in Fig. 9. Convolution neural network have been used here for training and classification of the data which gives the precise outcome in determining the concentration level of various biomolecule under test. The device is portable and able to provide real time monitoring for diagnosis and treatment without the need of spectrometer. Metasurface based photonic biosensors are bringing lot of attention due to its various advantages compared to conventional optical sensors in terms of its ability to control and manipulate light, portability and versatility. A highly sensitive plasmonic photonic crystal fiber has been proposed for identification of multi analytes in the visible to near infrared region [81]. For detection of various cancerous cells biological analytes need to be investigated. In this work [82], a two-dimensional photonic crystal-based device has been used with a very high-quality factor and sensitivity with low level of detection limit. The sensor is further integrated using ML techniques for optimization and accurate detection of cancer. Support vector machine and multiple regression method has been used and compared with good R^2 value obtained. Aptamer immobilized three-dimensional

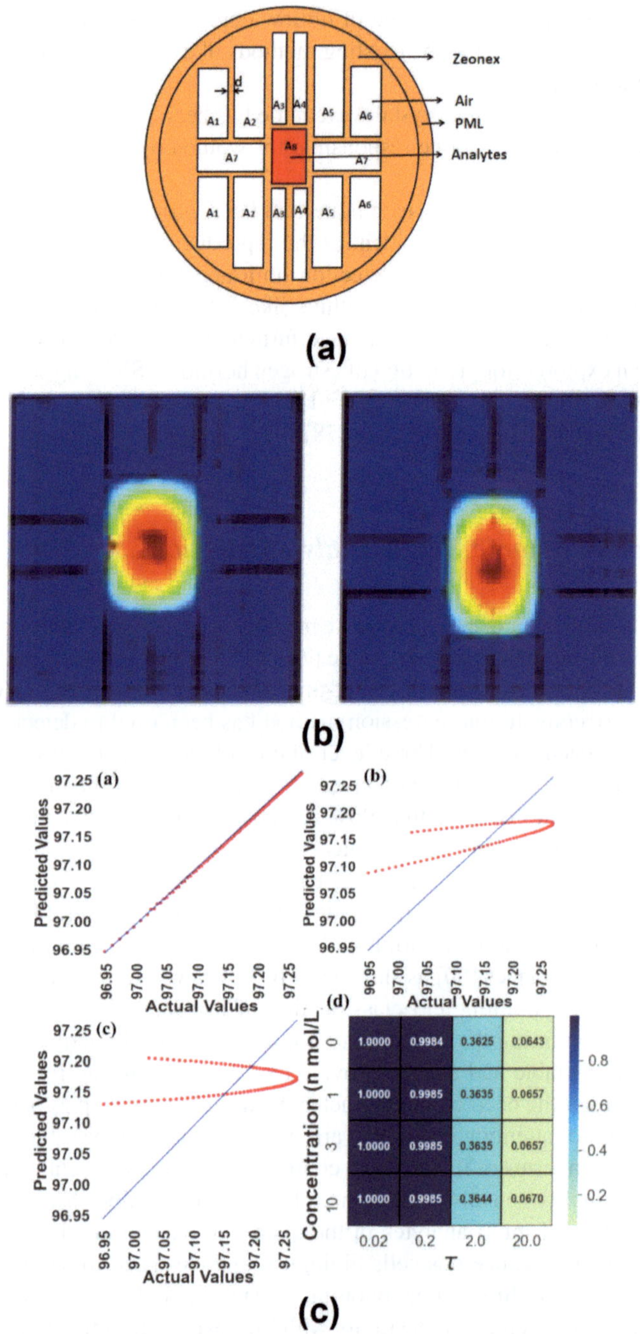

Fig. 8 **a** Hollow core photonic crystal fiber women reproductive hormone sensor. **b** Distribution of electric field energy in the x and y polarization direction at terahertz frequency of 1.35 THz. **c** Application of locally weighted linear regression ML method for comparing the actual and predicted values of relative sensitivity of the device for progesterone hormone [75]

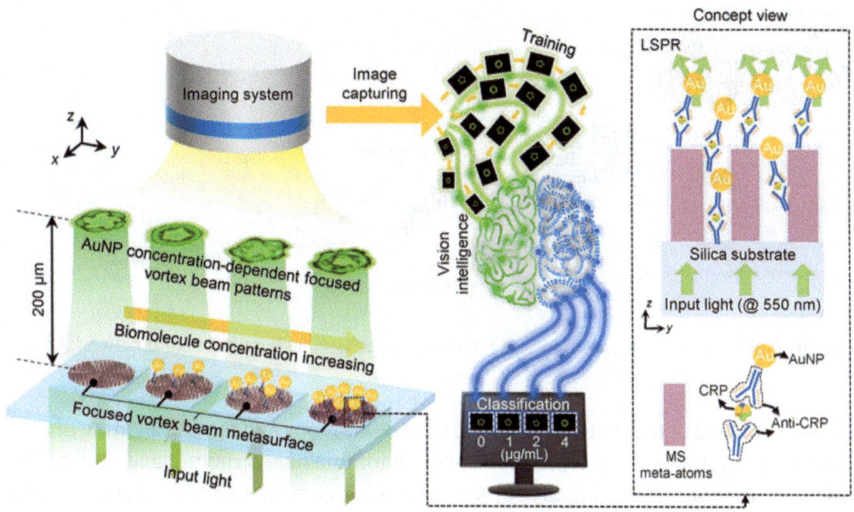

Fig. 9 AI integrated biosensor for diagnosis of biomolecules [80]

photonic crystal-based device has been investigated [83] for the identification of aflatoxin B1 which is categorized as group 1 carcinogen and mainly responsible for liver cancer in living beings. AI based optical analyte refractive index sensor has been studied for prediction of various cancerous cells with accuracy of more than 80% [84].

4.3 Photonic and AI Enabled Glucose Monitoring System

Non-invasive glucose monitoring system is very significant and important for diabetes control. Photonic based non-invasive glucose monitoring system based on optical coherence tomography, microwave spectroscopy, Raman spectroscopy is very convenient as it provides high resolution, fast and accurate method. In this study authors used laser light to record the speckle pattern from the finger and combined the ML and deep neural network methods for data processing and detection of glucose level with high accuracy shown in Fig. 10 [85]. A long short-term memory network is trained to detect the presence of glucose levels. Problem of overfitting has been solved by keeping 30% of data unused during training process. One dimensional photonic crystal based optical biosensor is proposed for measuring glucose level in blood and urine [86]. To determine the sensitivity transfer matrix method has been used to observe the absorption peak in the spectrum depends on the levels of glucose concentration. Also, optimization technique has been implemented to further enhance the sensitivity. Fiber Bragg grating glucose monitoring device is demonstrated with deep learning enabled method to improve the data and enhance

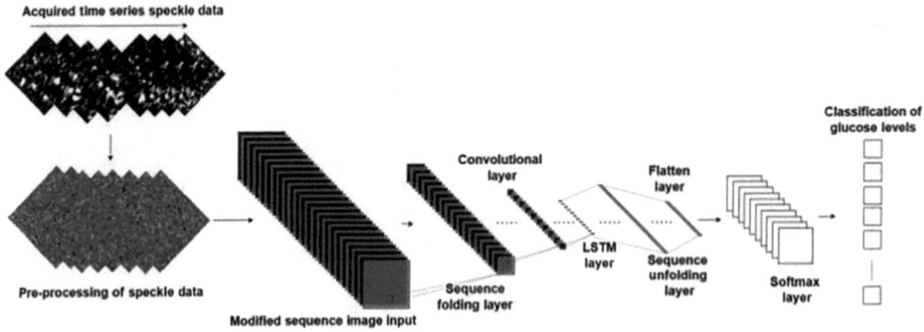

Fig. 10 Non-invasive glucose monitoring and data processing using deep neural network [85]

the performance of the system [87]. Real time continuous and remote locating glucose monitoring system has been proposed using photonic nanostructure and optical fiber based phenylboronic acid hydrogel sensor [88]. Fast response time along with low order of detection limit and sensitivity of 1.2 μW/mM has been reported. To make it portable smartphone has been deployed to observe the change in optical power which correlates with the concentration level. Photonic crystal fiber based biosensor has been studied for detection of glucose in haemoglobin [89].

Various modes of the fiber has been analysed to observe the performance in terms of its change in refractive index, optical spectrum, coupling length between the cores and sensitivity. To correlate the relation between fiber design parameters with its sensitivity the model is enabled with deep learning-based regression model. Further explainable AI and Shapley additive explanation techniques has been employed for feature selection and prediction. One dimensional photonic crystal nanostructure using Si/TiN material to monitor the level of sugar in blood with a very high sensitivity of more than 4000 nm/RIU is proposed [90] which is very useful for identification of any biological sensing. Optical spectroscopy based non-invasive glucose detection is advancing rapidly. Further ensemble ML models have been deployed to determine the concentration level of glucose with more than 90% of accuracy [91].

4.4 AI Enhanced Optical Biosensors for Pathogen Detection

Various optical based biosensors have been developed for detection of *E. coli* pathogen in food and water using surface enhanced Raman spectroscopy, surface plasmon resonance based colorimetric and fluorescence based optical techniques. Further the output generated by this sensor are processed using ML based techniques for developing next generation smart medical and biological sensing system for fast, rapid and accurate detection of various diseases. AI based biosensor network has been proposed for detection of pathogen in liquid food or water [92]. The output data from the optical based fluorescent scanning method is used to identify the bacterial infection and further classified and processed using AI based deep learning model. The AI model was trained for lab cultured bacteria and able to detect the real time untrained data from the agricultural water samples with accuracy of more than 90%. An integrated optical microsystem has been proposed using rib waveguide and micro fluid channel for detection of bacteria in fluid samples [93]. Identification of *S. typhi* bacterial infection has been diagnosed using the optical based immunosensor with nanoparticles [94]. In this study interaction between titanium dioxide nanoparticles and biological analyte is discussed which changes the characteristic intensity of the photoluminescence and shifting the peak in the infrared range and thereby offers highly sensitive device. The integration of AI with modern high-speed technology in nanomaterial based biosensing device has been explored for identification of pathogen detection in plants [95]. In pharmaceutical products detection of pathogen with low concentration level is very important. A SERS based technique [96] has been employed for identification of pathogens like *E. coli* and *S. typhi*. ML based softmax classifier algorithm has been used for improving the accuracy of the detection upto 100% (Table 3).

5 Advancements in Nanophotonic/Bio-photonic Sensors

Recent advances in nanophotonic and biophotonic sensors have made it possible to create sensing devices that are extremely sensitive, small, and effective for industrial applications, environmental monitoring, and medical diagnostics [107]. Nanophotonic sensors improve light-matter interactions at the nanoscale [108], resulting in increased detection sensitivity and miniaturization by employing nanostructures like plasmonic nanoparticles [109], photonic crystals, and metasurfaces. These sensors are very useful for early disease diagnosis and biomarker detection since they can identify even the smallest changes in biological material. In the meantime, real-time, non-invasive medical imaging and biosensing applications have been made possible by biophotonic sensors, which take use of light-based interactions with biological tissues and molecules. Raman spectroscopy, optical coherence tomography (OCT), fluorescence-based detection, and surface plasmon resonance (SPR) have all significantly increased their accuracy and range of applications.

The combination of artificial intelligence (AI) and machine learning (ML) considerably expands the capabilities of these sensors. Furthermore, developments in flexible photonic materials and lab-on-a-chip technology are opening the door for implantable and wearable biosensors that provide ongoing health monitoring [110]. For broad sense issues including cost-effectiveness, real-world integration, and fabrication complexity still need to be resolved. It is anticipated that as research advances, nanophotonic and biophotonic sensors will become increasingly important in precision medicine, smart sensing applications, and next-generation healthcare. Recent advances in AI enabled photonic based biosensors are reported in [111, 112] for early diagnosis of cancer and predicting it accurately using ML based method for personalized healthcare applications shown in Fig. 11. Personalized medical treatment using next generation sensing method by exploring nanomaterial with ML

Table 3 Various AI and ML integrated photonic sensors

References	Type of sensor	AI/ML methodology used	Outcome
[97]	Photonic crystal-based sensor using the concept of photonic bandgap	Support vector machine (SVM) and random forest (RF) classifiers has been used for optimizing and evaluation of performance. ANN is used for predicting the densities of the gases	100% accuracy obtained for identification of greenhouse gases
[98]	Twin core photonic crystal fiber	ML and neural network integrated with system	To increase the sensitivity by identifying optimal design features for various environment changes
[72]	Fiber optic based plasmonic sensor	Gaussian process regression model has been used to enhance accuracy in terms of figure of merit	Model is trained with 459 dataset and predicted figure of merit is compared with the actual value and obtained better result with R^2 close to one
[29]	Multi-dimensional photonic sensor for chemical analysis	Wavelength based hybrid feature selection Dimensionality reduction techniques like feature extraction and feature selection	Execution time is improved by 61% with accurate detection. Detection of ethanol is applied for real time application with reduced energy consumption
[99]	Nanoparticle doped fiber	AI and ML integrated with system for multiple disturbance tracking along the fiber length based on transmission reflection analysis	Accuracy of more than 99% obtained The method is useful for developing low cost distributing system method which can be helpful for remote health monitoring of patient with smart decision taking device
[100]	Ring resonator sensor	Neural network model for predicting and analysing mixed compound of chemicals to develop data driven intelligent sensor	High sensitivity and accuracy of 95% with low root mean square value for analysing multiple chemicals
[101]	Photonic crystal fiber SPR sensor	KNN regression method has been used to determine the loss in the presence of bend. Also compared with ANN and linear least square regression method	Improved response time and resolution. Faster than existing simulation techniques like FEM to optimize and design parameters of the sensor

(continued)

Table 3 (continued)

References	Type of sensor	AI/ML methodology used	Outcome
[102]	Nanophotonic waveguide sensors	CNN and multilayer perceptron algorithm have been utilized for analysing mixture of gases in the mid infrared region	Accuracy of 94% obtained in classification of 19 various gas mixture. AI enabled photonic system developed for analysing multiple organic gas mixtures for future healthcare monitoring
[103]	Photonic sensor	Deep learning and CNN for verification of seed quality for agriculture	Automated classification between good and low quality obtained with accuracy more than 97%
[65]	Photonic crystal optical biosensor	AI based CNN for lab on chip auto detection of bacterial contamination in water	Based on output spectrum obtained from sensor and trained data a prediction accuracy of more than 95% obtained
[104]	Photonic crystal fiber based SPR sensor	ML based support vector machine is used to investigate the sensor performance	Identification of malaria disease sensitivity analysis of the sensor with low mean square and very less error obtained
[105]	Waveguide Bragg grating sensor	Neural network model has been used to predict early cancer detection from the samples	Higher accuracy eliminating the need of higher resolution interrogator devices
[106]	Microwave photonic sensors	ML and deep learning integrated with sensor for better performance and miniaturization	Widening sensing range, multi analyte detection, enhanced signal to noise ratio

based optimization and predicting clinical outcomes using AI and further advancement with quantum technology for precise, fast and accurate diagnosis is reported [113].

6 Challenges and Future Directions in AI-Based Photonic Sensing

AI-based photonic sensing has become a transformative technology for biological and medicinal applications. However, with various advantages there are several obstacles which prevent its broad use as shown in Fig. 12. Since AI models need sizable, high-quality datasets for precise training and validation, one of the main issues is data quality and standardization. The dependability of AI-based sensors may

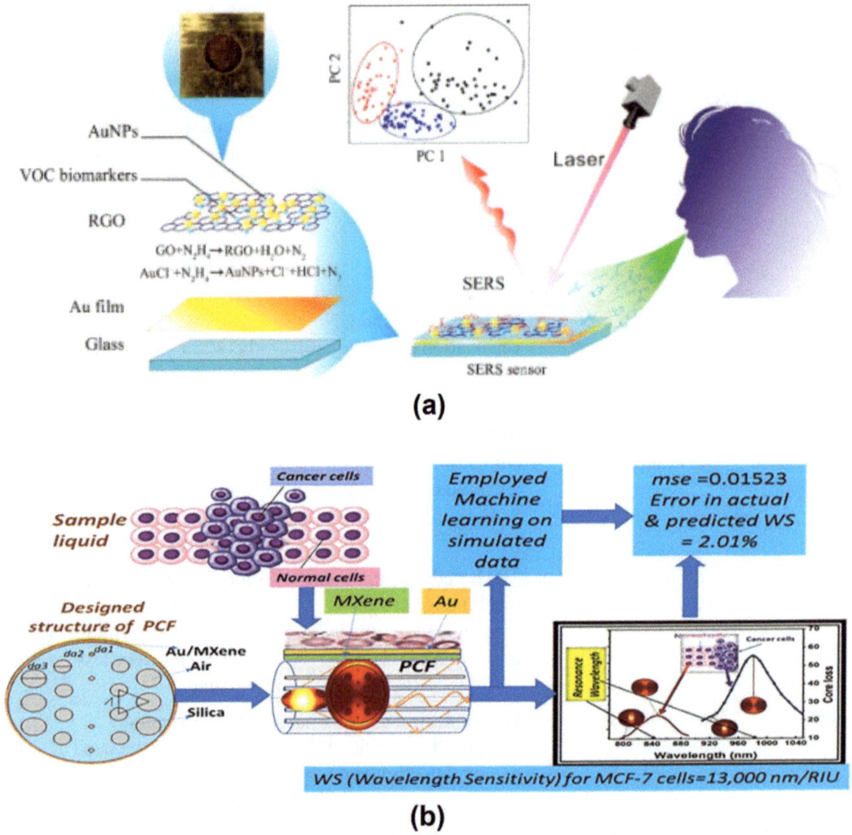

Fig. 11 Advances in AI enabled biosensors. **a** Surface enhanced Raman scattering based biosensor for diagnosis of gastric cancer using non-invasive method by analysing breath [111]. **b** 2D material based photonic crystal fiber for predicting cancer using machine learning approach [112]

be impacted by inconsistencies introduced by differences in data collection techniques, patient variability, and environmental circumstances. Another major challenge is the computational complexity of AI-based photonic devices. Large volumes of data are produced by these sensors, necessitating sophisticated hardware and algorithms for real-time processing, which raises expenses and power consumption. The critical challenge is the integration of AI-based photonic sensors with existing healthcare infrastructure. Due to compatibility problems and strict regulatory requirements, seamless interoperability with medical equipment and electronic health records (EHRs) is crucial but sometimes challenging. Furthermore, two significant issues with AI-integrated photonic sensors are still their high cost and scalability. The development of these systems necessitates sophisticated AI frameworks and specific materials, which makes cost and large-scale manufacture difficult. Additionally, ethical and regulatory issues provide obstacles to clinical implementation. Strict regulations must be followed by AI-driven medical technology to guarantee

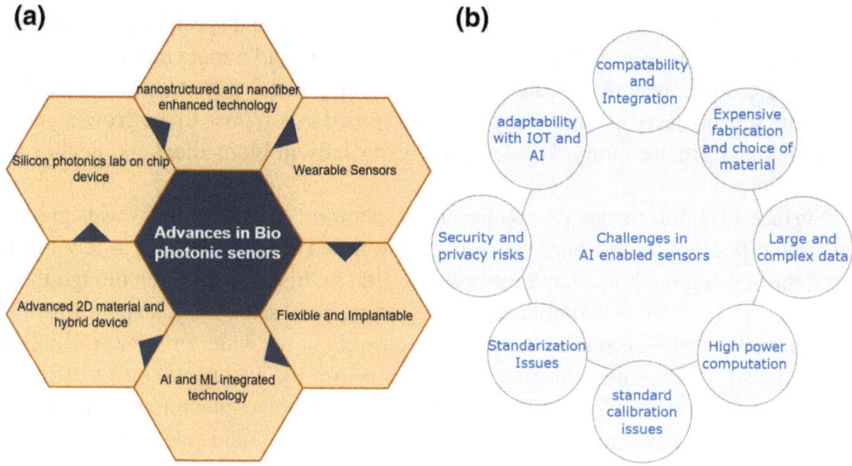

Fig. 12 **a** Advancement and **b** challenges in AI enabled photonic sensors

patient safety, data privacy, and decision-making openness. Gaining clinical trust also requires addressing the problem of bias in AI algorithms.

AI-based photonic sensing has promising future directions. The sensitivity, compactness, and accuracy of medical diagnostics will be improved by developments in nanophotonic and biophotonic sensor technologies. By integrating edge AI for real-time data processing, cloud computing reliance may be decreased, allowing for quicker and more effective diagnoses in wearable and point-of-care applications. The integration of photonic sensors with other sensing technologies such as electrochemical and acoustic sensors to enhance diagnostic accuracy is another fascinating approach that combines multi modal sensing with AI driven data fusion. Future studies should concentrate on low-cost production methods like 3D printing and flexible photonic materials to improve accessibility and affordability. Standardization and broad use of these technologies will be made possible by promoting cooperative research and open-source AI models, which will further speed innovation. All things considered, even if there are still difficulties with AI-based photonic sensing, ongoing developments in AI, photonics, and healthcare integration will open the door for next-generation intelligent sensing systems that will transform medical diagnosis and care.

7 Conclusion and Future Perspectives

The use of photonic sensing devices is quickly expanding in several domains, such as industry, agriculture, early disease detection, healthcare, medicine, and environmental monitoring [114]. They are very useful for high-precision, real-time sensing because they can identify changes in optical characteristics brought on by certain physical, chemical, or biological factors. High sensitivity and specificity, quick reaction times, non-invasive and non-destructive operation, and remote monitoring capabilities are some benefits of these sensors. Miniaturized, portable, and wearable photonic sensor-based systems have been developed in response to the growing need for quick and precise biomarker detection, especially in biomedical and healthcare applications.

Surface plasmon resonance, optical fiber, photonic crystal, optical waveguides, integrated photonics, and microwave photonic-based sensors are only a few of the photonic sensing methods that have been covered in this chapter. Even though these technologies have many benefits, issues including cost, scalability, standardization, and interaction with current systems must be resolved for wider use. By facilitating very precise, quick, and real-time data monitoring, the incorporation of artificial intelligence (AI) is transforming photonic sensing. AI-powered photonic sensors use deep learning and machine learning algorithms to identify and categorize recurring data patterns, enabling the creation of predictive models for early illness diagnosis. However, with AI-based photonic sensing systems, privacy and data security continue to be major issues.

The future of intelligent photonic sensors seems bright despite these obstacles. Their capabilities will be further enhanced by developments in biophotonic sensors, nanophotonics, and AI-driven data analytics, making them more effective and available for a range of applications [115]. AI-integrated photonic sensors have the potential to completely transform industrial automation, environmental monitoring, and medical diagnostics as research into these technologies' advances. These sensors will be essential in determining the direction of healthcare and beyond by resolving present issues and enhancing real-world integration.

References

1. P. Kakria, N.K. Tripathi, P. Kitipawang, A real-time health monitoring system for remote cardiac patients using smartphone and wearable sensors. Int. J. Telemed. Appl. **2015**, 373474 (2015). https://doi.org/10.1155/2015/373474. Epub 15 Dec 2015. PMID: 26788055; PMCID: PMC4692989
2. N. Bhalla, P. Jolly, N. Formisano, P. Estrela, Introduction to biosensors. Essays Biochem. **60**(1), 1–8 (2016). https://doi.org/10.1042/EBC20150001. PMID: 27365030; PMCID: PMC4986445
3. M. Mal, Noninvasive metabolic profiling for painless diagnosis of human diseases and disorders. Future Sci. OA **2**(2), FSO106. https://doi.org/10.4155/fsoa-2015-0014. PMID: 28031956; PMCID: PMC5137983
4. E.D. Evans, C. Duvallet, N.D. Chu et al., Predicting human health from biofluid-based metabolomics using machine learning. Sci. Rep. **10**, 17635 (2020). https://doi.org/10.1038/s41598-020-74823-1
5. Z. Wu, Z. Qiao, S. Chen et al., Interstitial fluid-based wearable biosensors for minimally invasive healthcare and biomedical applications. Commun. Mater. **5**, 33 (2024). https://doi.org/10.1038/s43246-024-00468-6
6. B. Katey, I. Voiculescu, A.N. Penkova, A. Untaroiu, A review of biosensors and their applications. ASME Open J. Eng. **2**, 020201 (2023). https://doi.org/10.1115/1.4063500

7. V. Naresh, N. Lee, A review on biosensors and recent development of nanostructured materials-enabled biosensors. Sensors **21**(4), 1109 (2021). https://doi.org/10.3390/s21041109. PMID: 33562639; PMCID: PMC7915135
8. C. Wang, T. He, H. Zhou et al., Artificial intelligence enhanced sensors—enabling technologies to next-generation healthcare and biomedical platform. Bioelectron. Med. **9**, 17 (2023). https://doi.org/10.1186/s42234-023-00118-1
9. M. Chen et al., Artificial intelligence-based medical sensors for healthcare system. Adv. Sens. Res. **3**(3) (2024). https://doi.org/10.1002/adsr.202300009
10. T. Wasilewski, W. Kamysz, J. Gębicki, AI-assisted detection of biomarkers by sensors and biosensors for early diagnosis and monitoring. Biosensors **14**(7), 356 (2024). https://doi.org/10.3390/bios14070356. PMID: 39056632; PMCID: PMC11274923
11. M. Javaid, A. Haleem, R.P. Singh, R. Suman, S. Rab, Significance of machine learning in healthcare: features, pillars and applications. Int. J. Intell. Netw. **3**, 58–73 (2022). ISSN 2666-6030. https://doi.org/10.1016/j.ijin.2022.05.002
12. G. Backiyalakshmi, U. Snekhalatha, A.L. Salvador, Recent advancements in non-invasive wearable electrochemical biosensors for biomarker analysis—a review. Anal. Biochem. **692**, 115578 (2024). ISSN 0003-2697. https://doi.org/10.1016/j.ab.2024.115578
13. S. Ye, S. Feng, L. Huang, S. Bian, Recent progress in wearable biosensors: from healthcare monitoring to sports analytics. Biosensors **10**(12), 205 (2020). https://doi.org/10.3390/bios10120205
14. M.K. Sahu, R. Yadav, S.P. Tiwari, Recent advances in nanotechnology. Int. J. Nanomater. Nanotechnol. Nnomed. **9**(2), 015–023 (2023). https://doi.org/10.17352/2455-3492.000053
15. H. Altug, S.H. Oh, S.A. Maier et al., Advances and applications of nanophotonic biosensors. Nat. Nanotechnol. **17**, 5–16 (2022). https://doi.org/10.1038/s41565-021-01045-5
16. J. Kim, W.R. de Araujo, I.A. Samek, A.J. Bandodkar, W. Jia, B. Brunetti, T.R.L.C. Paixão, J. Wang, Wearable temporary tattoo sensor for real-time trace metal monitoring in human sweat. Electrochem. Commun. **51**, 41–45 (2015). ISSN 1388-2481. https://doi.org/10.1016/j.elecom.2014.11.024
17. A.J. Bandodkar et al., Tattoo-based noninvasive glucose monitoring: a proof-of-concept study. Anal. Chem. **87**(1), 394–398 (2015). https://doi.org/10.1021/ac504300n
18. S. Emaminejad et al., Autonomous sweat extraction and analysis applied to cystic fibrosis and glucose monitoring using a fully integrated wearable platform. Biol. Sci. **114**(18), 4625–4630 (2017). https://doi.org/10.1073/pnas.1701740114
19. J. Qin et al., Photonic hydrogel sensing system for wearable and noninvasive cortisol monitoring. ACS Appl. Polym. Mater. **5**(9) (2023). https://doi.org/10.1021/acsapm.3c01119
20. Z. Liu et al., Explainable deep-learning-assisted sweat assessment via a programmable colorimetric chip. Anal. Chem. **94**, 15864–15872 (2022)
21. Y. Han et al., Sweat sensor based on wearable Janus textiles for sweat collection and microstructured optical fiber for surface-enhanced Raman scattering analysis. ACS Sens. **8**(12) (2023)
22. P. Ray et al., Label-free optical detection of multiple biomarkers in sweat, plasma, urine, and saliva. ACS Sens. **4**(5) (2019)
23. N. Sultana et al., Optical detection of pH changes in artificial sweat using near-infrared fluorescent nanomaterials. Sens. Diagn. **1**, 1189–1197 (2022)
24. X. Yue et al., Flexible wearable optical sensor based on optical microfiber Bragg grating. J. Lightwave Technol. **41**(6), 1858–1864 (2023). https://doi.org/10.1109/JLT.2022.3227186
25. M.A. Butt et al., A highly sensitive design of subwavelength grating double-slot waveguide microring resonator. Laser Phys. Lett. **17**, 076201 (2020). https://doi.org/10.1088/1612-202X/ab8faa
26. C. Chen et al., Protein analysis by Mach–Zehnder interferometers with a hybrid plasmonic waveguide with nano-slots. Opt. Express **25**(25), 31294 (2017)
27. X. Jiang et al., High-sensitivity optical biosensor based on cascaded Mach–Zehnder interferometer and ring resonator using Vernier effect. Opt. Lett. **39**(22), 6363–6366 (2014)

28. P. Ramirez-Priego et al., Label-free and real-time detection of tuberculosis in human urine samples using a nanophotonic point-of-care platform. ACS Sens. **3**(10), 2079–2086 (2018)
29. R. Marino, S. Quintero, A. Otero, J.M. Lanza-Gutierrez, M. Holgado, A machine learning-based methodology for in-process fluid characterization with photonic sensors. IEEE Sens. J. **21**(22), 26059–26073 (2021). https://doi.org/10.1109/JSEN.2021.3118490
30. Y. Fang et al., Ambulatory cardiovascular monitoring via a machine-learning-assisted textile triboelectric sensor. Adv. Mater. **33**(41), 2104178 (2021)
31. S. Sharma, L. Tharani, Use of AI techniques on photonic crystal sensing for the detection of tumor. J. Electron. Electromed. Eng. Med. Inform. **4**(2), 62–69 (2022)
32. L. Avellar, C. Stefano Filho, G. Delgado et al., AI-enabled photonic smart garment for movement analysis. Sci. Rep. **12**, 4067 (2022). https://doi.org/10.1038/s41598-022-08048-9
33. K. Amsalu, S. Palani, A review on photonics and its applications. Mater. Today Proc. **33**(Part 7), 3372–3377 (2020). ISSN 2214-7853. https://doi.org/10.1016/j.matpr.2020.05.184
34. M.A. Butt, N.L. Kazanskiy, S.N. Khonina, G.S. Voronkov, E.P. Grakhova, R.V. Kutluyarov, A review on photonic sensing technologies: status and outlook. Biosensors **13**(5), 568 (2023). https://doi.org/10.3390/bios13050568
35. M. Mancuso, J.M. Goddard, D. Erickson, Nanoporous polymer ring resonators for biosensing. Opt. Express **20**, 245–255 (2012)
36. Q. Ouyang, S. Zeng, L. Jiang, L. Hong, G. Xu, X.Q. Dinh, J. Qian, S. He, J. Qu, P. Coquet, K.T. Yong, Sensitivity enhancement of transition metal dichalcogenides/silicon nanostructure-based surface plasmon resonance biosensor. Sci. Rep. **16**(6), 28190 (2016). https://doi.org/10.1038/srep28190
37. Y. Khan, M.A. Butt, N.L. Kazanskiy, S.N. Khonina, Numerical study of fabrication-related effects of the structural-profile on the performance of a dielectric photonic crystal-based fluid sensor. Materials **15**, 3277 (2022). https://doi.org/10.3390/ma15093277
38. F. Esposito, Chemical sensors based on long period fiber gratings: a review. Results Opt. **5**, 100196 (2021). https://doi.org/10.1016/j.rio.2021.100196
39. R. Arunkumar, T. Suaganya, S. Robinson, Design and analysis of 2D photonic crystal based biosensor to detect different blood components. Photonic Sens. **9**, 69–77 (2019). https://doi.org/10.1007/s13320-018-0479-8
40. S. Robinson, N. Dhanlaksmi, Photonic crystal based biosensor for the detection of glucose concentration in urine. Photonic Sens. **7**, 11–19 (2017). https://doi.org/10.1007/s13320-016-0347-3
41. A. Sinibaldi, Cancer biomarker detection with photonic crystals-based biosensors: an overview. J. Lightwave Technol. **39**(12), 3871–3881 (2021). https://doi.org/10.1109/JLT.2021.3056225
42. M. Lee, P.M. Fauchet, Two-dimensional silicon photonic crystal based biosensing platform for protein detection. Opt. Express **15**, 4530–4535 (2007)
43. A. Asuvaran, G. Elatharasan, Design of two-dimensional photonic crystal-based biosensor for abnormal tissue analysis. SILICON **14**, 7203–7210 (2022). https://doi.org/10.1007/s12633-021-01442-4
44. F. Khozeymeh, F. Melli, S. Capodaglio, R. Corradini, F. Benabid, L. Vincetti, A. Cucinotta, Hollow-core fiber-based biosensor: a platform for lab-in-fiber optical biosensors for DNA detection. Sensors **22**(14), 5144 (2022). https://doi.org/10.3390/s22145144
45. S. Ko, S.A. Grant, A novel FRET-based optical fiber biosensor for rapid detection of *Salmonella typhimurium*. Biosens. Bioelectron. **21**(7), 1283–1290 (2006). ISSN 0956-5663. https://doi.org/10.1016/j.bios.2005.05.017
46. P. Hua, Z. Ding, K. Liu, H. Guo, M. Pan, T. Zhang, S. Li, J. Jiang, T. Liu, Distributed optical fiber biosensor based on optical frequency domain reflectometry. Biosens. Bioelectron. **228**, 115184 (2023). ISSN 0956-5663. https://doi.org/10.1016/j.bios.2023.115184
47. B. Krishnamoorthi, B. Elizabeth Caroline, M. Michael et al., A novel rhombic shaped photonic crystal bio-sensor for identifying disorders in the blood samples. Opt. Quant. Electron. **55**, 312 (2023). https://doi.org/10.1007/s11082-023-04584-4

48. H. Sharifi, S. Eskandari et al., Sensing blood components and cancer cells with photonic crystal resonator biosensor. Results Opt. **14**, 100593 (2024). https://doi.org/10.1016/j.rio.2023.100593
49. N.A. Mohammed, O.E. Khedr, E.S.M. El-Rabaie et al., Early detection of brain cancers biomedical sensor with low losses and high sensitivity in the terahertz regime based on photonic crystal fiber technology. Opt. Quant. Electron. **55**, 230 (2023). https://doi.org/10.1007/s11082-022-04515-9
50. A. Densmore et al., A silicon-on-insulator photonic wire based evanescent field sensor. IEEE Photonics Technol. Lett. **18**(23), 2520–2522 (2006). https://doi.org/10.1109/LPT.2006.887374
51. H. Pourhassan, S.O. Abdol, S.S. Nahaei, M.A. Bavil, Theoretical study of biophotonics sensor based on one-dimensional photonics crystal for the detection of dissolved oxygen in haemoglobin. Opt. Eng. **63**(6), 067101 (2024). https://doi.org/10.1117/1.OE.63.6.067101
52. D. Ignatyeva, G. Knyazev, P. Kapralov et al., Magneto-optical plasmonic heterostructure with ultranarrow resonance for sensing applications. Sci. Rep. **6**, 28077 (2016). https://doi.org/10.1038/srep28077
53. S. Khani, M. Hayati, Optical biosensors using plasmonic and photonic crystal band-gap structures for the detection of basal cell cancer. Sci. Rep. **12**, 5246 (2022). https://doi.org/10.1038/s41598-022-09213-w
54. F. Liu, W. Zhang, X. Lang, X. Liu, R. Singh, G. Li, Y. Xie, B. Zhang, S. Kumar, Development of taper-in-taper-based optical fiber sensors for chemical and biological sensing. Photonics **10**, 567 (2023). https://doi.org/10.3390/photonics10050567
55. H. Shafiee, E. Lidstone, M. Jahangir et al., Nanostructured optical photonic crystal biosensor for HIV viral load measurement. Sci. Rep. **4**, 4116 (2014). https://doi.org/10.1038/srep04116
56. N. Jahan et al., Photonic crystal fiber based biosensor for Pseudomonas bacteria detection: a simulation study. IEEE Access **9**, 42206–42215 (2021). https://doi.org/10.1109/ACCESS.2021.3063691
57. A. Gupta, T. Singh, R.K. Singh et al., Numerical analysis of coronavirus detection using photonic crystal fibre-based SPR sensor. Plasmonics **18**, 577–585 (2023). https://doi.org/10.1007/s11468-022-01761-1
58. G. Ramírez-García, L. Wang, A.K. Yetisen, E. Morales-Narváez, Photonic solutions for challenges in sensing. ACS Omega **9**(24), 25415–25420 (2024). https://doi.org/10.1021/acsomega.4c01953
59. F. Cui et al., Advancing biosensors with machine learning. ACS Sens. **5**(11), 3346–3364 (2020). https://doi.org/10.1021/acssensors.0c01424
60. M. Ramalingam, A. Jaisankar, L. Cheng et al., Impact of nanotechnology on conventional and artificial intelligence-based biosensing strategies for the detection of viruses. Discov. Nano **18**, 58 (2023). https://doi.org/10.1186/s11671-023-03842-4
61. K. Yadav, S. Bidnyk, A. Balakrishnan, Artificial intelligence and machine learning in optics: tutorial. J. Opt. Soc. Am. B **41**, 1739–1753 (2024)
62. T.M. Rawson, B. Hernandez, L.S.P. Moore, O. Blandy, P. Herrero, M. Gilchrist, A. Gordon, C. Toumazou, S. Sriskandan, P. Georgiou, A.H. Holmes, Supervised machine learning for the prediction of infection on admission to hospital: a prospective observational cohort study. J. Antimicrob. Chemother.Antimicrob. Chemother. **74**(4), 1108–1115 (2019). https://doi.org/10.1093/jac/dky514
63. A.H. Agbaria, A. Salman, G. Beck, I. Lapidot, D.H. Rich, J. Kapelushnik, M. Huleihel, S. Mordechai, Potential of bacterial infection diagnosis using infrared spectroscopy of WBC and machine learning algorithms, in *Clinical and Preclinical Optical Diagnostics II, Vol. EB101 of SPIE Proceedings*, paper 11073_40 (Optica Publishing Group, 2019)
64. C.S. Ho, N. Jean, C.A. Hogan et al., Rapid identification of pathogenic bacteria using Raman spectroscopy and deep learning. Nat. Commun. **10**, 4927 (2019). https://doi.org/10.1038/s41467-019-12898-9
65. S.A. Nehal, D. Roy, M. Devi et al., Highly sensitive lab-on-chip with deep learning AI for detection of bacteria in water. Int. J. Inf. Technol. **12**, 495–501 (2020). https://doi.org/10.1007/s41870-019-00363-1

66. C. Brown et al., Automated, cost-effective optical system for accelerated antimicrobial susceptibility testing (AST) using deep learning. ACS Photonics (2020). https://doi.org/10.48550/arXiv.2005.11454
67. J. Wang, B. Xu, L. Shi, L. Zhu, X. Wei, Prospects and challenges of AI and neural network algorithms in MEMS microcantilever biosensors. Processes **10**, 1658 (2022). https://doi.org/10.3390/pr10081658
68. D.L. Mancera-Zapata et al., AI-assisted real-time immunoassay improves clinical sensitivity and specificity. Anal. Chem. **96**(34) (2024). https://doi.org/10.1021/acs.analchem.4c00764
69. E.M. Materón et al., Recent advances in machine learning for electrochemical, optical, and gas sensors, in *Machine Learning for Advanced Functional Materials*, ed. by N. Joshi, V. Kushvaha, P. Madhushri (Springer, Singapore, 2023). https://doi.org/10.1007/978-981-99-0393-1_6
70. B.A. Taha et al., Synergizing nanomaterials and artificial intelligence in advanced optical biosensors for precision antimicrobial resistance diagnosis. ACS Synth. Biol. **13**(6), 1600–1620 (2024). https://doi.org/10.1021/acssynbio.4c00070
71. H. Shin et al., Early-stage lung cancer diagnosis by deep learning-based spectroscopic analysis of circulating exosomes. ACS Nano **14**(5), 5435–5444 (2020). https://doi.org/10.1021/acsnano.9b09119
72. Y.S. Dwivedi, R. Singh, A.K. Sharma, A.K. Sharma, Enhancing the performance of photonic sensor using machine-learning approach. IEEE Sens. J. **23**(3), 2320–2327 (2023). https://doi.org/10.1109/JSEN.2022.3225858
73. T.S. Hui, D.T.K. Xiang, H.-S. Loh, T. Maul, M.T.T. Tan, Graphene/TiO_2 nanocomposite based electrochemical biosensor enhanced by support vector machine classification model to detect different DENV serotype IgG, in *2023 IEEE Sensors*, Vienna, Austria (2023), pp. 1–4. https://doi.org/10.1109/SENSORS56945.2023.10325156
74. Mishra, D. Gupta, Comparative analysis of a bioelectric cell biosensor dataset employing machine learning classifiers for reliable listeria monocytogenes identification, in *2024 15th International Conference on Computing Communication and Networking Technologies (ICCCNT)*, Kamand, India (2024), pp. 1–6. https://doi.org/10.1109/ICCCNT61001.2024.10724850
75. A.N. Nithish et al., Terahertz women reproductive hormones sensor using photonic crystal fiber with behavior prediction using machine learning. IEEE Access **11**, 75424–75433 (2023). https://doi.org/10.1109/ACCESS.2023.3277955
76. N.R. Ramanujam et al., Numerical characterization of 1D-photonic crystal waveguide for female reproductive hormones sensing applications. Phys. B **639**, 414011 (2022). https://doi.org/10.1016/j.physb.2022.414011
77. J. Wekalao et al., Design and analysis of high-sensitivity hormone sensor with KNN behavior prediction for healthcare and biomedical applications. Measurement **242**(Part D), 116172 (2025). https://doi.org/10.1016/j.measurement.2024.116172
78. L. Arya, L. Singh, S. Yadav et al., Investigation of machine learning algorithms and plasmonic waveguide-based fano resonance sensor for diagnosis of estrogen. Plasmonics (2024). https://doi.org/10.1007/s11468-024-02680-z
79. J. Zhou et al., Artificial-intelligence-enhanced mid-infrared lab-on-a-chip for mixture spectroscopy analysis. Laser Photonics Rev. **19**(1), 2400754 (2025). https://doi.org/10.1002/lpor.202400754
80. J. Li et al., LSPR-susceptible metasurface platform for spectrometer-less and AI-empowered diagnostic biomolecule detection. Anal. Chim. Acta **1326**, 343094 (2024). https://doi.org/10.1016/j.aca.2024.343094
81. Q.M. Kamrunnahar, F. Haider, R.A. Aoni, J.R. Mou, S. Shifa, F. Begum, H.A. Abdul-Rashid, R. Ahmed, Plasmonic micro-channel assisted photonic crystal fiber based highly sensitive sensor for multi-analyte detection. Nanomaterials **12**(9), 1444 (2022). https://doi.org/10.3390/nano12091444
82. V.R. Balaji et al., Machine learning enabled 2D photonic crystal biosensor for early cancer detection. Measurement **224**, 113858 (2024). https://doi.org/10.1016/j.measurement.2023.113858

83. M. Jia et al., An aptamer-functionalized photonic crystal sensor for ultrasensitive and label-free detection of aflatoxin B1. Talanta **260**, 124638 (2023). https://doi.org/10.1016/j.talanta.2023.124638
84. R.K. Pathak, S. Mishra, P. Sharan, Design of optical sensor for cancer prognosis prediction using artificial intelligence. J. Opt. **53**, 1009–1017 (2024). https://doi.org/10.1007/s12596-023-01281-z
85. D. Pal, A. Kumar, N. Avraham, Y. Eisenbach, Y. Beiderman, S. Agdarov, Y. Beiderman, Z. Zalevsky, Noninvasive blood glucose sensing by secondary speckle pattern artificial intelligence analyses. J. Biomed. Opt. **28**(8), 087001 (2023). https://doi.org/10.1117/1.JBO.28.8.087001
86. R. Mahalakshmi et al., Optimization of optical biosensor based on 1D photonic crystals with metaheuristic algorithms for measuring glucose concentration. ECS J. Solid State Sci. Technol. **13**, 077012 (2024). https://doi.org/10.1149/2162-8777/ad6637
87. H. Tiwari, Y.S. Dwivedi, R. Singh, A.K. Sharma, A.K. Sharma, R. Krishna, N.S. Singha, Y.K. Prajapati, C. Marques, Deep learning-enabled de-noising of fiber Bragg grating-based glucose sensor: improving sensing accuracy of experimental data. Photonics **11**, 1058 (2024). https://doi.org/10.3390/photonics11111058
88. I. Ahmed et al., Photonic hydrogel for continuous glucose monitoring using smartphone readout. Mater. Des. **231**, 112065 (2023). https://doi.org/10.1016/j.matdes.2023.112065
89. M. Vijayan, S.S. Sridhar, D. Vijayalakshmi, A deep learning regression model for photonic crystal fiber sensor with XAI feature selection and analysis. IEEE Trans. NanoBiosci. **22**(3), 590–596 (2023). https://doi.org/10.1109/TNB.2022.3221104
90. M.G. Daher, Supersensitive biosensor based on a photonic crystal nanostructure for blood sugar level monitoring with ultra-high-quality factor and low detection limit. Optik **275**, 170581 (2023). https://doi.org/10.1016/j.ijleo.2023.170581
91. A. Aloraynan, S. Rassel, C. Xu, D. Ban, A single wavelength mid-infrared photoacoustic spectroscopy for noninvasive glucose detection using machine learning. Biosensors **12**(3), 166 (2022). https://doi.org/10.3390/bios12030166
92. J. Yi et al., AI-enabled biosensing for rapid pathogen detection: from liquid food to agricultural water. Water Res. Res. **242**, 120258 (2023). https://doi.org/10.1016/j.watres.2023.120258
93. D. Petrovszki et al., An integrated electro-optical biosensor system for rapid, low-cost detection of bacteria. Microelectron. Eng. **239–240**, 111523 (2021). https://doi.org/10.1016/j.mee.2021.111523
94. R. Viter et al., Toward development of optical biosensors based on photoluminescence of TiO_2 nanoparticles for the detection of Salmonella. Sens. Actuators B Chem. **252**, 95–102 (2017). https://doi.org/10.1016/j.snb.2017.05.139
95. Sonu, V. Chaudhary, A paradigm of internet-of-nano-things inspired intelligent plant pathogen-diagnostic biosensors. ECS Sens. Plus **1**, 031401 (2022). https://doi.org/10.1149/2754-2726/ac92ed
96. L. Jin et al., An aptamer-based SERS method for rapid screening and identification of pathogens assisted by machine learning technique with robustness evaluation. Sens. Actuators B Chem. **405**, 135356 (2024). https://doi.org/10.1016/j.snb.2024.135356
97. A. Ghosh, A. Pal, N.R. Das, An approach to design photonic crystal gas sensor using machine learning. Optik **208**, 163997 (2020). https://doi.org/10.1016/j.ijleo.2019.163997
98. S. Sharma, L. Tharani, Photonics for AI and AI for photonics integration: materials and characteristics. J. Inf. Optim. Sci. **45**(3), 805–816 (2024). ISSN 0252-2667 (print), ISSN 2169-0103 (online). https://doi.org/10.47974/JIOS-1579
99. L. Avellar, A. Frizera, H. Rocha, M. Silveira, C. Díaz, W. Blanc, C. Marques, A. Leal-Junior, Machine learning-based analysis of multiple simultaneous disturbances applied on a transmission-reflection analysis based distributed sensor using a nanoparticle-doped fiber. Photon. Res. **11**, 364–372 (2023)
100. Z. Li et al., Smart ring resonator-based sensor for multicomponent chemical analysis via machine learning. Photon. Res. **9**(2) (2020). https://doi.org/10.1364/PRJ.411825

101. C. Kalyoncu et al., Machine learning methods for estimating bent photonic crystal fiber based SPR sensor properties. Heliyon **8**(11), e11582 (2022). https://doi.org/10.1016/j.heliyon.2022.e11582
102. X. Liu et al., Artificial intelligence-enhanced waveguide "photonic nose"—augmented sensing platform for VOC gases in mid-infrared. Nano Micro Small **22**(34), 2400035 (2024). https://doi.org/10.1002/smll.202400035
103. S. Thakur et al., Deep transfer learning based photonics sensor for assessment of seed-quality. Comput. Electron. Agric. **196**, 106891 (2022). https://doi.org/10.1016/j.compag.2022.106891
104. K. Amit et al., Machine learning approach to surface plasmon resonance sensor based on MXene coated PCF for malaria disease detection in RBCs. Optik **274**, 170549 (2023). https://doi.org/10.1016/j.ijleo.2023.170549
105. N.P. Vishwaraj et al., Machine learning assisted cancer cell detection using strip waveguide Bragg gratings. Optik **284**, 170947 (2023). https://doi.org/10.1016/j.ijleo.2023.170947
106. X. Yi, X. Tian, L. Zhou, L. Li, L. Nguyen, R. Minasian, Integrated microresonator-based microwave photonic sensors assisted by machine learning. J. Lightwave Technol. **42**(12), 4271–4280 (2024). https://doi.org/10.1109/JLT.2024.3372015
107. Z.K.S. Mcotshana et al., Recent advances in the development of multiplexed nanophotonic biosensors. Photodiagn. Photodyn. Ther. **48**, 104246 (2024). https://doi.org/10.1016/j.pdpdt.2024.104246
108. A. González-Tudela, A. Reiserer, J.J. García-Ripoll et al., Light–matter interactions in quantum nanophotonic devices. Nat. Rev. Phys. **6**, 166–179 (2024). https://doi.org/10.1038/s42254-023-00681-1
109. L. Guo et al., Strategies for enhancing the sensitivity of plasmonic nanosensors. Nano Today **10**(2), 213–239 (2015). https://doi.org/10.1016/j.nantod.2015SS.02.007
110. Z. Song, S. Zhou, Y. Qin, X. Xia, Y. Sun, G. Han, T. Shu, L. Hu, Q. Zhang, Flexible and wearable biosensors for monitoring health conditions. Biosensors **13**, 630 (2023). https://doi.org/10.3390/bios13060630
111. Y. Chen et al., Breath analysis based on surface-enhanced Raman scattering sensors distinguishes early and advanced gastric cancer patients from healthy persons. ACS Nano **10**(9), 8169–8179 (2016). https://doi.org/10.1021/acsnano.6b01441
112. A. Kumar et al., Surface plasmon resonance sensor based on MXene coated PCF for detecting the cancer cells with machine learning approach. Microelectron. Eng. **267–268**, 111897 (2023). https://doi.org/10.1016/j.mee.2022.111897
113. H. Mazumdar et al., Artificial intelligence for personalized nanomedicine; from material selection to patient outcomes. Expert Opin. Drug Deliv. **22**(1), 85–108 (2024). https://doi.org/10.1080/17425247.2024.2440618
114. M. Shahbaz, M.A. Butt, R. Piramidowicz, A concise review of the progress in photonic sensing devices. Photonics **10**, 698 (2023). https://doi.org/10.3390/photonics10060698
115. A. Molani, F. Pennati, S. Ravazzani, A. Scarpellini, F.M. Storti, G. Vegetali, C. Paganelli, A. Aliverti, Advances in portable optical microscopy using cloud technologies and artificial intelligence for medical applications. Sensors **24**, 6682 (2024). https://doi.org/10.3390/s24206682

Generative AI in Multi-analytes Sensing

Kawsar Ahmed, Md. Mamun Ali, Nilanjana Basak, Francis M. Bui, and Santosh Kumar

Abstract Photonic Crystal Fiber (PCF) has demonstrated a prospective use in several areas of contemporary optics because of its superior transmission properties. But when it comes to building PCF architectures and describing pulse propagation in PCFs, traditional numerical solution techniques have drawbacks including high computing effort along with low efficiency. With the development of Artificial Intelligence (AI) technology, several intelligent computing approaches particularly machine learning (ML) algorithms and deep learning (DL) algorithms offer novel approaches to address the challenges. Numerous exciting applications have resulted from the design of multi-analyte sensor. With the growing need for an intelligent world, intelligent photonics has advanced quickly in recent years. The intelligent evolution of optical fiber sensors has been significantly aided by AI. Through the introduction of creative ways to problem-solving, its influence goes beyond improving sensor performance. This chapter's objective is to explain the properties that are acquired by integrating AI in multi-analyte sensing.

K. Ahmed (✉) · F. M. Bui
Department of Electrical and Computer Engineering, University of Saskatchewan, Saskatoon, SK, Canada
e-mail: k.ahmed@usask.ca

F. M. Bui
e-mail: francis.bui@usask.ca

K. Ahmed · N. Basak
Group of Bio-photomatiχ, Information and Communication Technology, Mawlana Bhashani Science and Technology University, Santosh, Tangail, Bangladesh
e-mail: nilanjanabasak96@gmail.com

Md. Mamun Ali
Division of Biomedical Engineering, University of Saskatchewan, Saskatoon, SK, Canada
e-mail: m.ali@usask.ca

S. Kumar
Department of Electronics and Communication Engineering, Centre of Excellence for Nanotechnology, Koneru Lakshmaiah Education Foundation, Vaddeswaram, Andhra Pradesh, India
e-mail: santosh@kluniversity.in

Y. Vasimalla and S. Kumar (eds.), *Generative AI for Photonic Sensing*, Progress in Optical Science and Photonics 36, https://doi.org/10.1007/978-981-95-1561-5_9

Keywords Artificial Intelligence · Biosensor · Deep learning · Machine learning · Multi-analyte sensing · Photonic crystal fiber

1 Introduction

The advent of muti-analyte sensors marks a significant milestone in the evolution of optical sensing, boasting extraordinary sensitivity and specificity for detecting chemical and biological substances. Plasmonic sensors mark a significant advancement in optical sensing technology due to their remarkable sensitivity and specificity in identifying chemical and biological components. PCF sensor devices are beneficial for constructing high rate, robust sensors for a variety of sensing applications because of their lightweight, compact architecture, and high signal-to-noise ratio. Additionally, when compared to conventional bulky sensor configuration approaches, their remote sensing capabilities and electromagnetic interference resistivity rank among the best alternative sensor devices [1]. The development of these sensors was prompted by the need for easier monitoring and confirmation of various natural conditions and elements. Additionally, sensors have been created to track other health-related parameters, including heart rate, sleep apnea, and glucose levels. Sensor development has advanced quickly in many areas of knowledge where movement, vibrations, and other factors are tracked. These sensors are now quicker, cheaper, and more suited to adjust surrounding requirements.

Nevertheless, new developments in software (like machine learning) and hardware (like optical fiber) have altered the design of these sensors. In order to decrease the complexity of the proposed structure and, consequently, computation time, machine learning techniques can generally be applied to optics and photonics systems. Moreover, crucial design elements like cladding geometry as well as material can be mapped using machine learning. Proper machine learning modeling can be used to prepare for the prediction of crucial propagation properties such confinement loss, sensitivity, birefringence, and dispersion.

For the purpose of identifying and tracking physical and sensory components, a variety of sensors have been created. The development of many kinds of sensors has led to advancements in the detection of temperature, movement, fire, and water, among other things. The developed systems use these sensors to detect irregularities in the input data they receive [2]. Features that increase operational precision have been implemented in sensor design [3]. To prevent a wide range of problems related to nature, health, architecture and roads, sensors must be designed, produced, and used in a variety of situations [4].

Since optical fiber has special characteristics and features that make signal manipulation easier, it is frequently employed in sensor design. As a result, faster, cleaner signals are produced by sensors that are more sensitive [5]. Sensor accuracy as well as efficiency can be increased by artificial intelligence approaches [6]. The goal of current research is to build more sensitive sensors and enhance their performance by utilizing a variety of machine learning approaches [7]. This makes it possible to classify and measure better samples [8]. Moreover, models analyzing the correlation among various optical spectrum parameters have been created using machine learning approaches [9–11].

2 Fundamentals of AI in Sensing Technology

Photonic sensors, such those found in spectroscopy, LiDAR, and optical communication, produce enormous volumes of data at a very high sampling rate. AI methods are needed for this in order to process the data effectively, extract relevant features, and make decisions in real time. By providing comprehensive data analytics capabilities that enhance sensor reliability and efficiency in complex, high-speed contexts, artificial intelligence (AI) greatly aids in overcoming these difficulties.

Whether due to sensor faults or external influences, photonic sensors are frequently susceptible to noise. The sensor data may be distorted by this noise, making it challenging to extract useful information. By using techniques like noise filtering, denoising autoencoders, or using effective machine learning models that are less susceptible to outliers, AI can reduce noise.

Thousands of features can be found in several photonic datasets, including hyperspectral images. These datasets are manageable without losing important information when neural networks are paired with dimensionality reduction techniques (e.g., Principal Component Analysis, t-SNE). Transmission and storage resources are strained by large data quantities. These procedures are optimized by AI algorithms, which eliminate duplication while maintaining significant patterns. To manage high-speed sensor inputs while preserving accuracy and low latency, requires for the models and the computing infrastructure to be optimized.

Machine learning techniques have become more and more popular in optical devices and fibres in recent years. However, due to their capacity to automatically fit to the problem, the majority of solutions focus on using Artificial Neural Networks (ANN) and other ML algorithms [3, 12–18]. Table 1 shows the surface plasmon resonance (SPR) based PCF biosensors where AI technology was applied to optimized or predict the biosensors.

3 AI Techniques for Photonic Sensor Optimization

The most common numerical simulation methods for PCF include the finite element method (FEM) [33, 34], beam propagation method [35, 36], finite difference time-domain method [37, 38], block-iterative frequency domain approaches [39], multipole method [40], and plane wave implementation approach [41]. FEM's high precision makes it a popular choice for numerical simulation of PCF optical characteristics. Computational modelling and performance optimization of PCF structures rely on time-consuming electromagnetic simulations carried out with the commercial FEM

Table 1 Brief description of SPR based PCF biosensors with application area, applied methods, input parameters and outcomes

References	Application area	Predicted parameters	Input parameters	Applied methods	Data set (training, testing)	Published year
Verma et al. [19]	PCF-SPR based biosensor (breast cancer cells detection)	Effective RI of core mode, SPP mode and CL	5 (RI, WL, pitch, gold thickness and TiO_2 thickness)	MLP regressor employed on ANN	207 (186, 21)	2022
Kumar et al. [20]	PCF-SPR based biosensor (malaria disease detection)	Effective RI of core mode and CL	5 (gold thickness, pitch, RI of the sample, WL and diameter)	ANN	556 (90%, 10%)	2022
Das et al. [21]	Prediction of PCF parameters	Real and imaginary (x-direction, y-direction) values of effective RI	CL, core radius, cladding radius, RI of analyte, and pitch	LR, DT, RF	NM	2022
Kumar et al. [22]	PCF-SPR based biosensor (cancer cells detection)	CL	5 (WL, pitch, RI, diameter, gold thickness)	MLP regressor employed on ANN	563 (556, 7)	2023
Chowdhury et al. [23]	PCF-SPR based biosensor (RI of 1.26–1.36)	CL, core effective and imaginary RI, core and total power, EML	6 (plasmonics materials: gold and TiO_2, RI of 2 analytes, WL, thickness)	ANN	5527 (4421, 9.06)	2023
Islam et al. [24]	PCF-SPR based biosensor (RI of 1.33–1.40)	CL	6 (RI of analyte, WL, real and imaginary components of the RI for both x and y axis)	ANN	NM	2023

(continued)

Table 1 (continued)

References	Application area	Predicted parameters	Input parameters	Applied methods	Data set (training, testing)	Published year
Ahmed et al. [25]	PCF based biosensor (RI of 1.33–1.35)	Effective indices: real and imaginary, core and cladding power profiles, effective area	5 (WL, 2 types of diameters, RI of analyte, pitch)	LSR, LASSO, ENet BRR	1402 (70%, 30%)	2023
Kaziz et al. [26]	PCF-SPR based biosensor (RI of 1.29–1.36)	Pick CL and WL, SNR	5 (3 types diameters, pitch, gold layer thickness)	PSO, MLP-ANN, genetic algorithm	NM	2024
Ehyaee et al. [27]	PCF-SPR based biosensor (RI of 1.31–1.4)	CL, sensitivity	4 (WL, RI of analyte, img. of RI, radius of gold ring)	ANN	1254 (885, 369)	2024
Al Mahfuz et al. [28]	PCF-SPR based biosensor (RI of 1.33–1.42)	RI	2 (WL, CL)	ANN	NM	2024
Guedri-Knani et al. [29]	PCF-SPR based biosensor (RI of 1.28–1.44)	CL	5 (3 types diameters, analyte and gold layer thickness)	MLR, MLP-ANN	NM	2024
Saha et al. [30]	PCF-SPR based biosensor (RI of 1.34–1.41)	CL	5 (RI of the analyte, WL, CL, real and imaginary parts of the RI)	MLR, SVR, and RF regression	1580 (NM)	2024

(continued)

Table 1 (continued)

References	Application area	Predicted parameters	Input parameters	Applied methods	Data set (training, testing)	Published year
Islam et al. [31]	PCF-SPR based biosensor (RI of 1.33–1.40)	CL	8 (RI of analyte, WL, pitch, 2 types diameters, open channels radius, and both real and imaginary components of the RI)	GAN, ANN	3240 (90%, 10%)	2024
Dogan et al. [32]	PCF SPR refractive index sensor	Sensitivity	4 (2 types diameters, pitch, gold layer thickness)	XGBoost, RF, NN	192 (174, 18)	2025

RI refractive index, *WL* wavelength, *CL* confinement loss, *SPP* surface plasmon, *EML* effective material loss, *MLP* multi layer perceptron, *LR* linear regression, *DT* decision tree, *RF* random forest, *LSR* least squares regression, *LASSO* Least Absolute Shrinkage and Selection Operator, *ENet* Elastic-Net, *BRR* Bayesian ridge regression, *SNR* signal to noise ratio, *PSO* Particle Swarm Optimization, *XGBoost* gradient boosting, *NM* not mentioned

package COMSOL Multiphysics. However, because PCFs require a lot of parameters tuning, using FEM can be computationally time-consuming. Machine learning is currently starting to address these problems and is the most effective method. Machine learning algorithms can become more capable of producing extremely precise predictions if they are trained on a balanced dataset.

The mechanical and optical characteristics of polymeric fibres were investigated by Omar et al. They analyzed digitally modified micro-interferograms using deep learning methods [42] and CNN [18]. With the use of ANN, the optical properties of PCF have been predicted. By employing ANN, Rodrguez-Esquerre et al. effectively created an indicator in 2013 that directed the chromatic dispersion of microstructured fibre [43]. In 2019 [44], Chugh et al. presented an ANN with three hidden layers, each with 50 neurons. This ANN predicts neff, effective mode area (Ae), dispersion (D), and αc all at once. 12 optical properties of a silica-based PCF were instantaneously determined by Jabin et al. in 2022 using artificial neural networks [45]. In regression applications, SVM may eventually replace ANN due to its better approximation capabilities and noticeably faster rate of convergence when using the sparse solution technique [46]. Moreover, GPR is a reliable and efficient learning technique for modelling complex and nonlinear functional mappings [47, 48]. The GPR model demonstrated comparatively low prediction errors, outperforming more robust nonlinear modelling techniques like the ANN and the LSSVM [49]. In addition to ANN, Abdul Kaium et al. [50] suggested using SVM, GPR, KNNR, and other machine learning models to calculate the optical properties of PCF, such as diameter and effective index. The data distribution properties of the models result in poor forecasts for loss. They suggested using logarithmic transformation techniques to alter the data distribution in order to resolve the problem.

There are three commonly used models for the application of optical fiber sensors: CNN, RNN, and ANN. ANNs are also known as feedforward NNs because the input only goes through unidirectional forward processing, which is another term multi-layer perception. The input, hidden, and output layer constituents of the ANN. The output layer is in charge of producing the result, the hidden layer is made up of one or more layers to process the input features in accordance with the requirements of the task, and the input layer receives the input and typically consists of one layer. Weight training is required for every layer. Each neuron's output is the total of multiplying the weight by the input value.

Loop connections are added to the hidden layer by RNNs, allowing them to extract sequential information from incoming data. Gated recurrent units (GRUs) better capture dependencies with longer time intervals in time series by managing the information flow through learnable gates. By employing storage units to efficiently filter redundant information, long short-term memory (LSTM) improves information extraction efficiency and circumvents the issue of disappearing or expanding gradients in RNNs.

Due to their ability to efficiently capture and identify features in images, CNNs are the most often used in image processing. CNN applies convolution operations on a sliding window over the input image in order to generate relevant features from that image, which is what convolutional layers are for.

4 Applications of AI-Integrated Photonic Sensors

Utilizing light-based technology, photonic sensors have become highly effective instruments in a variety of fields for the detection and measurement of physical, chemical, and biological phenomena. The incorporation of artificial intelligence (AI) is revolutionizing these sensors by offering improved features like predictive modeling, adaptable functioning, and real-time data processing. This chapter examines the various uses of AI-integrated photonic sensors, describing how they are affecting industries like smart infrastructure, industrial automation, healthcare, and environmental monitoring. Figure 1 represents the application of multi-analyte sensor integrated with AI.

Significant obstacles are presented by bacterial infections, biofilm formation, and drug resistance, highlighting the necessity of creative approaches to detection, monitoring, and treatment. Because multidrug-resistant bacterial strains are becoming more prevalent, conventional methods are frequently less successful and require more time and resources. By shielding bacteria from immune reactions and medications, biofilms increase complexity and cause persistent infections. By enabling

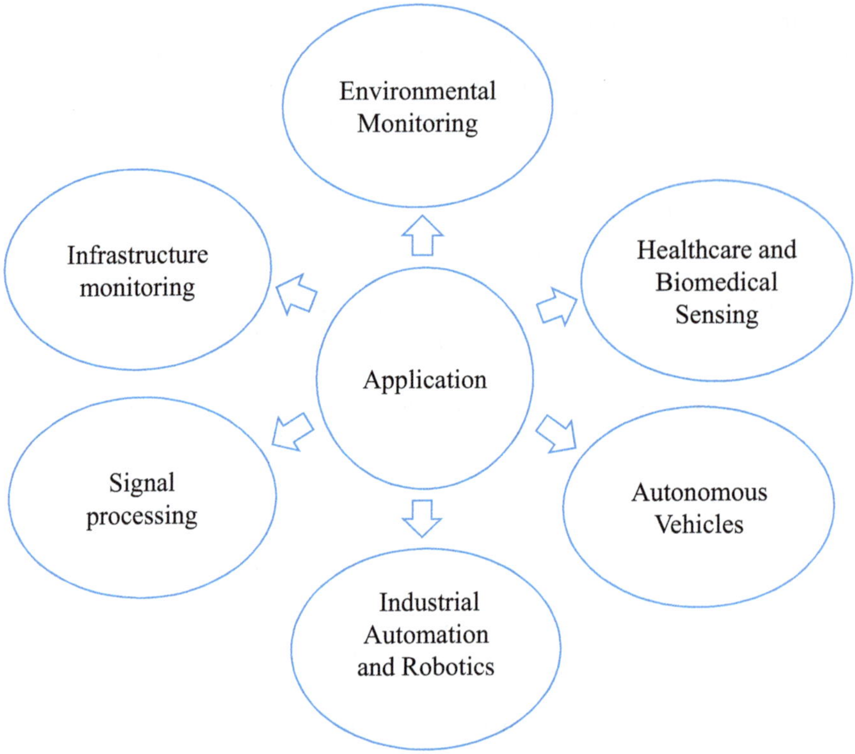

Fig. 1 Application of multi-analyte sensor integrated with AI

quick, sensitive, and precise pathogen recognition, advanced biosensor technologies offer possible solutions that enable prompt and focused therapy [51, 52]. To improve the accuracy and sensitivity of medical imaging, a PCF-based sensor has been developed that is combined with DL to improve tumour detection [53].

A new approach to sports medicine that uses optical sensors and quantum computers to evaluate players' health state was recently proposed. A quantum convolutional learning-adapted recurrent neural network (ResNet) (NResNetRNN_QCL) is used to interpret the data obtained from optical sensors [54]. This further work combines an optical sensor with a ML approach to evaluate musical signals in a health monitoring system. The work uses a Markov propagation theory that includes a reinforcement gradient vector to evaluate the collected data based on optical sensors (RGVMP) and quantum photonics to analyze musical signals [55].

5 Technical Implementation: Integrating AI with Photonic Sensors

After the PCF's collection of optical data from the samples, CNN and DL algorithms are applied. A PCF sensor's sensitivity is increased and DL's capacity to recognize intricate patterns in sensor data is increased when DL and CNNs are combined. AI, notably ML and NN is rapidly being used for improving photonics by providing novel capabilities in the structure, optimization, and deployment of photonic devices and systems [56]. AI systems can determine the best designs for certain applications by simulating different configurations and evaluating performance results. This results in fibers with customized properties for improved sensing performance. Technical Implementation of AI with multi-analyte sensor is illustrated in Fig. 2.

5.1 AI-Powered Data Processing and Analysis Integration

The incorporation of AI, specifically through NN and ML algorithms, allows for real-time evaluation of data gathered from the fiber's sensing potential. By spotting

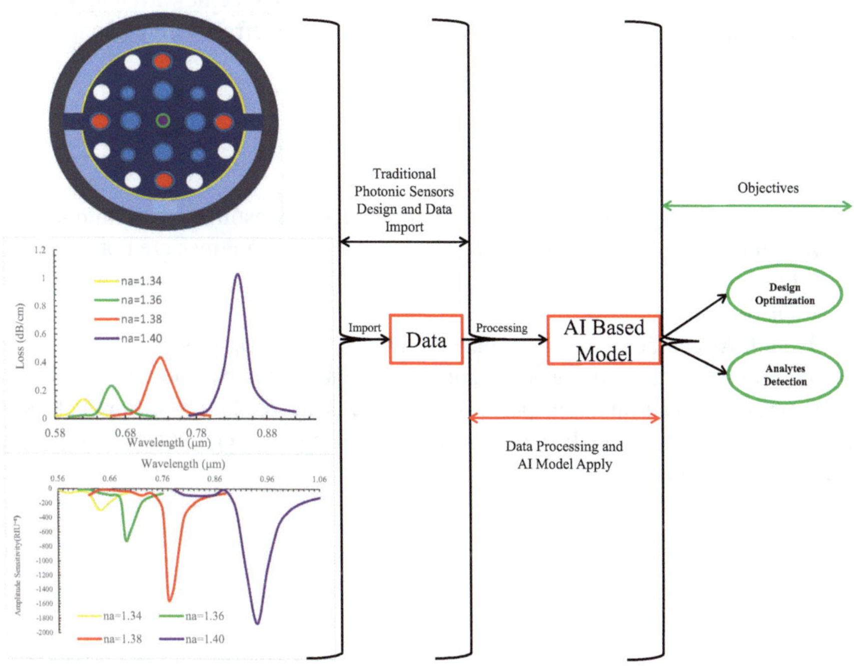

Fig. 2 Implementation of multi-analyte sensor with AI [57]

patterns or irregularities in the data, AI can provide rapid and precise detection of changes in the environment or the possible presence of particular compounds [58].

5.2 *Theoretical Background and Problem Formulation for Optimization of Design*

The loss in PCF sensors is mainly influenced by factors like confinement loss, bending loss, absorption by materials, and scattering effects. The overall loss in PCF can be represented as:

$$\mathcal{L}_{\text{total}} = \mathcal{L}_{\text{conf}} + \mathcal{L}_{\text{bend}} + \mathcal{L}_{\text{mat}} + \mathcal{L}_{\text{scat}}$$

where $\mathcal{L}_{\text{conf}}$ represents the confinement loss, which produced because of light leakage, $\mathcal{L}_{\text{bend}}$ represents the bending loss, which is induced by curvature effects, $\mathcal{L}_{\text{mat}}$ describes the material loss induced by absorption and dispersion, and the loss due to imperfection is called scattering loss and represented by $\mathcal{L}_{\text{scat}}$. Finally, $\mathcal{L}_{\text{total}}$ represents the total loss influenced by all types of loss in PCF sensors. In terms of PCF sensitivity performance, confinement loss variation plays a vital role, while other loss can be ignored because of low variation over the specific wavelength range. For confinement loss, the imaginary part of the effective refractive index n_{eff} is used to compute the loss:

$$\mathcal{L}_{\text{conf}} = \frac{40\pi}{\lambda \ln(10)} \text{Im}(n_{\text{eff}})$$

where λ represents the operating wavelength. Numerical estimation techniques like FEM estimate n_{eff} and imaginary part. However, these techniques are not computationally feasible. AI can predict the loss function $\mathcal{L}_{\text{total}}$ and $\mathcal{L}_{\text{conf}}$ for designing new PCF sensors.

Overall, AI can be used to optimize the PCF sensor design itself. AI algorithms can determine the best designs for particular applications by modelling different configurations and evaluating performance results by predicting different losses. This method results in PCF sensors with customized characteristics for improved sensing performance.

5.3 *Maintenance Prediction*

AI can evaluate data transmission to predict and prevent problems like fiber failure or signal deterioration, guaranteeing dependable and continuous operation. This fiber design sets the path for future advancements in photonics technology by integrating innovative photonic design with AI features, promising improved performance, intelligence, and versatility in its applications [59]. Figure 3 shows the applications of AI models in optimization or detection of anaytes of the sensors for different types of input data like simulated numerical data, optical images data or optical time series data etc.

6 Current Limitations and Solutions

There are several obstacles to integrating AI with PCF, including problems with data quality, design accuracy, and practical implementation. Explainable AI, transfer learning, sophisticated data preprocessing, and model optimization for scalable, reliable, and real-time applications are some of the solutions. These methods improve the efficacy and reliability of PCF-AI systems in a variety of fields.

Although ML-based spectrum demodulation does not require extra hardware, it is hindered by the lack of labelled data, which necessitates costly and skilled annotation. The shortage of labelled data not only adds to the complexity of models, but it also restricts their ability to generalize, making it harder for them to correctly recognize and categorize spectral patterns using small datasets. ML requires powerful resources like GPUs due to its high processing needs, caused by its reliance on big, labelled datasets. By keeping important information and boosting effectiveness, methods like dimensionality reduction and feature selection improve in computation optimization. It can be difficult for ML algorithms to recognize and deal with outliers, which are prevalent in spectra and can affect the precision of analysis. To solve these problems, preprocessing usually needs prior knowledge. A major obstacle for AI in spectroscopy is feature extraction from high-dimensional spectra, which necessitates efficiently handling substantial volumes of intricate data in order to perform precise analysis. Advanced features can be automatically learned from data using DL algorithms, doing away with the requirement for manual feature extraction and domain-specific knowledge. One important avenue for developing ML applications in fiber optic sensors is this self-learning capacity.

7 Future Research Directions

To enhance AI integration, future photonics and AI research should investigate cutting-edge technologies like quantum photonics and neuromorphic computing. Real-time data fusion, versatile AI models, and resolving ethical issues like transparency and privacy are important areas for development. Reducing the computational as well as energy requirements for broad adoption while preserving cost-effectiveness when scaling generative AI models for large-scale, multi-analyte sensing systems. AI-enhanced photonic sensing will improve more quickly if new researchers are encouraged to collaborate and work on open-source projects. The

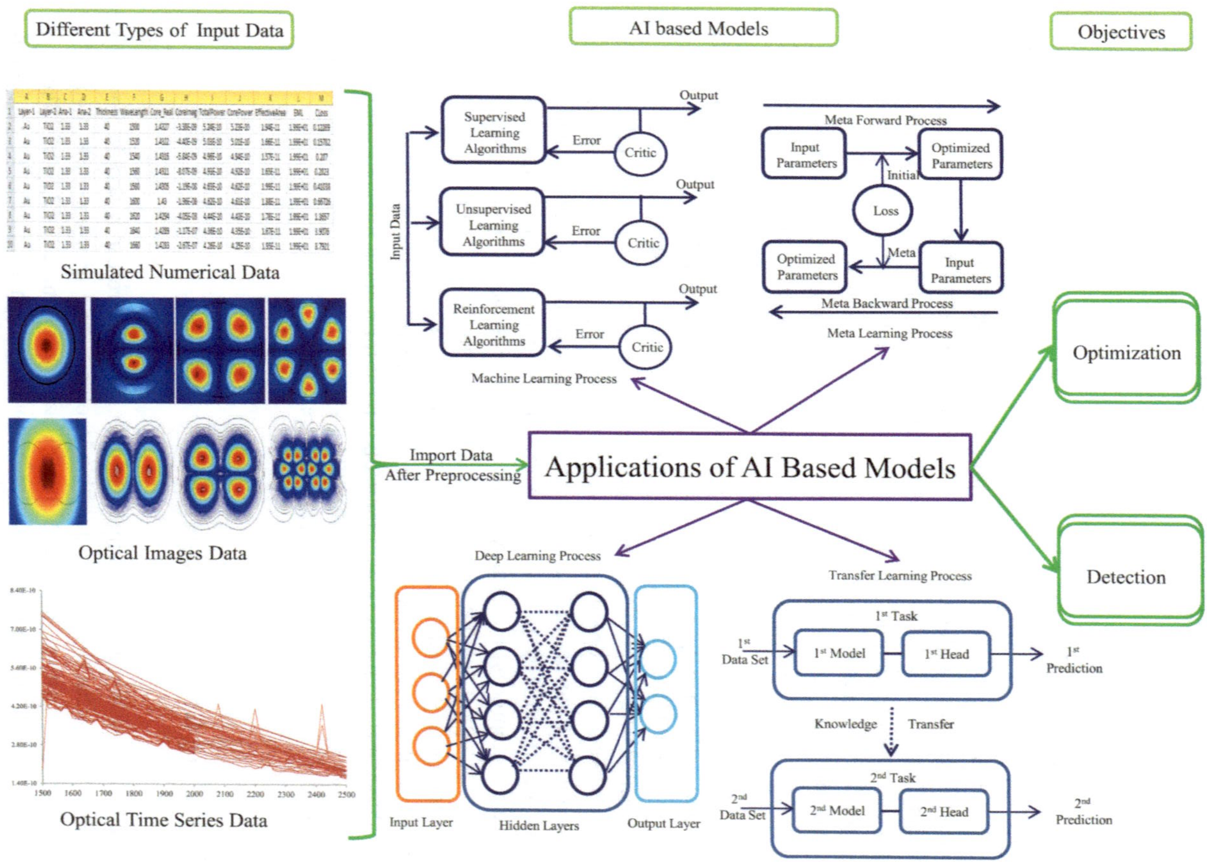

Fig. 3 Potential workflow diagram of optimization or detection of anaytes of the PCF sensors applying AI models for different types of input data like simulated numerical data, optical images data or optical time series data etc.

usefulness and dependability of AI in spectrum demodulation will be improved by developments in data collection, feature extraction, selecting models, and conceptual ML techniques. AI technology will soon reach its full potential through the integration of hybrid optical systems, opening the door to a smarter, greener future. Some of the potentials advanced AI techniques are mentioned in the following sections along with their application for future research.

7.1 Ensemble Machine Learning for Loss Prediction in PCF Sensors

Ensemble Machine Learning (EML) provides an effective solution by integrating several machine learning models to enhance predictive accuracy and generalization [60, 61]. Utilizing ensemble methods like Bagging, Boosting, and Stacking, EML can improve the resilience of PCF loss forecasting models. This method not only speeds up loss forecasting but also enhances dependability when compared to standalone models. Figure 4 depicts a potential framework for ensemble learning model developing and application for PCF sensor design and optimization through predicting the loss and other related parameters.

Bagging (Bootstrap Aggregating)

Bagging is a method that enhances the stability and precision of forecasts by training several models on various bootstrap samples from the dataset and averaging their results [62]. Let $\boldsymbol{X}$ is an input parameter vector, which has been found from PCF sensor design. Mathematically, given M base model and $f_m(\boldsymbol{X})$ is the prediction

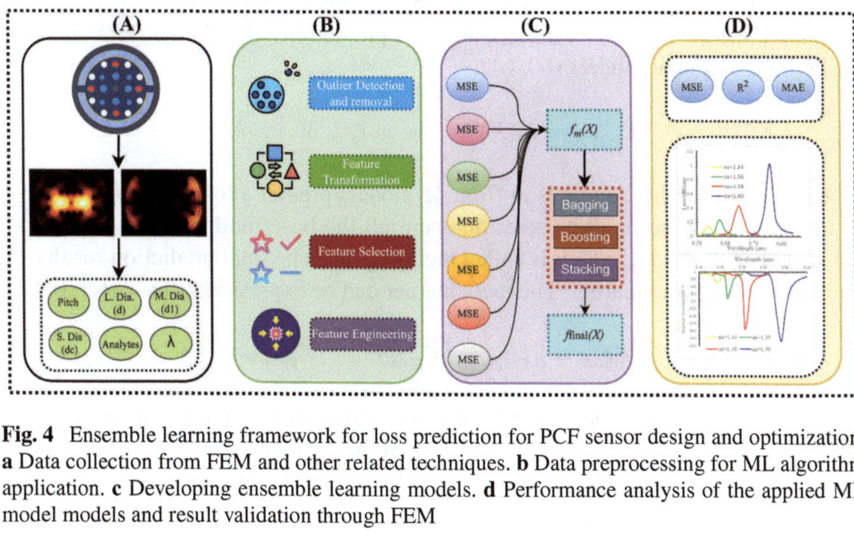

Fig. 4 Ensemble learning framework for loss prediction for PCF sensor design and optimization. **a** Data collection from FEM and other related techniques. **b** Data preprocessing for ML algorithm application. **c** Developing ensemble learning models. **d** Performance analysis of the applied ML model models and result validation through FEM

from each of the M models. The bagging technique for this data and models can be represented as follows [63]:

$$\mathcal{L}_{\text{pred}} = \frac{1}{M}\sum_{m=1}^{M} f_m(\boldsymbol{X}) \tag{1}$$

Random Forest is an example of bagging-based algorithm, which consists of multiple decision trees trained on different subsets of the data. However, a new types of ensemble model can be designed combining multiple ML algorithms, which are best fit for the PCF sensor optimization datasets. It will combine strengths of multiple algorithms and provide a better outcome for loss prediction.

Boosting

Boosting trains weak models in sequence, with each new model addressing the mistakes made by its predecessors [64]. The ultimate forecast is achieved by merging all weak learners. The mathematical representation for boosting is as follows [65]:

$$\mathcal{L}_{\text{pred}} = \frac{1}{M}\sum_{m=1}^{M} \alpha_m f_m(\boldsymbol{X}) \tag{2}$$

where α_m represents the weight of the mth model, which is calculated based on its accuracy. $f_m(X)$ represents each of the weak model's prediction result.

Some of the popular bosting algorithms are gradient boosting machine, Extreme gradient boosting, LightGBM, AdaBoost and so on.

Stacking Method

Stacking integrates several machine learning models by employing a meta-model to determine the best method for blending their predictions. The initial tier comprises several foundational models [60, 61]:

$$\mathcal{L}_{pred,i} = f_m(X), \qquad m = 1, 2, 3, \ldots, M \tag{3}$$

where each of the base models m from M models predict a loss value $f_m(X)$ for $\boldsymbol{X}$ input vectors. These prediction results from all the base models are used as input for another ML model, which is called meta-learner. The final prediction results are found from the meta-learner. The meta learner can be expressed as follows [60, 61]:

$$\mathcal{L}_{\text{final}} = g\left(\mathcal{L}_{\text{pred},1}, \mathcal{L}_{\text{pred},2}, \ldots, \mathcal{L}_{\text{pred},M}\right) \tag{4}$$

where g represents meta-learner, which is an individual ML algorithm. In this case, it will be regression or neural network algorithm for regression analysis. Stacking often outperforms individual models by leveraging their complementary strengths.

Ensemble learning offers a robust data-driven strategy to enhance and expedite PCF sensor design, lessening the dependence on costly numerical techniques.

7.2 Transfer Learning for Loss Prediction in PCF Sensors

Transfer learning is an approach in machine learning where a model created for a specific task is utilized as the foundation for a similar task [66]. This method is especially beneficial when labeled data is limited, which is frequently true in PCF sensor development. Utilizing pre-trained models, transfer learning can greatly decrease the time and resources needed for training [67]. Models that have been pre-trained, like convolutional neural networks (CNNs) and recurrent neural networks (RNNs), can be adjusted to predict loss values in PCF sensors. For instance, a CNN that has been pre-trained on a dataset related to surface plasmon resonance (SPR) sensors can be modified to predict confinement loss in PCF sensors by refining the model using a more limited dataset of PCF sensor setups. Domain adaptation is a crucial element of transfer learning, in which a model developed for one domain (e.g., SPR sensors) is modified to excel in a different yet related domain (e.g., PCF sensors). This is especially beneficial in the optimization of photonic sensors, as various sensor types have comparable foundational principles. A potential framework for loss prediction using transfer learning methods for PCF sensor design and optimization has been visualized on Fig. 5.

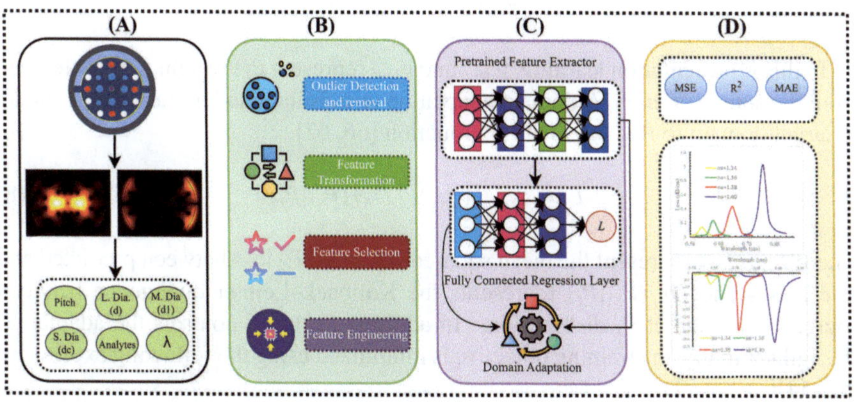

Fig. 5 Transfer learning framework for loss prediction for PCF sensor design and optimization. **a** Data collection from FEM and other related techniques. **b** Data preprocessing for transfer learning algorithm application. **c** Developing transfer learning models. **d** Performance analysis of the applied transfer learning model and result validation through FEM

7.2.1 Transfer Learning Framework for Loss Prediction

Transfer learning consists of training a deep learning model on an extensive dataset of established PCF losses and subsequently adjusting it for a new target dataset that has few labeled examples. The structure of the model includes [68]:

1. **Pre-trained Feature Extractor**: A convolutional neural network (CNN) or transformer model trained on a large dataset of PCF parameters and loss values.
2. **Fully Connected Regression Layers**: Custom layers that map the extracted features to predicted loss values.

Let's have PCF design parameter vector $\boldsymbol{X}$, the pre-trained model f_θ, parameterized by θ, learn a mapping as follows [66]:

$$f_\theta = \boldsymbol{X} \rightarrow \mathbf{Z} \tag{5}$$

where the latent representation of the PCF features is denoted by $\mathbf{Z}$. A new function for prediction loss denoted by g_ϕ is trained and it can be represented as follows [66]:

$$\mathcal{L}_{\text{pred}} = g_\phi(\mathbf{Z}) \tag{6}$$

where the fine-tuned parameters of this model are represented by ϕ. The feature representation of PCF sensors in transfer learning can be defined as following expression [67]:

$$Z = \psi(X) = W_{\text{feat}}X + b_{\text{feat}} \tag{7}$$

In this representation learning, w_{feat} and b_{feat} represent the weight metrics learned from the source domain. A domain adaptation term is included in training the model in target domain to find the final loss function [66, 67]:

$$\mathcal{L} = \mathcal{L}_{regression} + \lambda D_{KL}(P_s||P_t) \tag{8}$$

where $\mathcal{L}_{regression}$ represent the mean squared error (MSE) loss between predicted and true losses and $D_{KL}(P_s||P_t)$ represents the Kullback–Leibler divergence between source P_s and target P_t distributions. In addition to this, λ controls the adaptation strengths. Finally, the training objective is minimized using the following expression [52–54]:

$$\mathcal{L} = \sum_{i=1}^{N}\left(\mathcal{L}_{\text{true},i} - \mathcal{L}_{\text{pred},i}\right)^2 + \lambda D_{KL}(P_s||P_t) \tag{9}$$

where the true and predicted losses are denoted by $\mathcal{L}_{\text{true}}$ and $\mathcal{L}_{\text{pred}}$ respectively.

The transfer learning approach gains a high correlation between predicted and actual loss values, with a MSE of less than **0.01 dB/km**. Compared to traditional deep learning methods, which are trained from scratch, transfer learning significantly can be trained with a few amounts of training data while maintaining high accuracy. The trained model generalizes well to unseen PCF designs, demonstrating robustness in loss prediction.

7.3 Meta-Learning for Loss Prediction in PCF Sensors

Meta-learning is a subfield of ML. It is also called “learning to learn”, which aims to build ML models capable of adapting to new tasks with minimal amount of data [69]. Conventional ML methods require a high volume of data and computational resources. On the other hand, meta-learning does not depend on large amounts of data and computational resources [69]. Because it learns how to learn. As a result, it adapts to new tasks with a minimal amount of data using previous knowledge. This is especially beneficial in optimizing photonic crystal fiber (PCF) sensors, as gathering data can be both time-intensive and costly.

Application for PCF Sensors

In the context of PCF sensors, meta-learning can be employed to predict loss metrics such as confinement loss and effective material loss. Through exposure to diverse sensor setups and materials, meta-learning models are able to rapidly adjust to new sensor designs using only a small amount of new data. For instance, a meta-learning model developed using a dataset of PCF sensors featuring varied air-hole shapes and plasmonic materials can accurately forecast losses for a novel sensor design.

Meta-Learning Framework

Figure 6 depicts a potential meta-learning framework for PCF sensor design and optimization. A distribution of tasks is required to employ meta-learning in PCF sensors design optimization predicting different loss values related to design PCF sensors [70]. Meta-learning can be applied for various purposes like classification, regression, and reinforcement learning. Since predicting loss value is regression problem, meta-learning can be applied to solve regression problems. In this case, let we have $p(\mathcal{T})$ distributions of tasks, and a model denoted by f_θ parameterized by θ that map observations x to outputs a. During meta-learning, the model is trained to be able to adapt to a large or infinite number of tasks. A task $\mathcal{T}_i$ can be represented as follows [69]:

$$\mathcal{T}_i = \{\mathcal{L}(x_1, a_1, \ldots x_H, a_H), q(x_1), q(x_{t+1}|x_t, a_t), H\} \quad (10)$$

Formally, each of the tasks $\mathcal{T}_i$ consists of a loss function $\mathcal{L}$, a distribution over initial observation $q(x_1)$, a transition distribution $q(x_{t+1}|x_t, a_t)$, an episode length H (in i.i.d. supervised learning problems, the length $H = 1$).

Meta-learning is a bilevel optimization-based framework, which optimizes meta-learning algorithm in two levels such as outer level optimization (optimization across

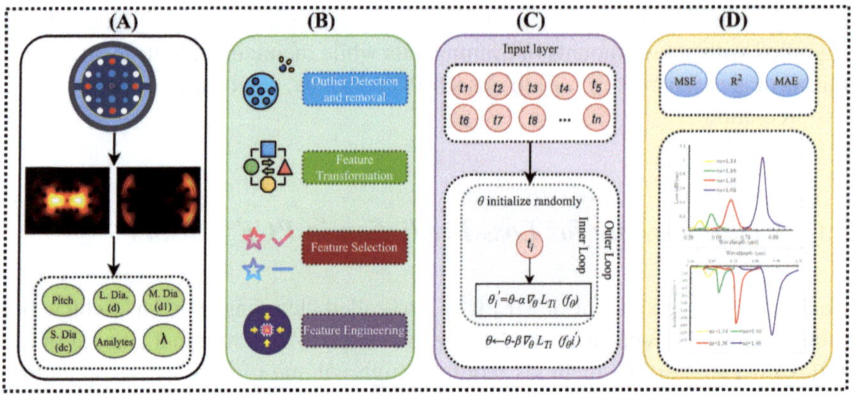

Fig. 6 Meta-learning framework for loss prediction for PCF sensor design and optimization. **a** Data collection from FEM and other related techniques. **b** Data preprocessing for meta-learning algorithm application. **c** Developing meta-learning models. **d** Performance analysis of the applied meta-learning model and result validation through FEM

all the tasks) and inner level optimization (optimization for each task). The model $f(\cdot)$ initialized by θ parameters will be optimized in inner level to get the optimal parameters θ_i' for each task $\mathcal{T}_i$ from $p(\mathcal{T})$ distribution over tasks using the following expression [58]:

$$\theta_i' = \theta - \alpha \nabla_\theta \mathcal{L}_{\mathcal{T}_i}(f_\theta) \tag{11}$$

where α is the step size of model in inner and $\mathcal{L}_{t_i}$ presents the loss value of the model f_θ by $\mathcal{L}_{\mathcal{T}_i}(f_\theta)$. Based on the loss value of each task, another loss value is calculated to update the initial parameters θ to train the model for the next unseen tasks. which will generalize all the tasks of the model. The meta-optimization across tasks is performed via stochastic gradient descent (SGD), such that the model parameter θ are updated as follows [71]:

$$\theta \leftarrow \theta - \beta \nabla_\theta \sum_{\mathcal{T}_i \sim p(\mathcal{T})} \mathcal{L}_{\mathcal{T}_i}\left(f_{\theta_i'}\right) \tag{12}$$

where β is step size of the outer level optimization. The process for meta-learning algorithms will be as follows [69, 71]:

Algorithm 1: Algorithm for meta-learning model training and updating the parameters

Step	Description
	Require: $p(\mathcal{T})$: distribution over tasks
	Require: α, β: step size hyperparameters

(continued)

(continued)

Step	Description		
1	Randomly initialize θ		
2	**While** not done **do**		
3		Sample batch of tasks $\mathcal{T}_i \sim p(\mathcal{T})$	
4		**for all** $\mathcal{T}_i$ **do**	
5			Evaluate $\nabla_\theta \mathcal{L}_{\mathcal{T}_i}(f_\theta)$ with respect to k samples
6			Compute adapted parameters with gradient descent: $\theta_i' = \theta - \alpha \nabla_\theta \mathcal{L}_{\mathcal{T}_i}(f_\theta)$
7		end for	
8		Update $\theta \leftarrow \theta - \beta \nabla_\theta \sum_{\mathcal{T}_i \sim p(\mathcal{T})} \mathcal{L}_{\mathcal{T}_i}\left(f_{\theta_i'}\right)$	
9	**end while**		

7.4 *Physics Informed Neural Network (PINN) for Loss Prediction*

Physics-Informed Neural Networks (PINNs) represent a type of neural network that integrates physical laws during their training, guaranteeing that the model adheres to essential governing equations [72]. In the realm of PCF sensors, predicting loss entails intricate physical interactions, such as confinement loss, bending loss, material absorption, and scattering effects. Conventional numerical techniques like the finite element method (FEM) and beam propagation method (BPM) yield precise forecasts but require significant computational resources. PINNs present a viable alternative by incorporating Maxwell's equations and loss constraints straight into the neural network, resulting in a model that is both data-efficient and physically coherent.

Physics-Informed Neural Network (PINN) Framework

An input vector $\boldsymbol{X}$ (such as fiber shape and material qualities) is mapped to an output $\mathcal{L}_{\text{total}}$ by a function f_θ that is learned by a typical neural network. To guarantee adherence to Maxwell's equations, we incorporate physical constraints into the loss function of a PINN rather than training it solely on data.

Governing Equations: Maxwell's Equations

Maxwell's equations are the main mathematical formula, which governs electromagnetic wave propagation in PCF. The formula is as follows [73]:

$$\nabla \times \mathbf{E} = -\frac{\partial \mathbf{B}}{\partial t} \tag{13}$$

$$\nabla \times \mathbf{H} = \mathbf{J} - \frac{\partial \mathbf{D}}{\partial t} \tag{14}$$

In these equations, the electric and magnetic field vectors are denoted **E** and **J**, and **B** and **D** represent the magnetic and electric flux densities. The current density has been denoted by **J**.

In the context of PCFs, the **Helmholtz equation** derived from Maxwell's equations controls the propagation of modes [74]:

$$\nabla^2 E + k^2 n^2 E = 0 \tag{15}$$

where $k = \frac{2\pi}{\lambda}$ represent the wave number and the refractive index profile is denoted by n.

The effective refractive index n_{eff} is found by solving this equation for given PCF structures.

Loss Function in PINN

In PINN, a hybrid loss function is defined, which includes data loss, physics loss, and boundary condition loss. Data ensures agreement with known data samples. Physics loss enforces Maxwell's and Helmholtz equations, where boundary condition loss confirms accurate field behavior at PCF interfaces. The total loss can be expressed as follows [75]:

$$\mathcal{L} = \mathcal{L}_{\text{data}} + \lambda_1 \mathcal{L}_{\text{physics}} + \lambda_2 \mathcal{L}_{\text{boundary}} \tag{16}$$

where $\mathcal{L}_{\text{data}}$ represent the MSE represents the true and predicted loss:

$$\mathcal{L}_{\text{data}} = \sum_{i=1}^{N} \left(\mathcal{L}_{\text{true},i} - \mathcal{L}_{\text{pred},i}\right)^2 \tag{17}$$

Besides, $\mathcal{L}_{\text{physics}}$ represents physics loss and it can be calculated using following expression:

$$\mathcal{L}_{\text{physics}} = \sum_{i=1}^{N} \left\| \nabla^2 E + k^2 n^2 E \right\|^2 \tag{18}$$

Boundary condition loss $\mathcal{L}_{\text{boundary}}$ ensures the field continuity at PCF core-cladding interfaces:

$$\mathcal{L}_{\text{boundary}} = \sum_{i=1}^{N} \left\| E_{\text{core}} - E_{\text{cladding}} \right\|^2 \tag{19}$$

In this case, the contribution of the physics and boundary constraints are controlled by λ_1 and λ_2.

PINN Architecture and Training Strategy

Figure 7 represents a potential framework for the application of the PINN model for PCF sensor design and optimization. The PINN consists of an input layer to input design parameters (e.g., air-hole diameter, pitch, core radius), hidden layer with activation function and an output layer to predict the loss value. Though this design is similar to traditional neural network model, a physics-informed residual block is included to enforce physical constraints at each layer in PINN.

During training, the designed model is first trained with loss values estimated using FEM simulations. Physics-Informed fine-tuning is applied to estimate the loss ensuring compliance with Maxwell's equations. A hybrid optimizer (Adam + L-BFGS) is applied to minimize both data and physics loss. The designed model is trained by minimizing loss by the following expression [72–75]:

$$\mathcal{L} = \sum_{i=1}^{N} \left(\mathcal{L}_{\text{true},i} - \mathcal{L}_{\text{pred},i}\right)^2 + \lambda_1 \sum_{i=1}^{N} \left\| \nabla^2 E + k^2 n^2 E \right\|^2 + \lambda_2 \sum_{i=1}^{N} \left\| E_{\text{core}} - E_{\text{cladding}} \right\|^2 \quad (20)$$

This above explanation of PINN demonstrates the effectiveness of PINNs for loss prediction in PCF sensors by integrating Maxwell's equations into neural network models. PINNs significantly reduce computational costs while maintaining high accuracy. By incorporating physics knowledge directly into machine learning, PINNs

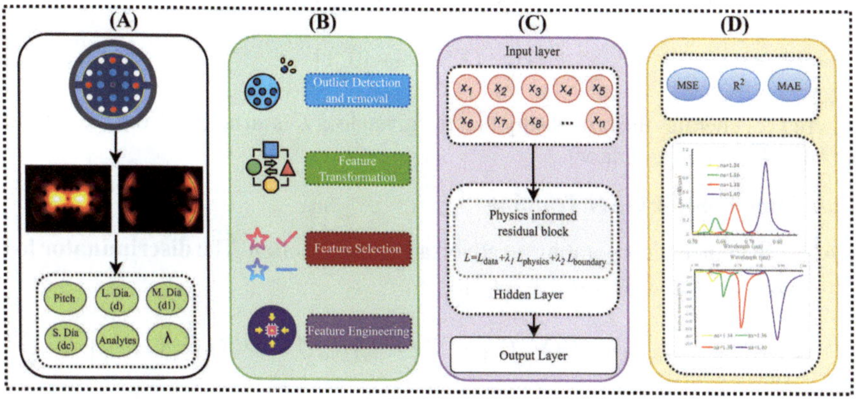

Fig. 7 Physics informed neural network framework for loss prediction for PCF sensor design and optimization. **a** Data collection from FEM and other related techniques. **b** Data preprocessing for PINN algorithm application. **c** Developing PINN models. **d** Performance analysis of the applied PINN model and result validation through FEM

provide a powerful tool for fast and accurate loss prediction in PCF sensors, paving the way for real-time optical system optimization.

7.5 Generative Adversarial Networks (GANs) for Loss Prediction

Generative Adversarial Networks (GANs), a generative model based on deep learning, offer a robust alternative by understanding the underlying loss distribution and producing realistic forecasts for new PCF designs [76]. GANs are made up of two rival neural networks: a generator that creates synthetic data and a discriminator that distinguishes between authentic and generated data [76, 77]. Through training GANs on a dataset containing PCF parameters and their associated loss values, the model is able to generalize unfamiliar PCF structures and efficiently forecast losses.

Generative Adversarial Network (GAN) Framework for Loss Prediction

Figure 8 mentions a potential GAN framework for PCF sensor design and optimization. A GAN has two different neural networks such as generator $G(\boldsymbol{X}, z)$, and discriminator $D(\boldsymbol{X}, \mathcal{L})$. Generator learns to make some synthetic loss values by conditioning on input PCF parameters $\boldsymbol{X}$ and discriminator differentiates between actual and generated loss values. The generator learns a function [76]:

$$G_\theta : (\boldsymbol{X}, z) \rightarrow \mathcal{L}_{\text{pred}} \tag{21}$$

where z represents a vector which is laten noise and $\mathcal{L}_{\text{pred}}$ represents the generated loss.

The discriminator learns a function as follows [76]:

$$D_\phi : (\boldsymbol{X}, \mathcal{L}) \rightarrow [0, 1] \tag{22}$$

where D_ϕ generates the probability that the given loss $\mathcal{L}$ is an actual (from the dataset) or fake (from the generator).

Loss Functions for GAN Training

GANs are trained incorporating an adversarial loss function. The **discriminator loss** can be represented as follows [62, 63]:

$$\mathcal{L}_D = \mathbb{E}_{(X,\mathcal{L})\sim p_{\text{data}}}\big[\log D(X, \mathcal{L})\big] - \mathbb{E}_{(X,z)\sim p_z}\big[\log(1 - D(X, G(X, z)))\big] \tag{23}$$

where the first portion of the equation maximizes the probability of the correctly predicting losses and the second part of the equation minimizes the probability of the wrong generated losses. The generator loss can be expressed as follows [76, 77]:

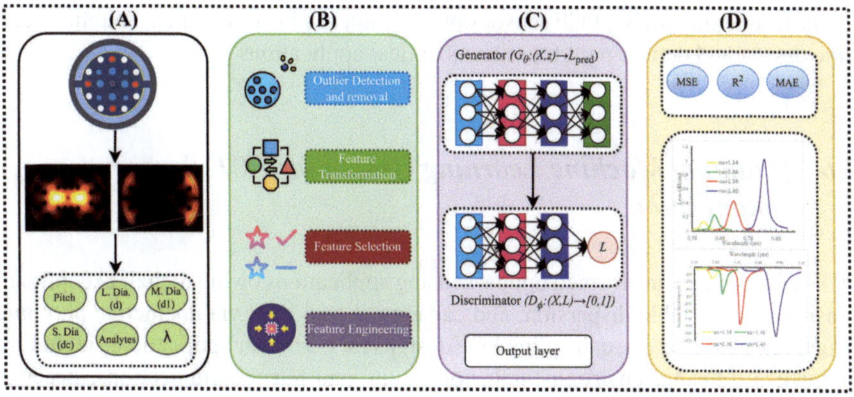

Fig. 8 GAN framework for loss prediction for PCF sensor design and optimization. **a** Data collection from FEM and other related techniques. **b** Data preprocessing for GAN algorithm application. **c** Developing GAN models. **d** Performance analysis of the applied GAN model and result validation through FEM

$$\mathcal{L}_G = \mathbb{E}_{(X,z)\sim p_z}\left[\log D(X, G(X, z))\right] \tag{24}$$

which forces the generator to produce loss values that are indistinguishable from real data. The overall GAN objective function can be presented as follows:

$$\min_G \max_D \mathbb{E}_{(X,\mathcal{L})\sim p_{\text{data}}}\left[\log D(X, \mathcal{L})\right] + \mathbb{E}_{(X,z)\sim p_z}\left[\log(1 - D(X, G(X, z)))\right] \tag{25}$$

which represents a minimax game between G and D.

Conditional GAN (cGAN) for Loss Prediction

Due to the reliance of loss prediction on particular PCF parameters, Conditional GANs (cGANs) are utilized. In a cGAN, the generator and discriminator are both dependent on the input X [78]:

$$G(X, z) \to \mathcal{L}_{\text{pred}} \tag{26}$$

$$D(X, \mathcal{L}) \to [0, 1] \tag{27}$$

The cGAN loss function is modified as:

$$\mathcal{L}_{cGAN} = \mathbb{E}_{(X,\mathcal{L})\sim p_{\text{data}}}\left[\log D(X, \mathcal{L})\right] + \mathbb{E}_{(X,z)\sim p_z}\left[\log(1 - D(X, G(X, z)))\right] \tag{28}$$

This conditioning improves training stability and ensures that generated loss values are physically meaningful.

By leveraging GANs, PCF sensor optimization can be accelerated, enabling real-time design and deployment for various optical applications.

7.6 *Quantum Machine Learning (QML) for PCF Sensor Optimization*

PCF sensors have transformed optical sensing applications owing to their exceptional sensitivity, adjustable dispersion, and capacity to trap light via distinctive photonic structures. Nonetheless, enhancing PCF designs for particular applications, including biomedical sensing, environmental monitoring, and telecommunications, necessitates computationally intensive simulations utilizing numerical techniques like FEM and BPM.

To speed up and improve the optimization process, Quantum Machine Learning (QML) has surfaced as a promising method. QML integrates quantum computing with traditional machine learning methods to effectively tackle intricate optimization challenges. Quantum algorithms, including Quantum Neural Networks (QNNs), Variational Quantum Circuits (VQCs), and Quantum Support Vector Machines (QSVMs), utilize the concepts of quantum mechanics (such as superposition and entanglement) to improve computational efficiency in the design of PCF sensors [79].

Quantum Machine Learning Framework for PCF Optimization

Quantum computing offers exponential acceleration for specific optimization issues using algorithms like Quantum Neural Networks (QNNs) and Quantum Variational Circuits (VQCs). The fundamental concept of QML for PCF optimization is to represent fiber design parameters as quantum states and apply quantum algorithms to speed up the learning process [80].

Quantum Feature Encoding

A set of feature vector $\boldsymbol{X}$, which are considered for classical PCF design, is converted into a quantum state $|\psi\boldsymbol{X}\rangle$ using quantum feature map (QFM) [79]:

$$|\psi\boldsymbol{X}\rangle = U_{\phi}(\boldsymbol{X})|0\rangle^{n} \tag{29}$$

where $U_{\phi}(\boldsymbol{X})$ is a unitary transformation encoding classical parameters into quantum states. A frequently utilized feature encoding technique is **Amplitude Encoding**, in which a normalized feature vector $\boldsymbol{X}$ is transformed into quantum amplitudes [80]:

$$|\psi\boldsymbol{X}\rangle = \sum_{i=1}^{N} \boldsymbol{X}_i|i\rangle \tag{30}$$

Another effective method is **Angle Encoding**, where each feature is encoded as a rotation in a qubit [67]:

$$|\psi X\rangle = \otimes_{i=1}^{n} R_y(X_i)|0\rangle \tag{31}$$

where $R_y(X_i)$ represents the **Y-axis rotation gate** applied to the ith qubit.

Quantum Neural Networks (QNNs) for Loss Prediction

A Quantum Neural Network (QNN) consists of different layers including Quantum Feature Encoding Layer to map PCF parameters to quantum states, Variational Quantum Circuit (VQC) (learnable quantum gates for transformation), and measurement layer to extract the loss values and gradients. A QNN function can be expressed as follows for predicting PCF loss value [79–81]:

$$\mathcal{L}_{\text{pred}} = f_\theta(|\psi X\rangle) \tag{32}$$

where f_θ denotes a trainable quantum circuit with the parameters θ. The Quantum Variational Circuit (VQC) is employed to transform the encoded state, and it can be expressed as follows:

$$|\psi_{\text{out}}\rangle = U(\theta)|\psi X\rangle \tag{33}$$

where $U(\theta)$ represents a unitary matrix and it is parameterized by trainable quantum gates. The quantum loss function is estimated using the following mathematical expression:

$$\mathcal{L}_{\text{quantum}} = \langle\psi_{\text{out}}|H|\psi_{\text{out}}\rangle \tag{34}$$

where H denotes the Hamiltonian, which encodes the loss function of PCF sensor.

Hybrid Quantum-Classical Optimization

Due to the limitations of current quantum hardware, hybrid quantum-classical models integrate QNNs with classical optimization techniques. The procedure includes quantum features encoding to transform PCF parameters into quantum states, quantum computation to apply QNN to approximate loss function, and classical optimizer to update parameters. The total hybrid loss function is as follows:

$$\mathcal{L}_{\text{total}} = \mathcal{L}_{\text{quantum}} + \lambda_1 \mathcal{L}_{\text{classical}} \tag{35}$$

where $\mathcal{L}_{\text{classical}}$ denotes the classical loss function and quantum and classical contributions are balanced by λ_1.

Overall, by employing quantum superposition and entanglement, QML provides a promising direction for next-generation PCF sensor optimization, enabling real-time optical design advancements.

7.7 Federated Learning (FL) for Collaborative PCF Sensor Optimization

Federated Learning (FL) offers a decentralized machine learning method in which various organizations jointly train a global model while keeping their raw data private [82]. FL is especially beneficial for the joint optimization of PCF sensors, allowing for privacy-protected and efficient learning from distributed datasets while ensuring data security. A number of PCF sensor fabrication labs or organizations can train a global model sharing their model design parameters and data to use in real life PCF sensor designing and optimization. However, since the parameters and data will come from different organizations, the parameters and the data privacy should be ensured. In this case, FL can play a vital role by training a decentralized global model ensuring their data privacy. It can be a collaborative approach among different organizations to have a single and efficient model to use. In this regard, FL is the most efficient method to apply. Figure 9 visualizes a potential framework for FL application in PCF sensor design and optimization.

Federated Learning Framework for PCF Sensor Optimization

In FL, several local clients (such as various fabrication labs or organizations) train a model using their private datasets and transmit only model updates (excluding raw data) to a central server. The federated optimization issue is expressed as [83]:

$$\min_{w} \sum_{k=1}^{K} p_k F_k(w) \tag{36}$$

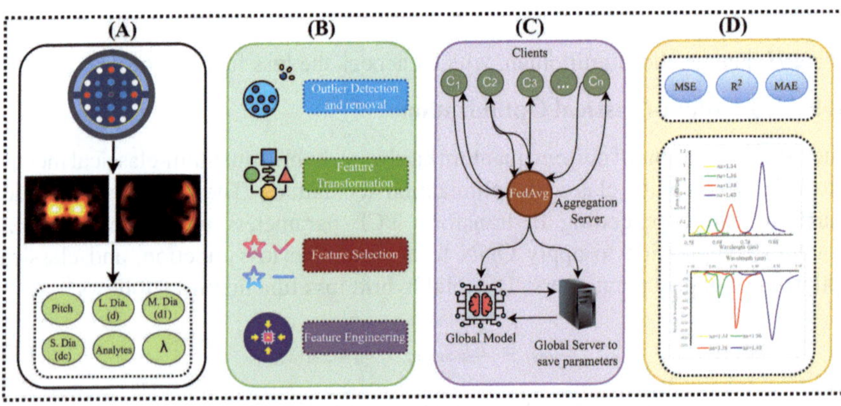

Fig. 9 Federated learning framework for loss prediction for PCF sensor design and optimization. **a** Data collection from FEM and other related techniques. **b** Data preprocessing for federated learning algorithm applications. **c** Developing federated learning models. **d** Performance analysis of the applied federated learning model and result validation through FEM

where K represents the number of clients (PCF sensor fabrication labs or organizations), p_k denotes the weight assigned to each client based on the size of the dataset, $F_k(w)$ represents loss function for each client k, and w mentions the parameters of the model. Each of the clients updates the parameters of the local model using Stochastic Gradient Descent (SGD) [83, 84]:

$$w_k^{t+1} = w_k^t - \eta \nabla F_k\left(w_k^t\right) \tag{37}$$

where the learning rate of the model is represented by η and $\nabla F_k\left(w_k^t\right)$ represents the gradients of the local loss function. The server aggregates the local updates using different methods. Federated Averaging (FedAvg) is one of the most popular methods to aggregate the local updates to get the parameters of the global model. The FedAvg can be expressed as follows [83, 84]:

$$w^{t+1} = \sum_{k=1}^{K} p_k w_k^{t+1}$$

where w^{t+1} is the most updated global model.

Loss Function for Federated PCF Optimization

Every client develops a deep learning model aimed at forecasting PCF sensor loss using structural parameters. The local loss function for client k is defined as [85]:

$$F_k(w) = \frac{1}{N_k} \sum_{i=1}^{N_k} \left(\mathcal{L}_{\text{true},i} - \mathcal{L}_{\text{pred},i}\right)^2 \tag{38}$$

where the number of samples of each client k is denoted by N_k and $\mathcal{L}_{\text{true},i}$ represents the actual PCF loss from FEM simulations, and $\mathcal{L}_{\text{pred},i}$ represents the predicted loss from the developed model. The global federated loss function is calculated using the following expression [85]:

$$F(w) = \sum_{k=1}^{K} p_k F_k(w) \tag{39}$$

This expression ensures that all client models are contributing to the update of the global model.

FL offers a distributed, efficient solution that protects privacy while optimizing PCF sensors in various research settings, enabling real-time enhancements in collaborative design.

Neuromorphic Computing for Real-Time Loss Prediction

Neuromorphic computing offers a biologically inspired method that replicates the human brain's parallel processing and energy efficiency [86]. Utilizing Spiking

Neural Networks (SNNs), neuromorphic computing enables real-time inference with minimal power usage, presenting it as a hopeful technology for predicting PCF sensor loss. This chapter investigates the ways neuromorphic hardware and SNN-based models can facilitate the acceleration of PCF loss estimation [87].

Spiking Neural Networks (SNNs)

SNNs serve as the basis for neuromorphic computing, replicating biological neurons through spikes rather than continuous activations. The behavior of an SNN neuron can be characterized by the Leaky Integrate-and-Fire (LIF) model [86, 87]:

$$\tau_m = \frac{dV(t)}{dt} = -(V(t) - V_{rest}) + I(t) \tag{40}$$

where $V(t)$ represents the membrane potential at time t, and the resting potential is denoted by V_{rest}. τ_m represents the constant membrane time and the synaptic input is denoted by $I(t)$. A neuron **hits a spike** when $V(t)$ exceeds a threshold V_{th}, after which the potential resets [86, 87].

$$V(t) \geq V_{th} \Rightarrow \text{Spike emited}, \;\; V(t) \rightarrow V_{\text{reset}} \tag{41}$$

SNNs outperformed for real-time processing because the information is processed by active neurons, and ideal for edge computing. Besides, maintaining biologically plausible learning rules is another important criterion for outperforming.

The neuromorphic computing model consists of three types of layers like any other neural network model. However, they have different computation techniques. The first layer is called the input layer, which extracts PCF parameters into spike protein. A number of hidden layers are also incorporated to apply SNN neuron, which is model via LIF or Hodgkin–Huxley equations. The last layer is called the output layer, which predicts the loss in PCF sensor optimization. The forward propagation is computed using the following expression [86, 87]:

$$S_i(t) = H(V_i(t) - V_{\text{th}}) \tag{42}$$

where $S_i(t)$ denotes the spike emitted by a neuron i at time t and $H(\cdot)$ the Heaviside step function. The final prediction is calculated through **rate coding**, where the firing rate of output neurons encodes the predicted loss [72, 73]:

$$\mathcal{L}_{\text{pred}} = \sum_{t=0}^{T} S_{\text{out}}(t) \tag{43}$$

Neuromorphic Learning Algorithms

Two major learning methods for training SNNs in neuromorphic computing are Spike Timing-Dependent Plasticity (STDP) and Surrogate Gradient Learning (SGL).

Spike Timing-Dependent Plasticity (STDP)

A biologically motivated principle in which synaptic weights are modified according to the timing differences between pre- and post-synaptic spikes [86, 87]:

$$\Delta w = A_+ e^{-\frac{\Delta t}{\tau_+}} \quad \text{if} \quad \Delta t > 0 \tag{44}$$

$$\Delta w = -A_- e^{-\frac{\Delta t}{\tau_-}} \quad \text{if} \quad \Delta t < 0 \tag{45}$$

where Δw denotes the weight change, and Δt represents the difference between spike time, and $A_+,\ A_-,\ \tau_+,\ \tau_-$ represents the learning parameters of the model.

Surrogate Gradient Learning (SGL)

Since SNNs feature non-differentiable spike functions, gradient-based training employs a smoothed approximation of the spike function $H(x)$:

$$\frac{dH(x)}{dx} \approx \sigma(x)(1 - \sigma(x)) \tag{46}$$

where $\sigma(x)$ is the sigmoid function of the neuromorphic computing model.

Overall, Neuromorphic computing offers a groundbreaking method for predicting optical loss in real-time, setting the stage for smart, energy-efficient PCF sensing systems.

8 Conclusion

Multi-analyte sensing combined with generative AI may allow for more accurate and dependable systems for industrial automation, environmental monitoring, and medical diagnostics. With its ability to create realistic data, extract significant features, and manage the complexity of multidimensional datasets, generative AI offers a revolutionary approach to multi-analyte sensing. It can greatly improve the accuracy and efficiency of sensing systems by adapting to a variety of sensing settings and analytes. By combining different aspects, the development of AI has enhanced sensor systems, producing gadgets with higher performance and more accurate outcomes. Generative AI has the potential to significantly influence the development of sophisticated sensing systems in the future by guaranteeing flexibility, affordability, and real-time deployment, which will ultimately propel advancement in a number of fields.

Acknowledgements This work was supported, in part, by funding from the Natural Sciences and Engineering Research Council of Canada (NSERC).

References

1. W.C. Wong, C.C. Chan, J.L. Boo, Z.Y. Teo, Z.Q. Tou, H.B. Yang, C.M. Li, K.C. Leong, Photonic crystal fiber surface plasmon resonance biosensor based on protein G immobilization. IEEE J. Sel. Top. Quantum Electron. **19**(3), 4602107–4602107 (2013)
2. A. Hassani, M. Skorobogatiy, Design of the microstructured optical fiber-based surface plasmon resonance sensors with enhanced microfluidics. Opt. Express **14**(24), 11616–11621 (2006)
3. I. Abdelaziz, F. AbdelMalek, S. Haxha, H. Ademgil, H. Bouchriha, Photonic crystal fiber with an ultrahigh birefringence and flattened dispersion by using genetic algorithms. J. Lightwave Technol. **31**(2), 343–348 (2013)
4. D. Vijayalakshmi, C.T. Manimegalai, N. Ayyanar, D. Vigneswaran, K. Kalimuthu, Detection of blood glucose with hemoglobin content using compact photonic crystal fiber. IEEE Trans. Nanobiosci. **20**(4), 436–443 (2021)
5. H. Feng, M. Wang, Z. Sha, L. Du, D. Qi, Research on weak vibration characteristics based on EMD and design of MZI classifier. J. Lightwave Technol. **41**(10), 3182–3191 (2023)
6. W. Naku, A.K. Nambisan, M. Roman, C. Zhu, R.E. Gerald, J. Huang, Identification of volatile organic liquids by combining an array of fiber-optic sensors and machine learning. ACS Omega **8**(5), 4597–4607 (2023)
7. Y.S. Dwivedi, R. Singh, A.K. Sharma, A.K. Sharma, Enhancing the performance of photonic sensor using machine-learning approach. IEEE Sens. J. **23**(3), 2320–2327 (2022)
8. M. Hassani-Marand, N. Fahimi-Kashani, M.R. Hormozi-Nezhad, Machine-learning assisted multiplex detection of catecholamine neurotransmitters with a colorimetric sensor array. Anal. Methods **15**(9), 1123–1134 (2023)
9. D. Schletz, M. Breidung, A. Fery, Validating and utilizing machine learning methods to investigate the impacts of synthesis parameters in gold nanoparticle synthesis. J. Phys. Chem. C **127**(2), 1117–1125 (2023)
10. G.A. Wellbrock, T.J. Xia, M.F. Huang, S. Han, Y. Chen, T. Wang, Y. Aono, Explore benefits of distributed fiber optic sensing for optical network service providers. J. Lightwave Technol. **41**(12), 3758–3766 (2023)
11. Y. Zhou, Y.N. Zhang, B. Han, L. Cheng, D. Li, W. Zheng, Y. Zhao, Biochemical sensor based on functional material assisted optical fiber surface plasmon resonance: a review. Measurement **207**, 112353 (2023)
12. G. Fornarelli, L. Mescia, F. Prudenzano, M. De Sario, F. Vacca, A neural network model of erbium-doped photonic crystal fibre amplifiers. Opt. Laser Technol. **41**(5), 580–585 (2009)
13. M.F.O. Hameed, S.S.A. Obayya, K. Al-Begain, A.M. Nasr, M.I. Abo el Maaty, Accurate radial basis function based neural network approach for analysis of photonic crystal fibers. Opt. Quant. Electron. **40**, 891–905 (2008)
14. P.D. McAtee, S.T. Bukkapatnam, A. Lakhtakia, Artificial neural network to estimate the refractive index of a liquid infiltrating a chiral sculptured thin film. J. Nanophotonics **13**(4), 046006 (2019)
15. A. Zelaci, A. Yasli, C. Kalyoncu, H. Ademgil, Generative adversarial neural networks model of photonic crystal fiber based surface plasmon resonance sensor. J. Lightwave Technol. **39**(5), 1515–1522 (2020)
16. Y. Kiarashinejad, M. Zandehshahvar, S. Abdollahramezani, O. Hemmatyar, R. Pourabolghasem, A. Adibi, Knowledge discovery in nanophotonics using geometric deep learning. Adv. Intell. Syst. **2**(2), 1900132 (2020)
17. T. Asano, S. Noda, Optimization of photonic crystal nanocavities based on deep learning. Opt. Express **26**(25), 32704–32717 (2018)
18. S. Chugh, A. Gulistan, S. Ghosh, B.M.A. Rahman, Machine learning approach for computing optical properties of a photonic crystal fiber. Opt. Express **27**(25), 36414–36425 (2019)
19. P. Verma, A. Kumar, P. Jindal, Machine learning approach for SPR based photonic crystal fiber sensor for breast cancer cells detection, in *2022 IEEE 7th Forum on Research and Technologies for Society and Industry Innovation (RTSI)*, Aug 2022 (IEEE, 2022), pp. 7–12

20. A. Kumar, P. Verma, P. Jindal, Machine learning approach to surface plasmon resonance sensor based on MXene coated PCF for malaria disease detection in RBCs. Optik **274**, 170549 (2023)
21. T. Das, M.N. Hossen, S.M. Rahman, T. Parvin, K. Ahmed, F.M. Bui, PrePCF_ML: prediction of photonic crystal fiber parameters using machine learning algorithms, in *2022 Second International Conference on Advances in Electrical, Computing, Communication and Sustainable Technologies (ICAECT)*, Apr 2022 (IEEE, 2022), pp. 1–5
22. A. Kumar, P. Verma, P. Jindal, Surface plasmon resonance sensor based on MXene coated PCF for detecting the cancer cells with machine learning approach. Microelectron. Eng. **267**, 111897 (2023)
23. S. Chowdhury, L.F. Abdulrazak, S.A. Mitu, K. Ahmed, F.M. Bui, L.K. Smirani, L. Chen, A highly sensitive multi-channel SPR-PCF based biosensor with deep learning prediction approach. Alex. Eng. J. **77**, 189–203 (2023)
24. N. Islam, I.H. Shibly, M.M.S. Hasan, M.N. Hasan, M.A. Yousuf, Computing optical properties of open–channels based plasmonic biosensor employing plasmonic materials with ML approach, in *2023 26th International Conference on Computer and Information Technology (ICCIT)*, Dec 2023 (IEEE, 2023), pp. 1–6
25. K. Ahmed, F.M. Bui, F.X. Wu, PreOBP_ML: machine learning algorithms for prediction of optical biosensor parameters. Micromachines **14**(6), 1174 (2023)
26. S. Kaziz, F. Echouchene, M.H. Gazzah, Optimizing PCF-SPR sensor design through Taguchi approach, machine learning, and genetic algorithms. Sci. Rep. **14**(1), 7837 (2024)
27. A. Ehyaee, A. Rahmati, A. Bosaghzadeh, S. Olyaee, Machine learning-enhanced surface plasmon resonance based photonic crystal fiber sensor. Opt. Express **32**(8), 13369–13383 (2024)
28. M. Al Mahfuz, S. Afroj, A. Rahman, M.A. Hossain, M.A. Hossain, M.S. Habib, Ultra-sensitive visible-IR range fiber based plasmonic sensor: a finite-element analysis and deep learning approach for RI prediction. IEEE Access (2024)
29. L. Guedri-Knani, S. Kaziz, C. Dridi, Optimizing and predicting performance of dual-side polished SPR photonic crystal fiber using MLR and ANN models. Plasmonics 1–12 (2024)
30. C. Saha, F. Haque, N. Islam, M.M. Hossain, M.E. Arafat, M.A. Yousuf, M.M. Rahman, Dual-core silver-coated plasmonic sensor modeling with machine learning. Heliyon 10(19) (2024)
31. N. Islam, M.M. Shoaib Hasan, I. Hossain Shibly, M.B. Rashid, M.A. Yousuf, F. Haider, R. Ahmmed Aoni, R. Ahmed, Plasmonic sensor using generative adversarial networks integration. Opt. Express **32**(20), 34184–34198 (2024)
32. Y. Dogan, R. Katirci, I. Erdogan, Machine learning-based optimization for D-shaped PCF SPR refractive index sensor. Opt. Commun. **575**, 131304 (2025)
33. A. Cucinotta, S. Selleri, L. Vincetti, M. Zoboli, Perturbation analysis of dispersion properties in photonic crystal fibers through the finite element method. J. Lightwave Technol. **20**(8), 1433 (2002)
34. F. Bréchet, J. Marcou, D. Pagnoux, P.J.O.F.T. Roy, Complete analysis of the characteristics of propagation into photonic crystal fibers, by the finite element method. Opt. Fiber Technol. **6**(2), 181–191 (2000)
35. K. Saitoh, M. Koshiba, Full-vectorial imaginary-distance beam propagation method based on a finite element scheme: application to photonic crystal fibers. IEEE J. Quantum Electron. **38**(7), 927–933 (2002)
36. Y.Z. He, F.G. Shi, Finite-difference imaginary-distance beam propagation method for modeling of the fundamental mode of photonic crystal fibers. Opt. Commun. **225**(1–3), 151–156 (2003)
37. N.H. Vu, I.K. Hwang, Y.H. Lee, Bending loss analyses of photonic crystal fibers based on the finite-difference time-domain method. Opt. Lett. **33**(2), 119–121 (2008)
38. O. Shulika, I. Guryev, N. Gurieva, I. Sukhoivanov, J.A. Andrade Lucio, Graphics processing unit–accelerated finite-difference time-domain method for characterization of photonic crystal fibers. Opt. Eng. **52**(12), 126109–126109 (2013)
39. F. Lin, G. Qiu, Y. Li, Analyzing point defect two-dimensional photonic crystals with transfer matrix and block-iterative frequency-domain method. Liangzi Dianzi Xuebao/Chin. J. Quantum Electron. (China) **20**(1), 35–41 (2003)

40. D.G. Popescu, P. Sterian, Photonic crystal fiber mode characterization with multipole method, in *UPB Scientific Bulletin, A Series* (University "Politehnica" of Bucharest, Bucharest, 2013), pp. 205–215
41. S. Guo, S. Albin, Simple plane wave implementation for photonic crystal calculations. Opt. Express **9**(2), 167–175 (2003)
42. G. Abo-Lila, T. Sokkar, E. Seisa, E. Omar, Adaptive investigation of the optical properties of polymer fibers from mixing noisy phase shifting microinterferograms using deep learning algorithms. Microsc. Res. Tech. **85**(2), 667–684 (2022)
43. M.A. Jabin, M.P. Fok, Prediction of 12 photonic crystal fiber optical properties using MLP in deep learning. IEEE Photonics Technol. Lett. **34**(7), 391–394 (2022)
44. T. Khan, C. Roy, Prediction of slot-position and slot-size of a microstrip antenna using support vector regression. Int. J. RF Microw. Comput. Aided Eng. **29**(3), e21623 (2019)
45. C. Roy, T. Khan, B.K. Kanaujia, Performance parameters prediction of slotted microstrip antennas with modified ground plane using support vector machine. Int. J. Microw. Wireless Technol. **9**(5), 1169–1177 (2017)
46. F. Güneş, N.T. Tokan, F. Gürgen, Support vector design of the microstrip lines. Int. J. RF Microw. Comput.-Aid. Eng. **18**(4), 326–336 (Center for Advanced Manufacturing and Packaging of Microwave, Optical, and Digital Electronics (CAMPmode) at the University of Colorado at Boulder, 2008)
47. J. Hu, J. Wang, Short-term wind speed prediction using empirical wavelet transform and Gaussian process regression. Energy **93**, 1456–1466 (2015)
48. L. Zhou, J. Chen, Z. Song, Recursive Gaussian process regression model for adaptive quality monitoring in batch processes. Math. Probl. Eng. **2015**(1), 761280 (2015)
49. N.D. Hoang, A.D. Pham, Q.L. Nguyen, Q.N. Pham, Estimating compressive strength of high performance concrete with Gaussian process regression model. Adv. Civ. Eng. **2016**(1), 2861380 (2016)
50. S.M. Abdul Kaium, M.A. Mollah, Optical properties estimation of photonic crystal fiber using Gaussian process regression. Opt. Contin. **3**(8), 1369–1388 (2024)
51. E.Z. Omar, T.Z.N. Sokkar, A.A. Hamza, In situ investigation and detection of opto-mechanical properties of polymeric fibres from their digital distorted microinterferograms using machine learning algorithms. Opt. Laser Technol. **129**, 106295 (2020)
52. H. Khan, Z. Jan, I. Ullah, A. Alwabli, F. Alharbi, S. Habib, M. Islam, B.J. Shin, M.Y. Lee, J. Koo, A deep dive into AI integration and advanced nanobiosensor technologies for enhanced bacterial infection monitoring. Nanotechnol. Rev. **13**(1), 20240056 (2024)
53. S. Sharma, S. Das, L. Tharani, Analyzing trends in medical imaging using intelligent photonics. Eng. Proc. **56**(1), 264 (2023)
54. W. Xu, Optical sensor based quantum computing in sports medicine for diagnosis and data analysis using machine learning model. Opt. Quant. Electron. **56**(4), 528 (2024)
55. C. Zhou, C. Huang, H. Zhang, W. Yang, F. Jiang, G. Chen, S. Liu, Y. Chen, Machine-learning-driven optical immunosensor based on microspheres-encoded signal transduction for the rapid and multiplexed detection of antibiotics in milk. Food Chem. **437**, 137740 (2024)
56. M. Soler, A. Scholtz, R. Zeto, A.M. Armani, Engineering photonics solutions for COVID-19. APL Photonics **5**(9) (2020)
57. K. Ahmed, R. Amin, F.M. Bui, L. Chen, N. Mohammadd, F.A. Al-Zahrani, S. Kumar, Design and analysis of multi-analyte detection based biosensor in the visible to near-infrared (VNIR) region. IEEE Trans. Nanobiosci. **23**(1), 42–50 (2023)
58. F.Y. Wu, Remote sensing image processing based on multi-scale geometric transformation algorithm. J. Discrete Math. Sci. Cryptogr. **20**(1), 309–321 (2017)
59. S. Das, L. Tharani, Analyzing trends in medical imaging using intelligent photonics (2023)
60. M.M. Ali, K. Ahmed, F.M. Bui, F.X. Wu, DrugEL: ensemble learning model for identification of druggable proteins, in *2024 IEEE International Conference on Bioinformatics and Biomedicine (BIBM)*, Dec 2024 (IEEE), pp. 168–171
61. M.M. Ali, K. Ahmed, F.M. Bui, L. Chen, STACKION: ion channel-modulating peptides identification using stacking-based ensemble machine learning, in *2023 IEEE Canadian Conference on Electrical and Computer Engineering (CCECE)*, Sept 2023 (IEEE), pp. 216–221

62. L. Breiman, Bagging predictors. Mach. Learn. **24**, 123–140 (1996)
63. P. Bühlmann, B. Yu, Analyzing bagging. Ann. Stat. **30**(4), 927–961 (2002)
64. A. Mayr, H. Binder, O. Gefeller, M. Schmid, The evolution of boosting algorithms. Methods Inf. Med. **53**(06), 419–427 (2014)
65. A.J. Ferreira, M.A. Figueiredo, Boosting algorithms: a review of methods, theory, and applications, in *Ensemble Machine Learning: Methods and Applications* (2012), pp. 35–85
66. K. Weiss, T.M. Khoshgoftaar, D. Wang, A survey of transfer learning. J. Big Data **3**, 1–40 (2016)
67. S.M. Salaken, A. Khosravi, T. Nguyen, S. Nahavandi, Seeded transfer learning for regression problems with deep learning. Expert Syst. Appl. **9**(5), 565–577 (2019)
68. K. Choi, G. Fazekas, M. Sandler, K. Cho, Transfer learning for music classification and regression tasks. arXiv preprint arXiv:1703.09179, 27 Mar 2017
69. T. Hospedales, A. Antoniou, P. Micaelli, A. Storkey, Meta-learning in neural networks: a survey. IEEE Trans. Pattern Anal. Mach. Intell. **44**(9), 5149–5169 (2021)
70. A. Abraham, Meta learning evolutionary artificial neural networks. Neurocomputing **56**, 1–38 (2004)
71. C. Finn, P. Abbeel, S. Levine, Model-agnostic meta-learning for fast adaptation of deep networks, in *International Conference on Machine Learning*, July 2017 (PMLR, 2017), pp. 1126–1135
72. S. Cuomo, V.S. Di Cola, F. Giampaolo, G. Rozza, M. Raissi, F. Piccialli, Scientific machine learning through physics–informed neural networks: where we are and what's next. J. Sci. Comput. **92**(3), 88 (2022)
73. M. Raissi, P. Perdikaris, G.E. Karniadakis, Physics-informed neural networks: a deep learning framework for solving forward and inverse problems involving nonlinear partial differential equations. J. Comput. Phys. **378**, 686–707 (2019)
74. G. Pang, L. Lu, G.E. Karniadakis, fPINNs: fractional physics-informed neural networks. SIAM J. Sci. Comput. **41**(4), A2603–A2626 (2019)
75. E. Zhang, M. Dao, G.E. Karniadakis, S. Suresh, Analyses of internal structures and defects in materials using physics-informed neural networks. Sci. Adv. **8**(7), abk0644 (2022)
76. I. Goodfellow, J. Pouget-Abadie, M. Mirza, B. Xu, D. Warde-Farley, S. Ozair, A. Courville, Y. Bengio, Generative adversarial networks. Commun. ACM **63**(9), 139–144 (2020)
77. A. Creswell, T. White, V. Dumoulin, K. Arulkumaran, B. Sengupta, A.A. Bharath, Generative adversarial networks: an overview. IEEE Signal Process. Mag. **35**(1), 53–65 (2018)
78. K. Wang, C. Gou, Y. Duan, Y. Lin, X. Zheng, F.Y. Wang, Generative adversarial networks: introduction and outlook. IEEE/CAA J. Autom. Sin. **4**(4), 588–598 (2017)
79. J. Biamonte, P. Wittek, N. Pancotti, P. Rebentrost, N. Wiebe, S. Lloyd, Quantum machine learning. Nature **549**(7671), 195–202 (2017)
80. M. Schuld, I. Sinayskiy, F. Petruccione, An introduction to quantum machine learning. Contemp. Phys. **56**(2), 172–185 (2015)
81. M. Cerezo, G. Verdon, H.Y. Huang, L. Cincio, P.J. Coles, Challenges and opportunities in quantum machine learning. Nat. Comput. Sci. **2**(9), 567–576 (2022)
82. L. Li, Y. Fan, M. Tse, K.Y. Lin, A review of applications in federated learning. Comput. Ind. Eng. **149**, 106854 (2020)
83. P.M. Mammen, Federated learning: opportunities and challenges. arXiv preprint arXiv:2101.05428 (2021)
84. C. Zhang, Y. Xie, H. Bai, B. Yu, W. Li, Y. Gao, A survey on federated learning. Knowl.-Based Syst. **216**, 106775 (2021)
85. P. Kairouz, H.B. McMahan, B. Avent, A. Bellet, M. Bennis, A.N. Bhagoji, K. Bonawitz, Z. Charles, G. Cormode, R. Cummings, R.G. D'Oliveira, Advances and open problems in federated learning. Found. Trends® Mach. Learn. **14**(1–2), 1–210 (2021)
86. C.D. Schuman, S.R. Kulkarni, M. Parsa, J.P. Mitchell, P. Date, B. Kay, Opportunities for neuromorphic computing algorithms and applications. Nat. Comput. Sci. **2**(1), 10–19 (2022)
87. C.D. Schuman, T.E. Potok, R.M. Patton, J.D. Birdwell, M.E. Dean, G.S. Rose, J.S. Plank, A survey of neuromorphic computing and neural networks in hardware. arXiv preprint arXiv:1705.06963 (2017)

Future Trends and Challenges in Generative AI for Photonic Sensing

Abhishek Chaurasia, Oindrila Banik, Santosh Kumar, and Earu Banoth

Abstract Artificial intelligence (AI), which has revolutionary potential for applications ranging from telecommunications to quantum computing, is redefining the design and operation of photonic systems. The traditional design of photonic structures relies heavily on physics-based techniques and iterative numerical simulations. Although these techniques have been effective, there have always been limitations in complex design areas, and the ability to reach optimal results has always been a challenge. AI techniques, especially those based on machine learning (ML) and Generative AI (Gen-AI), provide out-of-the-box solutions based on data-driven model simulations that are both forward and reverse. The "real-time" prediction generation of photonic devices and fast search in the design space are some benefits of implementing deep learning techniques. Like reinforcement learning, the design of photonic geometries can be the subject of efficient optimization methods that are enabled by Actor and critic Models. In contrast, PINNs, which are based on physics principles, use Maxwell equations for mode analysis. Due to the design and fabrication of grating couplers and metasurfaces, traditional optimization methods are often outperformed in terms of accuracy and speed. Generative AI oversees changes in photonic computing when photonics-based hardware technologies are developed to allow faster and more energy-efficient neural networks, which in turn can overshadow electrical computing. This combined AI-photonics approach is intended to change the face of things by providing a platform for the development of smart optical systems, holographic imaging, and quantum electronics. However, related issues are numerous. These problems involve quality assurance, data volume, and scalability. The primary solution is to expedite the development of generative AI algorithms and photonics hardware. The future promises photonics as the joining link between the

A. Chaurasia · O. Banik · E. Banoth (✉)
Department of Biotechnology and Medical Engineering, National Institute of Technology, Rourkela, Rourkela, Odisha, India
e-mail: banothe@nitrkl.ac.in

S. Kumar
Department of Electronics and Communication Engineering, Centre of Excellence for Nanotechnology, Koneru Lakshmaiah Education Foundation, Vaddeswaram, Andhra Pradesh, India
e-mail: santosh@kluniversity.in

Y. Vasimalla and S. Kumar (eds.), *Generative AI for Photonic Sensing*, Progress in Optical Science and Photonics 36, https://doi.org/10.1007/978-981-95-1561-5_10

real and digital worlds, and at the same time, the two systems will work together seamlessly, enabling the development of entirely new levels of creative output.

Keywords Generative artificial technology · Generative adversarial network · Photonic sensing · Deep learning · Metasurfaces

1 Introduction

The confluence of Generative Artificial Intelligence (Gen-AI) and photonics is set to revolutionize sensing technology, redefining how precision, efficiency, and adaptability are to be interpreted. Photonic sensing dwells in the amplification of the properties of light for measurement in diverse areas such as environmental monitoring, biomedical imaging, and quantum mechanical computation, offering a quantum leap due to the integration of AI. At the same time, Gen-AI, which synthesizes complex data to optimize intricate designs and push forward intelligent minimalistic processes, presents itself as a potential mode for the growth of photonics. This chapter examines these two frontiers and their emerging trends, unforeseen developments, and challenges to capitalize on the trend.

Vast advances in AI have been witnessed in photonics in the recent uplift of their importance in all areas, including optoelectronics, waveguides, metasurfaces, and multipolar filters. In optical communications, AI can be used to eliminate the signal processing, with the enormous real-time error correction and adaptive modulation for high-bandwidth data transmission. Gen-AI techniques such as deep generative models and reinforcement learning facilitate the design of new devices—the design of photonic devices with automation. One great example is the use of Generative Adversarial Network (GAN). Algorithms developed in one of the GAN groups may define the shape of diffraction gratings and holographic elements, thus increasing the generation speed of applications while reducing complexities during the production stage. Besides, deep learning-controlled holography is making 3D imaging rather lightning fast in terms of noise reduction, establishing its usefulness in real-time applications such as augmented reality (AR) and virtual reality (VR) systems. In one fell swoop, AI and photonic sensing could set a new horizon for technology change together.

Photonics will benefit from the introduction of AI if it manages to discard its long-accepted puzzles. Traditional photonic sensing systems have opportunities to perform noise suppression, efficiently interpret their data in more complicated scenarios, or have real-time adaptability, as AI makes way to do so by employing its personalized ways of intelligent sensing and automation protocols in the development of photonic components for specific application purposes. Speaking of other good things, sound knowledge of AI greatly helps with handling vast amounts of data, in contrast to the McGill High Energy Physics Working Group (HEP). Massive data storage capabilities will also support the optics insights to work efficiently around data, as in high-throughput biomedical imaging and large-environment monitoring. In this way,

AI contributes to more advanced research, minimizing the use of excess energy to make photo-netting and reducing waste in manufacturing.

The beginning of this section describes the basics of generative AI and photonic sensing, placing a foundation on the integration of the two. It discusses the latest developments in the generation of photonic design frameworks and AI-enhanced real-time holographic systems and describes a roadmap for the future of this interdisciplinary field. By shedding light on trends, challenges, and future directions, this book chapter stands as a manual for those researchers, engineers, and policymakers navigating through the intricate junction of AI and photonics, propelling innovation across industrial borders.

2 Fundamentals of Generative AI Photonic Sensor

In the developing world of technology, the combination of Gen-AI and photonic sensing offers intriguing opportunities for improving different sectors of the economy, from telecommunications to biomedical applications. In this chapter, an attempt has been made to explain the basic concepts of both Gen-AI and photonic sensing and also to demonstrate how they can be integrated to perform useful tasks using real-life examples and a mathematical explanation of Variational Autoencoders (VAEs).

2.1 *Photonic Sensing: A Brief Introduction*

It is a new technology involving photonics for sensing, where the peculiar properties of light are beautifully manipulated for sensing and measuring applications across a variety of physical, chemical, and biological phenomena, all with amazing precision. By incorporating the delicate characteristics of photons, like their wavelength, their intensity, and their phase, we would manipulate photonic sensing. This can raise the benefits in terms of accuracy, speed, and sensitivity, making it quite valuable in almost every application, transforming the manner of measurement while monitoring and understanding the environment. In medicine, for example, photonic sensor real-time detection of glucose in urine is very useful in tracking diabetes conditions and forms an important part of relevant and timely information that will lead to improved treatment and management processes or considerations. It also boosts changing approaches to chronic disease management by enhancing patient outcomes. This paper goes to the fundamentals of photonic sensing and its principles and innovations, with much emphasis on how this modern technology is fundamentally changing diverse fields through its diverse and impactful applications. The goal is to maximize the Evidence Lower Bound (ELBO), defined as follows (Fig. 1):

$$\mathrm{ELBO} = \mathbf{E}_{\mathrm{q}(z|x)}\left[\mathbf{logp}(\boldsymbol{x}|z)\right] - \mathbf{KL}(\mathbf{q}(z|\boldsymbol{x})||\mathbf{p}(z)) \tag{1}$$

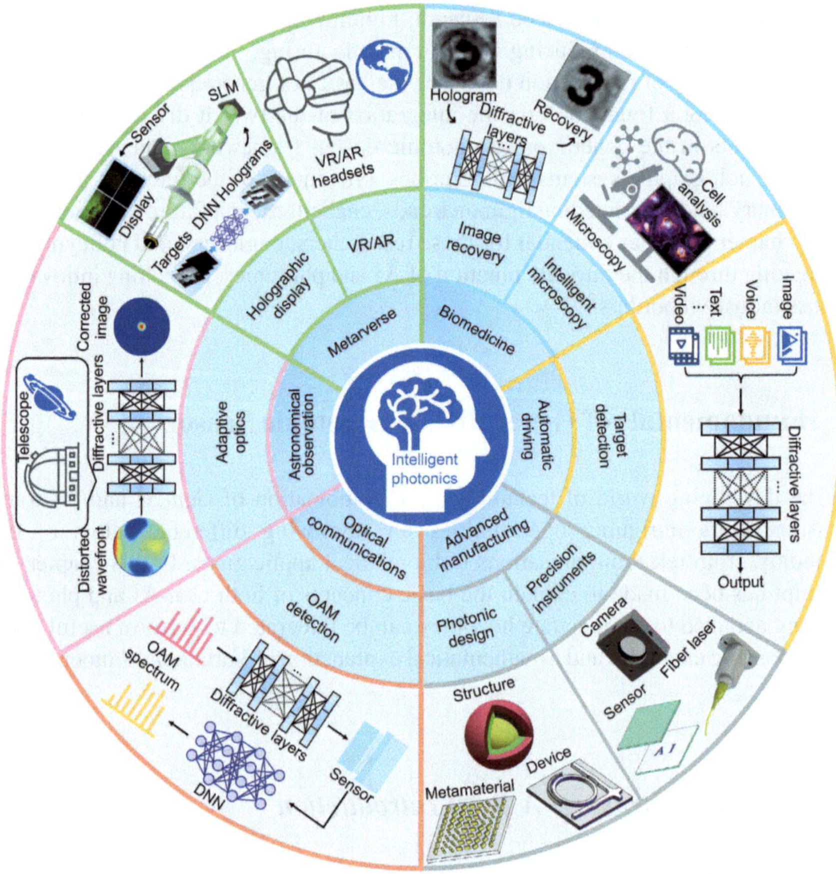

Fig. 1 An intelligent photonics-related application may find an entry in several domains starting from the metaverse, biomedicine, automatic driving, advanced manufacturing, optical communications, and astronomical observation [1]

The first term measures the likelihood that the data arise from the latent variables; the second term, the divergence Kullback–Leibler, regularizes the learned representation. Therefore, VAEs generate new data instances while ensuring such generated data is similar to the underlying distribution of the original dataset.

3 Role of Generative AI in Photonic Sensing

Gen-AI is rapidly becoming a turning point in photonic sensing with new methods for interpreting data, designing systems, and optimizing them. By utilizing the most recent algorithms, Gen-AI improves the modeling capacity of complex photonic systems so that researchers can predict and simulate sensor behavior under varied conditions. Such a technology can generate synthetic data to complement the training datasets, thus delivering machine-learning models in photonic applications leading to improvements in terms of accuracy and robustness. Further, generation assistants in the design of novel photonic materials and structures speed up the discovery process and help develop sensors with better sensitivities and specificities. With an ongoing integration of generative AI into scientists' workflows, its impact on the advancement of photonic sensing technologies becomes promising for industries like telecommunications, environmental monitoring, and healthcare. Here we review these exciting advances made in intelligent photonics from the viewpoints of deep learning, metaphotonics, and holography.

3.1 Deep Learning in Photonics

Deep learning has become a revolutionary type of information processing and statistical inference, based on innovations in network architecture, availability of giant training datasets, and the rapidly increasing amounts of computational power [2]. Mimicking the biological nervous system, deep neural networks (DNNs) comprise multilayer architectures with interconnected input, hidden, and output layers. These cascade operations can perform linear operations and nonlinear activations so that DNNs can approximate complex functions and apply them to a wide range of scientific and engineering problems [3].

The computational analysis in DNN involves neurons operated through weighted addition and non-linear activation. Mathematically, the outputs of neurons in the lth layer can be effectively represented as follows:

$$y = f_1(x) = \phi\left(\mathbf{w}_l^T \mathbf{x} + \mathbf{b}_l\right) \quad (2)$$

Here, $y = (\mathbf{y_1}, \mathbf{y_2}, \ldots \mathbf{y_m})^T$ is the output vector ($y \in \mathbf{R^{m\times 1}}$), $x = (x_1, x_2, \ldots x_n)^T$ is the input vector ($x \in R_{n\times 1}$), ϕ denotes the activation function, $\mathbf{w}_l$ represents the weight matrix ($w_l \in R^{n\times m}$), and b_I is the bias vector ($b_l \in \mathbf{R}_{m\times 1}$). Here, **n** and **m** denote the dimensions of the input and output vectors, respectively, while l specifies the layer index, and R denotes the field of real numbers. The goal of deep learning is to identify a mapping function f_1, $x \rightarrow y$ ($\mathbf{R}^n \rightarrow \mathbf{R}^m$). A DNN is formed by stacking several such single-layer networks. A DNN is formed by stacking several such single-layer networks into N-layer architectures, represented as $y_N(x) = f_N(f_{N-1}(\ldots f_2(f_1(x))))$ [1] (Fig. 2).

Deep learning exploits the intelligence of light—it gives light in photonics to solve the problems posed by traditional electronic hardware. Photonics, high speed, and parallelism are disseminated by novel paradigms for computing. An example is Photonic Neural Cellular Automata (PNCA), where sparse connectivity is employed to avoid scalability problems but remain efficient and robust. The experimental PNCA

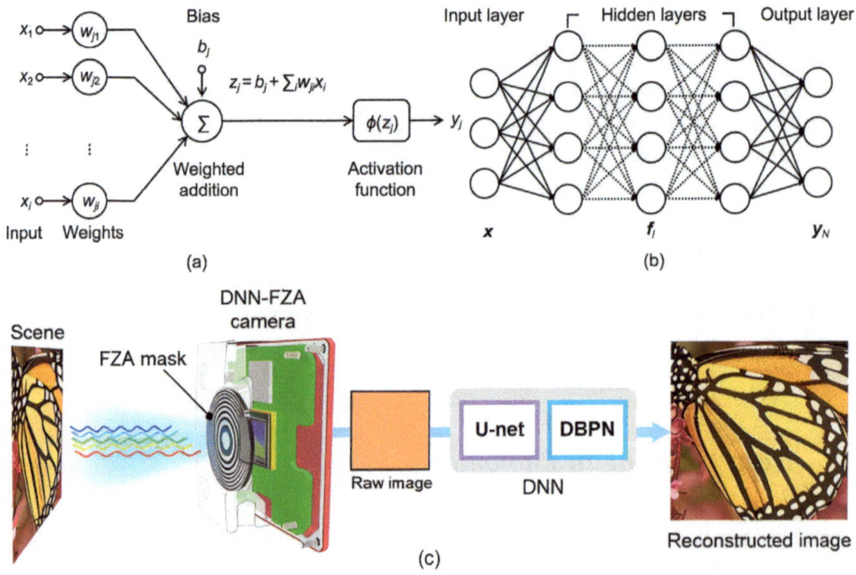

Fig. 2 **a** Exploding neurons, where *zj* is a linear transformation. **b** The general structure of the DNN model [1]. **c** Image acquisition and reconstruction principle of DNN-FZA camera [4]

setups demonstrate excellent accuracy with few photonic parameters used, making it a very attractive avenue for energy-efficient computation [5].

Silicon photonics (SiPh)-integrated circuits benefit from increased opportunities in photonic deep learning. These systems employ light-based interactions to go beyond the bandwidth and latency constraints associated with CMOS-based systems and result in performance improvements in energy cost and speed of computation. SiPh accelerators fit in versatile tasks related to deep learning besides providing a sustainable computational framework [6]. The integration of phase-change materials in silicon waveguide arrays permits fine-tuning of light-manipulating capabilities in the realization of unitary operators for reconfigurable optical systems and quantum simulations [7]. Optical deep learning networks exploit another milestone of non-linear interaction in the multimode fiber to realize ultrafast decomposition and identification of signals beyond the capacity of any traditional electronic systems [8]. Together, these promising paradigms turn photonic deep learning into one of the most important domains in developing computational methodologies to address energy-efficiency problems.

3.2 Generative AI in Metaphotonics

In a definitive sense, the emergence of generative AI is setting the tangent for metaphotonics through the building of precision-engineered optical systems and design optimization depths. Generative applications or models then extrapolate novel optical patterns and datasets, such as the digital framework of optical diffusion-based GANs, which underpins the manipulation of light for the high-tech applications. Generative models in general use phase-based encoding and encode information through free-space optics for energy-conscious physical programming. Recent work has further demonstrated the facility for optical generation models to tackle complex visual datasets like handwritten digits and natural scenes that hold and slightly exhibit performance concerning traditional neural networks, hence decreasing computational consumption [9].

The base of the metamaterials appears to delve into the early years of metamaterials research in 1968, achieving dire achievements marked by the birth of metasurfaces in 1996. The big-bang arrival had been thus far Capasso et al., when Capasso's team demonstrated plasmonic V-arranged horn antennas to imprint phase discontinuities on propagating light, enabling metamaterials to exact wavefront manipulation precisely. Alongside the creation crusade of such surfaces and other dissident developments, effects got multifunctional, such as polarization-multiplexing and wavelength-multiplexing designs, to attend to critical requirements of biomedical functional applications that encapsulate computational and quantum applications, among others [10].

With aid from programmable metasurfaces, Cui et al. put forward digital coding metasurfaces in 2014, with empathy denoting real-time modulation of electromagnetic waves. Within an ensemble with digital information, 'information metasurfaces' seamlessly open fresh layout avenues with spatial modulation, wireless communications, and countless other applications [11]. The splendid mix of AI and metamaterial metasurface constructs has stretched this field even further. Deep neural networks (DNNs) provide efficient, data-driven frameworks for simulating light-matter interactions, designing meta-atoms, and optimizing metasurface functionality, nevertheless in wide fields of view and highly aberration correction capacity [12]. Through these models, real-time imaging, intelligent target tracking, and communication enhancements offer a real platform with intelligent meta-photonic systems [5].

As another new trend, generative AI also provides neat ideas, such as photonic meta-neurons, which mimic biological neurons using metasurfaces. These high-speed, energy-efficient neurons may eventually pledge green and sustainable alternatives as opposed to conventional electronic models [13]. Together, the integration of generative AI into meta-photonics has impacted the design and functionality enhancements of photonic devices, offering to mark the beginning of a new era of scalable, high-performance optical systems and intelligent human–machine interfaces (Fig. 3).

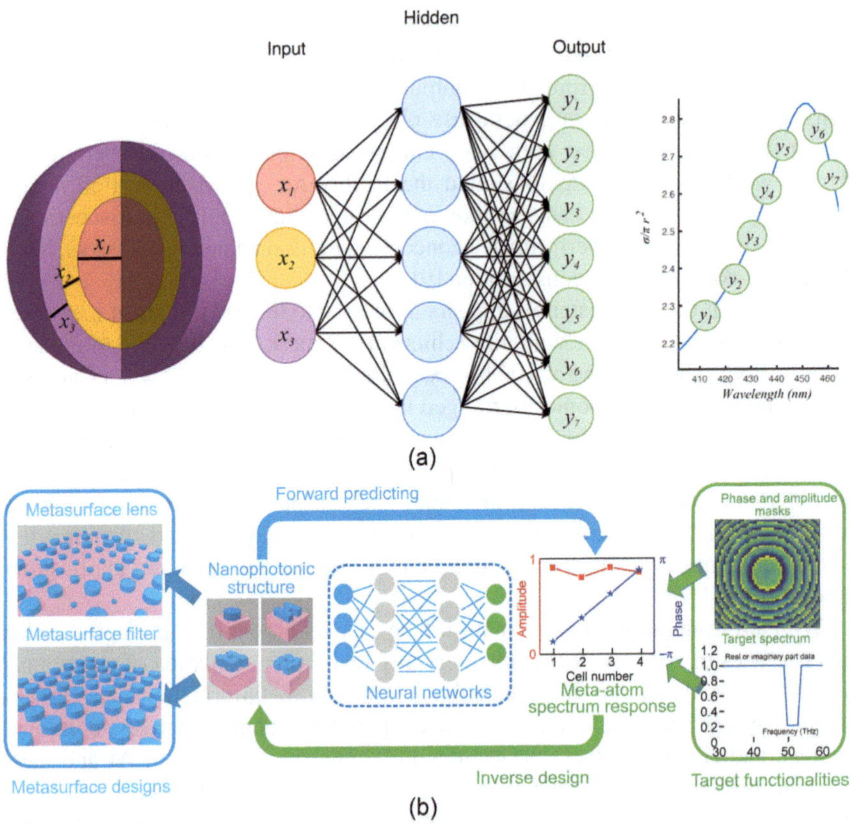

Fig. 3 **a** Artificial neural networks employed to predict and simulate scattering cross-section of nanoscale structures in multiple wavelengths of the scattering spectrum. **b** A means of deep learning towards different metasurface devices engineered over multiple wavelengths of the entire electromagnetic spectrum [1]

3.3 *Generative AI in Holography*

Generative AI is changing the landscape of "metaphotonics" for the generation of high-quality holograms with high efficiency and presents a solution to traditional hurdles such as concern with computational complexity and design limitations. Models such as the generative pre-trained transformer (GPT) and HoloGPT offer unprecedented speed and accuracy in generating phase masks for computer-generated holograms (CGHs). These models reduce computational overhead while ensuring output quality higher than that achieved by traditional algorithms, the Gerchberg–Saxton method, by speeding up the calculations. Real-time hologram generation is made easy by this method while successfully generalizing with different datasets for application areas such as 3D displays, immersive environments, and advanced visualization systems [16].

The P-Hologen framework stands as another breakthrough, combining vector quantized variational autoencoders with the angular spectrum method to solve POH complications. This end-to-end framework generates complex holographic content in scenarios without pre-existing images while enhancing computational efficiency and visual accuracy. Enabling semantic-aware holography-ready content generation allows P-Hologen to support transformative applications in entertainment, medical imaging, and scientific visualization, highlighting the potential of generative AI to redefine holography [17] (Fig. 4).

Another major thrust forward in holographic technologies lies in generative adversarial networks (GANs) that uplift the reconstruction quality of 3D scenes. These networks eliminate noise and further optimize the phase and amplitude distribution so that multilayered holographic content can be very accurately created. Such inventions extend the realm of possible applications into microscopy, virtual reality, and 3D visualization, thereby proving the transformative force of programming groups with GAN on metaphotonic systems [19].

AI-guided methods have equally served to democratize holography by simplifying workflows and limiting dependency on complex equipment and/or expertise. To further this promise, AI-generated prompts have come to the fore in creating holographic stereograms that push high-fidelity holography into the realm of possibility for applications in education, art, and technological applications [20]. By linking

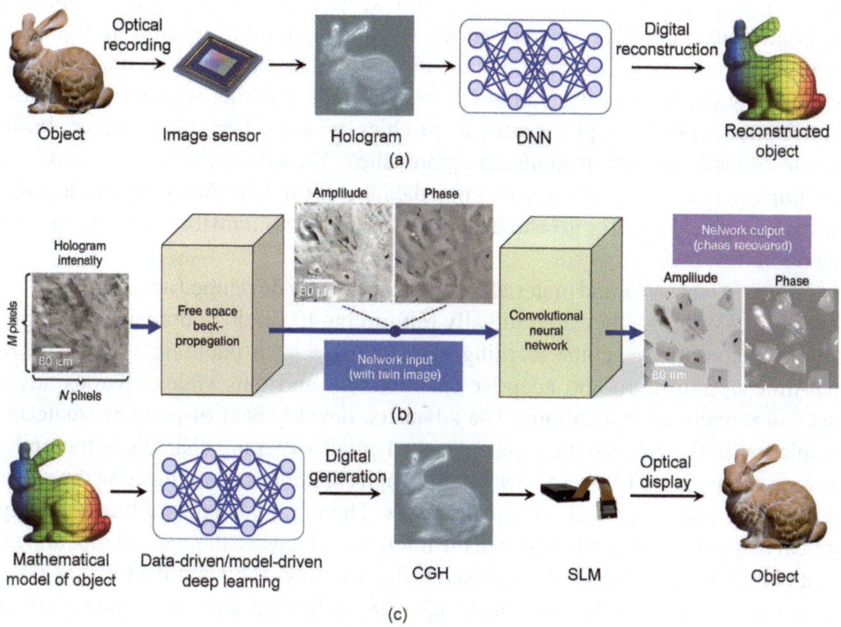

Fig. 4 **a** The principle of deep learning for digital holographic imaging. **b** A convolutional network-based methodology for rapid phase recovery and holographic image reconstruction. **c** Principle deep learning for computer-generated holography [1]

computational optimization with physical design in metaphotonics, generative AI continues to open unprecedented opportunities across scientific, medical, and creative domains.

4 Trends in AI-Driven Photonic Sensing

The integration of AI will radically impact applications from environmental monitoring to biomedical sciences and will open new vistas for creating remarkable advances in the challenges at the very frontiers of science, health, and industry. Photonic sensing has, until now, largely relied on conventional signal-processing techniques. While these techniques can show great reliability, they are often unable to provide the flexibility, scalability, and efficiency suited for the advanced processing of intricate, high-dimensional datasets. These methods were successful for many years but have not answered satisfactorily to the now-increasing demand for dynamic adaptability and preciseness.

Encouragingly, recent advances in artificial intelligence, particularly concerning complex neural network architectures and machine-learning frameworks, have considerably advanced the functionality and practicality of photonic sensors in real-world applications. Fully connected layers are instrumental in several AI applications for organizing and processing complex, high-dimensional data acquired by photonic sensors. Such layers are widely used in examining spectral data acquired from light emitters for material compositions to relate raw data meaningfully with useful insights. Nevertheless, the FCL system finds it harder with intrinsic optical data frameworks of complex spatial hierarchies, which is becoming a severe limitation in conventional implementation approaches. Therefore, CNNs were developed and implemented to learn and work upon the inherent spatial characteristics and hierarchies in optical data for greater analytical capacity and transformative advances in the photonics realm.

Photonic hardware and materials are synchronously developed and evolved synergistically. Therefore, they are mutually reinforcing technology domains. The application of complex machine learning algorithms in such photonic systems allows real-time optimization and adaptive functionality and thus yields obvious advantages in numerous applications. The advanced development of photonic materials, coupled with the AI analytics, was meant to entail sweeping changes in the optical domain and push for new remedies for the tough challenges touching upon the performance of scientific and technological sensors. The improvement is achieved through optical responses tailored to enhance the interaction between the sensing medium and light [21]. Similarly, dielectric multilayers and the core–shell architecture are attractive topics with outstanding prospects regarding enhanced sensitivity and specificity for photonic sensing applications. These advances take advantage of their peculiar optical properties to detect much smaller changes in biological processes, chemical composition, or environmental parameters [22] (Fig. 5).

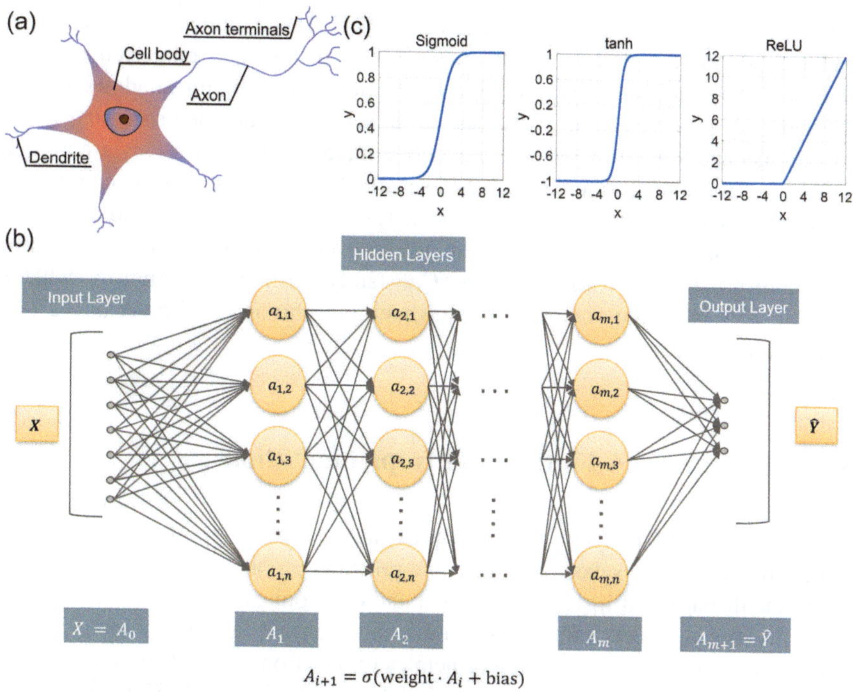

Fig. 5 **a** Neurons are represented biologically. **b** A neural network based on fully connected layers in which all the neurons in adjacent layers are interconnected. **c** Three common activation functions: Sigmoid, tanh, and ReLU [23]

Plasmonic nanostructures focus light energy at the nanoscale and are integrated into intricate photonic sensors that greatly enhance the interaction between light and matter, essentially enabling ultralow detection thresholds and extremely high precision metrology [24]. *Optical neural networks* stand at the revolutionary point where photonics is reinforced to artificial intelligence resulting in a pioneering integration that allows for the direct realization of tasks upon optical signals. In many respects, this advance in technology not only lowers the latency and computational burden but also elevates the throughput to such a degree that it transforms potentials across bold operational fronts [25]. The reprogrammable surfaces introduce a new dimension of adaptability, thereby affording unprecedented operability for sensing systems to obtain real-time modifications and adjustments, enabling the system to attain its task under rapidly changing or dynamic environmental conditions [26].

Other than this innovation pipeline, *3D printing of photonic structures* creates a less-discriminating fabrication environment for complex device fabrication; therefore, it promotes rapid and inexpensive prototyping for customization toward very specific sensing applications. 3D printing provides infinite opportunity for rapid iteration and development without exceedingly extensive and steep production costs and timelines to entertain creative experimentation [27]. The combination of these high-end technologies, including AI-driven analyses, metasurface-enhanced photonic sensors, ONNs, and cheaply accessible fabrication processes, foreshadows a future whereby any sensing system possesses higher sensitivity, unique processing speed, and unmatched versatility. It goes beyond a mere combination; it is a complete redesign of what has become possible in sensing technologies, thereby giving birth to new applications and solutions across a variety of dispersed multidimensional domains—from trace pollutant detection into innovative biomedical tools for smarter and more efficient industrial systems. Constantly pushing the limits of what is possible now has been achieved through merging photonics with AI into a new epoch of technological advancement.

5 Challenges in Adoption and Implementation

Deep learning combined with photonics presents interesting opportunities, but at the same time, it imposes severe scientific and practical problems. The main difficulty lies in the disparity intrinsic to the optical and electronic systems. Optical systems relate principles of continuous waveforms and physical light phenomena, which stand in stark contrast to the design principles for electronic circuits based on which deep-learning algorithms operate [28]. Digital computations need to be converted to optical processes, which may require some novel designs, most probably photonic neural networks [29]. Nonetheless, these technologies are still relatively nascent and not mature enough to be practically deployed.

Here are some challenges in AI integration in photonics.

Data Quality and Volume: A key challenge is to acquire large and high-quality datasets for the training of deep learning models. Bad data could halt the scalability of networks and their abilities to tackle the more complicated questions effectively [3].

Algorithm Training Difficulties: The very commonly used backpropagation algorithm for the training of neural networks is faced with challenges such as the dependence on digital training models, a lack of scalability, and issues that arise during training. Hence, other alternative learning methodologies have been suggested, including backpropagation-free methods and physics-informed training methods [30, 31].

Limited Manipulation of Optical Fields: The deep learning application in photonics is limited due to the factors of complex optical field control. All of the above limitations can hamper the efficiency of deep-learning architectures in the optimization of photonic systems [32].

Energy Consumption: Recently, a lot of concern has arisen regarding the amount of energy consumed in training large AI models. As per the paper, training a single large language model emits as much carbon dioxide as 125 round-trip flights between major cities; this opens a discussion regarding the feasibility of the current infrastructure for electronic computing with the fast growth of AI-enabled applications [33].

Computational Limitations of Classical Systems: Conventional electronic computational designs, particularly those based on the von Neumann architecture, face latency and increased power consumption owing to the decoupling of memory from processing units [34]. The architecture is now gradually failing to fulfill the high demands for training artificial intelligence [35].

Integration of Optical and Electronic Systems: Hybrid optical-electronic computing systems need to be devised to synergize the advantages enjoyed by both. Again, this is not very easy, with the accompanying technical hurdles, which include high speed, low power consumption, and high throughput in such systems [36].

Standardization and Paradigm Development: The domain of intelligent photonics shows a certain intrinsic requirement for well-defined paradigms related to the design and training of the networks. Absent such paradigms, the further design and implementation of efficient photonic systems intended for artificial intelligence will be hampered [32].

Interdisciplinary Collaboration: An urgent necessity exists for interdisciplinary approaches to encompass this challenge. Physics, engineering, and computer science have huge implications for the development of intelligent photonics [28, 37].

The management of thermal dynamics and energy efficiency is another added complexity in photonic systems. Even though photonics is described as energy-efficient for applications such as matrix multiplication, its implementation in the real world is afflicted by thermal limits during high-speed operations, which may affect the overall performance. Besides, error correction mechanisms, which are quite often exploited in conventional computation, remain underappreciated in photonics, thereby threatening paramount applications in the finance or aerospace industries where precision becomes highly critical [32]. The above results open several possibilities for fundamentally changing both the capabilities and application areas of AI from multiple orders of magnitude faster processing and higher energy efficiency advantages in deep photonics learning. With the advancement in these areas, true real-time applications of AI appear to be within the realm of possibilities for high-frequency financial trading or the personalization of medicine. If these hurdles are not crossed, they can hinder the pace of advancing innovative solutions, slow down the introduction of the next generation of AI systems, and constrain the abilities of the latter to act as catalysts for change. Such advanced work in these three realms—photonics, computer science, and materials engineering—will be required for the integration to overcome these hindrances.

6 Future Directions

The incorporation of Generative AI with photonic sensing systems opens up new transformative avenues while stretching the lines of accuracy and computational efficiency, also benefiting both. The primary neural network approaches would include fully connected layers (FCLs), convolutional neural networks (CNNs), and other models, optical neural networks (ONNs) in addressing this field. However, the future calls for critical improvements and innovations to tackle many persistent challenges while taking advantage of emerging opportunities.

Current-day FCLs are perfect for processing such types of data: they have nearly high dimensions but fall short of achieving the very complex spatial hierarchies that are mostly present in optical datasets. Therefore, future research should address this hybrid architecture that combines FCLs with attention mechanisms, catering to better non-linearity within the optical data patterns. Similarly, lengths could be added to CNNs, which modernized the spatial hierarchy use via graph neural networks (GNNs) to integrate complex, many non-Euclidean data structures present in an array of photonic sensing applications. ONNs, on the other hand, have proven to be of transformative potential by enabling straightforward computation on optical signals. Yet, still, major problems remain with scalability or blending these with current digital systems. Future efforts should focus on scalable ONN designs, which will further anchor the principles of quantum computing to raise computation efficiency and establish new paradigms for processing optical signals. Another promising feature is the co-development with plasmonic nanostructures, which obtain detection thresholds into regions previously unheard of through nanoscale light-matter interactions. The evolution of AI algorithms and advanced photonic materials presents a fertile field for innovation. The self-adaptive metasurfaces have shown great promise such that they can adapt themselves in real-time to changing conditions in dynamic environments. Current research will have to focus on the production of such metasurfaces that hold within them AI controllers for autonomously optimizing the metasurfaces, thereby injecting flexibility concerning environmental changes. Advances in new material architectures, such as core–shell structures and dielectric multilayers, will also be necessary to enhance sensitivity and specificity in applications such as biomedical diagnostics and environmental monitoring. 3D printing technologies will provide unprecedented customization and fast prototyping capacity for photonic structures. These AI-driven design tools could be implemented well within the GAN framework to facilitate an accelerated pathway toward an optimal timing specifically tailored to the requirements of specific types of sensing applications. Such an approach would democratize access to advanced sensing technologies to be used by important sectors of society.

7 Conclusion

Generatively trained AI is heralding the advance in photonic sensing technologies by enabling the synthesis of extremely advanced neural architectures such as fully connected layers, convolutional neural networks, and optical neural networks with some innovative photonic materials and systems. Such integration presents unmatched data processing, optimization, and adaptability in pushing the envelope of sensor accuracy and applicability in a variety of domains, including environmental monitoring, healthcare, and industrial automation. So-called photonic-enabled AI innovation creates new vistas for possibilities, such as reprogrammable metasurfaces, plasmonic nanostructures, and 3D-printed photonic devices. Generative models will create further improvements by delivering data synthesis, metasurface optimization, and the possibility of real-time adaptability in complex applications. Such advances may well address traditional limitations by offering scalable, efficient routes to transform sensing capabilities. This chapter elucidates the role of generative AI as a catalyst in the development of photonic sensing and demonstrates different ways this technology could help demand more complex solutions and open up new ground. The mesh between AI and photonics heralds a landmark in advanced sensing technology, bringing intelligent, speedy, and versatile systems capable of responding to the envisioned world, a world that is dynamic and deeply interconnected.

References

1. D. Xu et al., Intelligent photonics: a disruptive technology to shape the present and redefine the future. Engineering **58**, 1–2 (2024)
2. Y. LeCun et al., Deep learning. Nature **44** (2015)
3. I. Goodfellow et al., *Deep Learning* (MIT Press, USA, 2016), 781 pp
4. J. Wu, L. Cao, G. Barbastathis, DNN-FZA camera: a deep learning approach toward broadband FZA lensless imaging. Opt. Lett. **46**(1), 130–133 (2021). https://doi.org/10.1364/OL.411228
5. Z. Li et al., Retrieval augmented generation or long-context LLMs? A comprehensive study and hybrid approach. ACL Anthol. (2024)
6. M. Atwany et al., A review of emerging trends in photonic deep learning accelerators. Front. Phys. **24** (2024)
7. A. Radford et al., Improving language understanding by generative pre-training. Semantic Scholar 1–12 (2018)
8. Y. Tamir et al., Spatio-temporal dynamics of pulses in multimode fibers. Nonlinear Opt. Appl. **23** (2024)
9. X. Chen et al., A neural speech decoding framework leveraging deep learning and speech synthesis. Nat. Mach. Intell. **24** (2024)
10. X. Xing et al., Optimization design of a polarization-independent grating coupler on lithium-niobate-on-insulator. Appl. Opt. **25** (2024)
11. T.J. Cui et al., Coding metamaterials, digital metamaterials and programmable metamaterials. Light Sci. Appl. 37 (2014)
12. I.J. Goodfellow et al., Explaining and harnessing adversial examples, in *ICLR* (Semantic Scholar, Mountain View, CA, 2015), p. 11
13. B. Liu et al., Empowering large language models with optimal planning proficiency. arXiv preprint (2023), p. 35

14. J. Peurifoy, Y. Shen, L. Jing, Y. Yang, F. Cano-Renteria, B.G. DeLacy, J.D. Joannopoulos, M. Tegmark, M. Soljačić, Nanophotonic particle simulation and inverse design using artificial neural networks. Sci. Adv. **4**(6) (2018). https://doi.org/10.1126/sciadv.aar4206
15. S. An, C. Fowler, B. Zheng, M.Y. Shalaginov, H. Tang, H. Li, L. Zhou, J. Ding, A.M. Agarwal, C. Rivero-Baleine, K.A. Richardson, T. Gu, J. Hu, H. Zhang, A deep learning approach for objective-driven all-dielectric metasurface design. ACS Photon. **6**(12), 3196–3207 (2019). https://doi.org/10.1021/acsphotonics.9b00966
16. H. Wang et al., Machine learning potential for serpentines. Adv. Earth Space Sci. **35** (2024)
17. S. Park et al., A deep learning model of tumor cell architecture elucidates response and resistance to CDK4/6 inhibitors. Nat. Cancer **29** (2024)
18. Y. Rivenson, Y. Zhang, H. Günaydın, et al., Phase recovery and holographic image reconstruction using deep learning in neural networks. Light. Sci. Appl. **7**, 17141 (2018). https://doi.org/10.1038/lsa.2017.141
19. S.A. Kiriy et al., Object image reconstruction: method for reconstructing images from digital off-axis holograms using a generative adversarial network. Meas. Tech. **40** (2024)
20. P. Gentet et al., From text to hologram: creation of high-quality holographic stereograms using artificial intelligence. Photonics **9** (2024)
21. A.V. Kildishev et al., Planar photonics with metasurfaces. Science **339**, 35 (2013)
22. L. Zhang et al., Deep learning for sentiment analysis: a survey. Wiley Interdiscip. Rev. Data Min. Knowl. Discov. **44** (2018)
23. Y. Xu, X. Zhang, Y. Fu, Y. Liu, Interfacing photonics with artificial intelligence: an innovative design strategy for photonic structures and devices based on artificial neural networks. Photon. Res **9**(4), B135 (2021). https://doi.org/10.1364/PRJ.417693
24. G. Huang et al., Densely connected convolutional networks, in *Proceedings of the IEEE Conference on Computer Vision and Pattern Recognition* (IEEE, 2017), p. 30
25. T.W. Hughes et al., Adjoint method and inverse design for nonlinear nanophotonic devices. ACS Photonics **7** (2018)
26. P. Yu et al., Broadband metamaterial absorbers. Adv. Opt. Mater. **25** (2019)
27. N. Baker et al., Deep convolutional networks do not classify based on global object shape. PLoS Comput. Biol. **43** (2018)
28. B. Shastri et al., Photonics for artificial intelligence and neuromorphic computing. Nat. Photonics **14** (2021)
29. Huang et al., A photonics-inspired compact network: toward real-time AI processing in communication systems. IEEE J. Sel. Top. Quantum Electron. **13** (2022)
30. K. Liao et al., Integrated photonic neural networks: opportunities and challenges. ACS Photonics **14** (2023)
31. Lin et al., All-optical machine learning using diffractive deep neural. Science **34** (2018)
32. M. Miscuglio, V.J. Sorger, Photonic tensor cores for machine learning. Appl. Phys. Rev. **15** (2020)
33. E. Strubell et al., Energy and policy considerations for modern deep learning research. AAAI Conf. Artif. Intell. 14–15 (2020)
34. J. Backus, Can programming be liberated from the von Neumann style? A functional style and its algebra of programs. ACM Commun. 15–16 (1978)
35. C.S. Chai et al., University students' intentions to learn artificial intelligence: the roles of supportive environments and expectancy–value beliefs. Int. J. Educ. Technol. High. Educ. 23–24 (2023)
36. Shen et al., Deep learning with coherent nanophotonic circuits. Nat. Photonics 14–15 (2017)
37. G. Wetzstein et al., Inference in artificial intelligence with deep optics and photonics. Nature 13–14 (2020)

MIX
Papier aus verantwortungsvollen Quellen
Paper from responsible sources
FSC® C105338

If you have any concerns about our products,
you can contact us on
ProductSafety@springernature.com

In case Publisher is established outside the EU,
the EU authorized representative is:
Springer Nature Customer Service Center GmbH
Europaplatz 3, 69115 Heidelberg, Germany

Printed by Libri Plureos GmbH
in Hamburg, Germany